Capillary Electrophoresis of Proteins and Peptides

METHODS IN MOLECULAR BIOLOGY™

John M. Walker, SERIES EDITOR

METHODS IN MOLECULAR BIOLOGY™

Capillary Electrophoresis of Proteins and Peptides

Edited by

Mark A. Strege

Eli Lilly and Company, Indianapolis, IN

Avinash L. Lagu

Lilly Research Laboratories, Indianapolis, IN

HUMANA PRESS ✷ TOTOWA, NEW JERSEY

999 Riverview Drive, Suite 208
Totowa, New Jersey 07512

www.humanapress.com

This publication is printed on acid-free paper. ∞
ANSI Z39.48-1984 (American Standards Institute)

Permanence of Paper for Printed Library Materials.

Cover design by Patricia F. Cleary.

For additional copies, pricing for bulk purchases, and/or information about other Humana titles, contact Humana at the above address or at any of the following numbers: Tel.: 973-256-1699; Fax: 973-256-8341; E-mail: humana@humanapr.com; or visit our Website: www.humanapress.com

Printed in the United States of America. 10 9 8 7 6 5 4 3 2 1

E-ISBN 1-59259-798-X

Library of Congress Cataloging-in-Publication Data

Capillary electrophoresis of proteins and peptides / edited by Mark A. Strege, Avinash L. Lagu.
p. ; cm. -- (Methods in molecular biology, ISSN 1064-3745 ; v. 276)
Includes bibliographical references and index.
ISBN 1-58829-017-4 (alk. paper)
1. Capillary electrophoresis--Laboratory manuals. 2. Proteins--Separation--Laboratory manuals. 3. Peptides--Separation--Laboratory manuals. [DNLM: 1. Electrophoresis, Capillary--methods. 2. Proteins--analysis. QU 25 C2418 2004] I. Strege, Mark A. II. Lagu, Avinash L. III. Series: Methods in molecular biology (Clifton, N.J.) ; v. 276.
QP519.9.C36C355 2004
572'.636--dc22

2003028088

Preface

Throughout the more than 20 years that have followed the beginnings of capillary electrophoresis (CE), its application to the analysis of proteins and peptides has continued to be reliable, versatile, and productive. Over time, CE has matured to become a superb complement to HLPC, and in many cases has also evolved as an automated and quantitative replacement for conventional slab gel electrophoresis methods such as SDS-PAGE and isoelectric focusing.

Within *Capillary Electrophoresis of Proteins and Peptides,* we have assembled contributions from researchers who are applying state-of-the-art CE for protein and peptide analysis, including topics that we believe are of great potential both in the present and for the future.

In comparison to traditional separation methods, CE represents a miniaturized analysis technique (especially in its microchip-based format) that is highly dependent upon the basic fundamentals of effective sample recovery and high sensitivity detection. With these issues in mind, Chapters 1–4 describe recently developed approaches for both capillary coatings and analyte detection via laser-induced fluorescence.

Since the discipline of biotechnology has established itself as a primary platform for the application of CE to the analysis of proteins and peptides, Chapters 5–7 demonstrate a variety of examples of the specific techniques that have been applied for the development of biopharmaceuticals and their commercialization. The methods covered here include also the analysis of oligosaccharides from glycoproteins.

Studies of the association of proteins with other molecules can provide insight into the very heart of biological processes. Therefore, a major focus within both the pharmaceutical industry and academia is the utilization of CE for the characterization of protein interactions with ligands, other proteins, and large biopolymers. Chapters 8–11 describe in detail the most recent approaches for performing affinity capillary electrophoresis for the evaluation of protein binding, including the use of protein charge ladders.

CE and capillary isoelectric focusing have been providing rapid, high-resolution separations of proteins. When combined with electrospray mass spectrometry detection they constitute a powerful analysis system capable of supporting complex studies such as those associated with proteomics. Chapters 12–15 focus on the use of CE within this exciting field. The use of CE in microfluidics format is also presented here.

The objective of *Capillary Electrophoresis of Proteins and Peptides,* by its breadth, topicality, and forward focus, is to serve as a valuable guide for researchers across many disciplines. We look forward with great anticipation to the impact this collection will have, as researchers new to the field are carried forward in their work by the experts' step-by-step guidance and notes provided within these chapters.

Mark A. Strege
Avinash L. Lagu

Contents

Contributors

HOSSEIN AHMADZADEH • *Department of Chemistry, University of Minnesota, Minneapolis, MN*
JOHN AUSTIN • *Department of Chemistry, University of Nebraska, Lincoln, NE*
MARYAM AZAD • *Department of Chemistry and Biochemistry, California State University, Los Angeles, CA*
NICOLE E. BARYLA • *Eli Lilly Canada, Inc., Toronto, Ontario, Canada*
STACEY CUMMINS-BITZ • *Indianapolis Bulk Operations Technical Excellence Centre, Eli Lilly and Company, Lilly Corporate Center, Indianapolis, IN*
JEFFREY D. CARBECK • *Department of Chemical Engineering, Princeton University, Princeton, NJ*
GREGORY S. CLANTON • *Lilly Research Laboratories, Eli Lilly and Company, Lilly Corporate Center, Indianapolis, IN*
NORMAN J. DOVICHI • *Department of Chemistry, University of Washington, Seattle, WA*
PAUL L. DUBIN • *Department of Chemistry, Indiana University-Purdue University at Indianapolis, Indianapolis, IN*
RAEANN M. FIELDS • *Indianapolis Bulk Operations Technical Excellence Centre, Eli Lilly and Company, Lilly Corporate Center, Indianapolis, IN*
FRANK A. GOMEZ • *Department of Chemistry and Biochemistry, California State University, Los Angeles, CA*
DAVID L. GOOD • *Indianapolis Bulk Operations Technical Excellence Centre, Eli Lilly and Company, Lilly Corporate Center, Indianapolis, IN*
DAVID S. HAGE • *Department of Chemistry, University of Nebraska, Lincoln, NE*
JED HARRISON • *Department of Chemistry, University of Alberta, Edmonton, Alberta, Canada*
TOSHIAKI HATTORI • *Research Center for Chemometrics, Toyohashi University of Technology, Toyohashi, Japan*
LILI HERNANDEZ • *Department of Chemistry and Biochemistry, California State University, Los Angeles, CA*
TIEMIN HUANG • *Department of Chemistry, University of Waterloo, Waterloo, Ontario, Canada*
JOHN KADDIS • *Department of Chemistry and Biochemistry, California State University, Los Angeles, CA*
HEE SEUNG KIM • *Department of Chemistry, University of Nebraska, Lincoln, NE*

VADIM KLYUSHNICHENKO • *Altus Biologics Inc., Cambridge, MA*
SERGEY KRYLOV • *Department of Chemistry, York University, Toronto, Ontario, Canada*
AVINASH L. LAGU • *Lilly Research Laboratories, Eli Lilly and Company, Lilly Corporate Center, Indianapolis, IN*
JIANJUN LI • *Institute for Biological Sciences, Ottawa, Ontario, Canada*
CHARLES A. LUCY • *Department of Chemistry, Gunning-Lemieux Chemistry Centre, University of Alberta, Edmonton, Alberta, Canada*
SUZANA MARTINOVIĆ • *Environmental and Molecular Sciences Laboratory, Pacific Northwest National Laboratory, Richland, WA*
MEHDI MOINI • *Department of Chemistry and Biochemistry, University of Texas at Austin, Austin, Texas*
BRIAN K. NUNNALLY • *Vaccine Analytical Development, Wyeth Research, Wyeth, Sanford, NC; formerly, Indianapolis Bulk Operations Technical Excellence Centre, Eli Lilly and Company, Lilly Corporate Center, Indianapolis, IN*
LJILJANA PAŠA-TOLIĆ • *Environmental and Molecular Sciences Laboratory, Pacific Northwest National Laboratory, Richland, WA*
GABOR PATONAY • *Department of Chemistry, Georgia State University, University Plaza, Atlanta, GA*
JEFFREY S. PATRICK • *Lilly Research Laboratories, Eli Lilly and Company, Lilly Corporate Center, Indianapolis, IN*
JANUSZ PAWLISZYN • *Department of Chemistry, University of Waterloo, Waterloo, Ontario, Canada*
BRENDA P. RENER • *Lilly Research Laboratories, Eli Lilly and Company, Lilly Corporate Center, Indianapolis, IN*
JOZEF SALON • *Department of Chemistry, Georgia State University, University Plaza, Atlanta, GA*
EMEK SEYREK • *Department of Chemistry, Indiana University-Purdue University at Indianapolis, Indianapolis, IN*
UPMA SHARMA • *Department of Chemical Engineering, Princeton University, Princeton, NJ*
CATHERINE SILVERIO • *Department of Chemistry and Biochemistry, California State University, Los Angeles, CA*
RICHARD D. SMITH • *Environmental and Molecular Sciences Laboratory, Pacific Northwest National Laboratory, Richland, WA*
JOHN SOWELL • *Department of Chemistry, Georgia State University, University Plaza, Atlanta, GA*
LUCJAN STREKOWSKI • *Department of Chemistry, Georgia State University, University Plaza, Atlanta, GA*

PIERRE THIBAULT • *Institut de Recherche en Immunovirologie et Cancérologie, Université de Montréal, Quebec, Canada*
TAMMY-LYNN TREMBLAY • *Institute for Biological Sciences, Ottawa, Ontario, Canada*
VALERIE VILLAREAL • *Department of Chemistry and Biochemistry, California State University, Los Angeles, CA*
JIAQI WU • *Convergent Bioscience Ltd., Toronto, Ontario, Canada*
XING-ZHENG WU • *Department of Materials Science and Engineering, Fukui University, Fukui-shi, Japan*
KEN K.-C. YEUNG • *Departments of Chemistry and Biochemistry, University of Western Ontario, London, Ontario, Canada*

1

Surfactant-Based Methods for Prevention of Protein Adsorption in Capillary Electrophoresis

Charles A. Lucy, Nicole E. Baryla, and Ken K.-C. Yeung

Summary

Surfactants such as didodecyldimethyl ammonium bromide (DDAB) and 1,2-dilauroyl-*sn*-phosphatidylcholine (DLPC) form bilayers at the walls of bare silica capillaries. Once formed, these bilayers are stable in the absence of surfactant in the buffer. DDAB provides a cationic bilayer coating which yields a strong reversed EOF and is effective for separation of cationic proteins. DLPC provides a zwitterionic bilayer coating which is effective for both cationic and anionic proteins. The electroosmotic flow (EOF) is strongly suppressed in DLPC-coated capillaries, thus low mobility proteins are slow to elute, and so the coating is favored for separation of high mobility proteins.

Key Words

Bilayer coatings; capillary electrophoresis; DDAB; DLPC; double-chained surfactants; protein adsorption.

1. Introduction

Although capillary electrophoresis (CE) provides rapid, high-resolution separations of many analytes, there are situations in CE that demand altering the chemistry at the capillary wall to improve (and even allow) a separation. For example, protein analysis by CE has been limited because proteins adhere strongly to the negatively charged capillary wall. Mazzeo and Krull identified the four characteristics that an ideal coating should exhibit for the separation of proteins *(1)*.

1. Separation efficiency (in theory, this should approach 1–2 million plates/m).
2. Protein recovery (this should approach 100%).
3. Reproducibility of migration time from run to run and day to day.

From: *Methods in Molecular Biology, vol. 276: Capillary Electrophoresis of Proteins and Peptides*
Edited by: M. A. Strege and A. L. Lagu © Humana Press Inc., Totowa, NJ

4. Retention of the electroosmotic flow (EOF) so that cationic and anionic proteins can be separated in the same run.

   To these, we would add that the coating should also ideally be.

   a. Easy to generate.
   b. Inexpensive.
   c. Applicable over a wide range of buffer conditions.
   d. Should not interfere with detection (i.e., should be compatible with both optical and mass spectral detection).

Capillary coatings can be categorized as: permanent; dynamic, and semipermanent ***(2)***. Permanent coatings are discussed in Chapter 2, and typically involve covalent attachment of polymers such as polyethyleneimine or polyacrylamide to permanently shield the bulk solution from the silanol groups on the capillary wall ***(3–7)***. Such permanent coatings are used for both EOF suppression and prevention of protein adsorption. Despite the success of these capillaries, derivatization procedures can be lengthy, their lifetime may be short, they can be unstable outside a limited pH range, and reproducibility from capillary to capillary may be poor. For example, polyethyleneimine has been covalently attached to the capillary wall ***(5)***. The derivatization procedure took longer than 2 h to perform and the separations were only reproducible for 5 d. The stability of this coating was improved by crosslinking the polyethyleneimine to the wall ***(4,5)***; however the derivatization procedure took more than 12 h. Furthermore, the cost associated with permanently derivatized capillaries can be substantial (>$300 for commercially available permanently coated capillaries) and may not be practical for high-throughput analyses.

As an alternative, dynamic coatings are desirable because of their low cost and simplicity of application ***(8)***. Dynamic coatings are formed by adding an additive to the background electrolyte. The additive has a strong affinity for the capillary surface. Thus, it equilibrates between the bulk electrolyte and the capillary wall, and thus, competes with the protein for adsorption onto the capillary wall. Typical buffer additives used for dynamic coatings are: amines to oligoamines; neutral polymers ***(3)***; and single-chained surfactants ***(9–11)***. Near ideal separation efficiencies and protein recoveries have been achieved with such dynamic coatings ***(9,10)***. However, the additive must be present in the background electrolyte for the additive to be effective. This may interfere with the separation and/or subsequent detection of the proteins.

To overcome some of the problems associated with derivatized capillaries and dynamic coatings, additives that form semipermanent coatings have been investigated. With semipermanent coatings, a reagent is rinsed through the capillary to form a noncovalent wall coating. Cationic polymers such as Polybrene ***(12–14)***, neutral polymers such as hydroxyethylcellulose and polyvinyl alcohol ***(3)***, and double-chained surfactants (discussed herein) have been used for noncovalent capillary coatings. These additives adsorb strongly onto the capillary surface. After the capillary is coated, the excess additive is flushed out of the capillary. Capillaries can be coated and regenerated between runs to maintain reproducible EOF. Alternatively, if the capillary becomes contaminated, the coating can be removed and then regenerated. This makes semipermanent coated capillaries more cost-effective than permanently derivatized coatings. However, procedures for semipermanent coatings can be time consuming (up to 2 h *[13]*) and reproducibility can be poor. For instance, in a recent study of four poly-cation additives, the EOF decreased by as much as 10% over 25 runs performed after the initial coating procedure *(14)*. Thus, recoating of the capillary must be performed frequently to achieve adequate reproducibility.

The coatings described in this chapter use double-chained surfactants to form semipermanent micellar aggregates at the capillary wall which prevent protein adsorption. These coatings are rapidly formed (<20 min) and their characteristics can be easily modified by changing the surfactant used. The key characteristics of such semipermanent surfactant coatings are depicted in **Fig. 1**. To form a coating, the capillary is equilibrated with a solution containing the surfactant (*see* **Fig. 1A**). Generally, the concentration of surfactant is greater than the critical aggregation concentration. Thus, micelles or vesicles are present in solution and may participate in the actual coating mechanism.

Appropriate selection of the surfactant is critical to the success of the coating. Ideally, the geometry of the surfactant monomer should be cylindrical in nature such that the surfactant aggregates to form a bilayer *(15)*. If the surfactant geometry is conical in shape, such as with single-chained surfactants (e.g., cetyltrimethylammonium bromide-[CTAB]), the surfactant will aggregate to form spherical micelles which cannot provide complete surface coverage *(16)*. Therefore, these coatings are not as effective at preventing protein adsorption. The geometry of the surfactant is dictated by the area of the head group relative to the width and length of the hydrophobic tail. The head group area is governed by electrostatic repulsion between the headgroups of adjacent surfactants in the bilayer. Thus, the headgroup area is affected by the ionic strength as determined by the buffer concentration and the ion association as determined by the nature of the buffer.

When the capillary is equilibrated with the surfactant solution (~0.1 m*M*), surfactant aggregates onto the capillary wall and forms the bilayer (*see* **Fig. 1A**). The surfactant solution is then flushed out of the capillary, and replaced with the run buffer, which does not contain surfactant. Under these conditions, the coatings described herein are stable for about 1 h, and so are best viewed as semipermanent coatings. The stability of the bilayer coatings is enhanced by factors that lower the critical aggregation concentration (i.e., higher ionic strength, longer tail) *(17)*. To maximize reproducibility of migration times, the coating

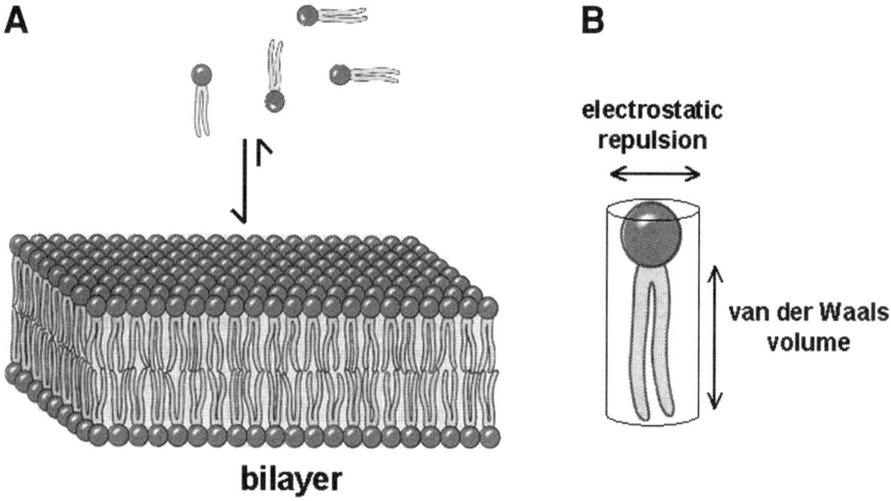

Fig. 1. Key attributes of surfactant-based wall coatings in CE. (**A**) Mechanism of surfactant aggregation and formation of a bilayer. (**B**) Aspects of surfactant geometry that affect structure of surfactant aggregate.

is refreshed before each run. Because surfactants are not present in the actual separation buffer, this method will be compatible with downstream mass spectral analysis. Ionization interference caused by surfactants was not evident in the previously reported CE fraction collection—offline MALDI–MS analysis ***(18)***.

Procedures for two semipermanent surfactant coatings used for protein separations in CE are described later. Didodecydimethyl ammonium bromide (DDAB) provides a cationic coating which yields a strong reversed EOF and is effective for separation of cationic proteins ***(15,16,18)***. 1,2-Dilauroyl-*sn*-phosphatidylcholine (DLPC) provides a zwitterionic coating which is effective for both cationic and anionic proteins ***(19)***. However, the EOF is strongly suppressed in DPLC-coated capillaries. Thus, low mobility proteins are slow to elute, and so the coating is favored for separation of high mobility proteins.

2. Materials

2.1. DDAB Cationic Semipermanent Coating

1. 18 *M*Ω Ultrapure water.
2. 0.1 *M* Sodium hydroxide (BDH, Darmstadt, Germany; *see* **Note 1**).
3. 25 m*M* phosphate buffer prepared from reagent grade ortho-phosphoric acid and adjusted to the desired pH with sodium hydroxide (*see* **Note 2**).

4. 0.1 m*M* DDAB (Aldrich, Milwaukee, WI) in 25 m*M* phosphate buffer (buffer prepared as in **item 3** above). DDAB solutions were stable and could be used for up to 1 mo when prepared and stored in Nalgeneware (*see* **Notes 3** and **4**).

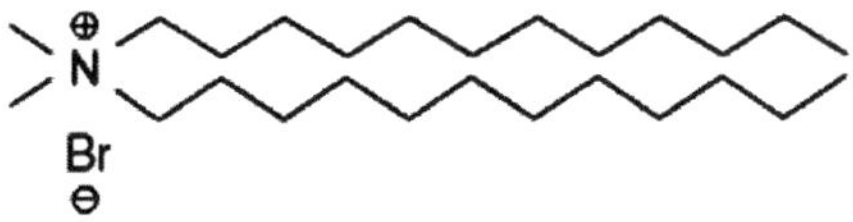

DDAB

5. 1 m*M* Mesityl oxide (Aldrich) in distilled water for EOF measurements (*see* **Subheading 3.3.** for procedure) (*see* **Note 5**).
6. Electrophoretic buffer or dilute (<10^{-3} *M*) HCl for electrokinetic fraction collection. (*see* **Note 6**).

2.2. Zwitterionic Semipermanent Coating Using DLPC

1. 18 *M*Ω Ultrapure water (Barnstead, Dubuque, IA).
2. 0.1 *M* Sodium hydroxide (BDH).
3. 20 m*M* Ultrapure Tris-HCl (Schwartz/Mann Biotech) adjusted to desired pH using 1 *M* HCl.
4. DLPC (Sigma) was used as received.

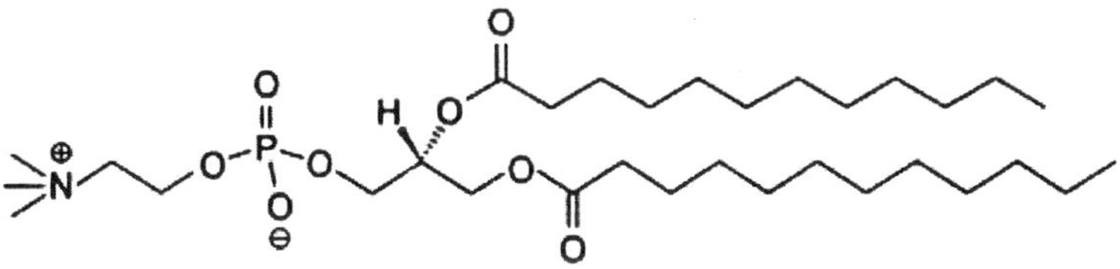

DLPC

5. Calcium chloride dihydrate (Molecular Biology grade, Sigma) added to Tris-HCl buffer to give a final concentration of 20 m*M*, as described in **Subheading 3.2.2.**
6. 1 m*M* Mesityl oxide (Aldrich) in distilled water for EOF measurements (*see* **Subheading 3.3.** for procedure).

2.3. Equipment

1. High-performance capillary electrophoresis instrument (e.g., Beckman P/ACE 5000 or MDQ (Fullerton, CA) or Agilent ^{3D}CE instrument (Palo Alto, CA).
2. Fused silica capillaries, 50 μm id and 360 μm od (Polymicro, Tucson, AZ; (*see* **Note 7**).
3. Detection: on-column ultraviolet (UV).
4. Polyethylene solution vials (*see* **Notes 1** and **4**).
5. Sonicator (Bransonic 220, Shelton, CT).
6. pH meter.

3. Methods

3.1. Cationic Semipermanent Coating Using DDAB

3.1.1. Capillary Pretreatment Steps (see **Note 8**)

1. 10 min (20 psi) with 0.1 *M* NaOH.
2. 10 min (20 psi) with water.
3. 5 min rinse (20 psi) with 0.1 m*M* DDAB in buffer.
4. 1 min rinse (20 psi) with the separation buffer to flush out excess surfactant (*see* **Note 9**).
5. 10 min of voltage application (at voltage used for the separation) to condition the coating, then repeat **steps 3** and **4** (*see* **Note 10**).
6. The resultant EOF is strongly reversed. **Table 1** indicates the EOF that has been observed with various buffers.

3.1.2. Separations With DDAB-Coated Capillaries

1. 2 min rinse (20 psi) with 0.1 m*M* DDAB in buffer.
2. 1 min rinse (20 psi) with separation buffer (*see* **Note 9**).
3. Sample injected hydrodynamically (0.5 psi) for 3.0 s.
4. Voltage is applied with a field strength of –200 V/cm using a 0.17-min rise time.
5. Data acquisition rate was 10 Hz, the detector rise time was 0.1 s, and direct UV detection was performed at 214 nm. An example electropherogram of cationic proteins is shown in **Fig. 2**.

3.2. Zwitterionic Semipermanent Coating Using DLPC

3.2.1. Capillary Pretreatment Steps (see **Note 8**)

1. 5 min (20 psi) rinse with 0.1 *M* NaOH.
2. 5 min (20 psi) rinse with Nanopure water.
3. 20 min (20 psi) rinse with 0.1 m*M* DLPC in 20 m*M* Tris-HCl, pH 7.4, buffer containing 20 m*M* $CaCl_2$ (prepared as described in **Subheading 3.2.2.**). At this point, the EOF is weakly reversed (-1.2×10^{-4} cm^2/Vs, as determined using the procedure in **Subheading 3.3.2.**).
4. 1 min (20 psi) rinse with 20 m*M* Tris-HCl, pH 7.4, buffer to flush excess DLPC and calcium from the capillary before protein samples were introduced. The EOF is strongly suppressed ($+0.2–0.5 \times 10^{-4}$ cm^2/Vs, as measured using the procedure in **Subheading 3.3.2.**).

3.2.2. Preparation of DLPC Coating Solution

1. Addition of 20 m*M* $CaCl_2$ to the buffer, followed by the addition of the DLPC. Solution preparation should take place in Nalgeneware (*see* **Note 11**).
2. The solution was sonicated for 10-min periods. Between each 10-min sonication period there was a 10-min "rest" interval where the solution was stirred at room temperature to cool (*see* **Note 12**).
3. Repeat the sonicate/stir cycle until the solution is clear. Generally, three sonicate/stir cycles are needed.

Table 1
Typical Electroosmotic Flow Mobilities Observed With Semipermanent Surfactant Coatings

Buffer	μ_{EOF} (× 10^{-4} cm^2/Vs)	Reference
DDAB Coatings		
10 m*M* Phosphate pH 7.2	–4.7	***(15)***
25 m*M* Phosphate pH 4.0	–6.3	***(18)***
25 m*M* Formate pH 4.0	–13.0	***(18)***
25 m*M* Acetate pH 4.0	–9.5	***(18)***
DLPC Coatings		
20 m*M* Tris-HCl pH 7.4 with 20 m*M* Ca^{2+}	–1.2	***(19)***
20 m*M* Tris-HCl pH 7.4	+0.2–0.5	***(19)***

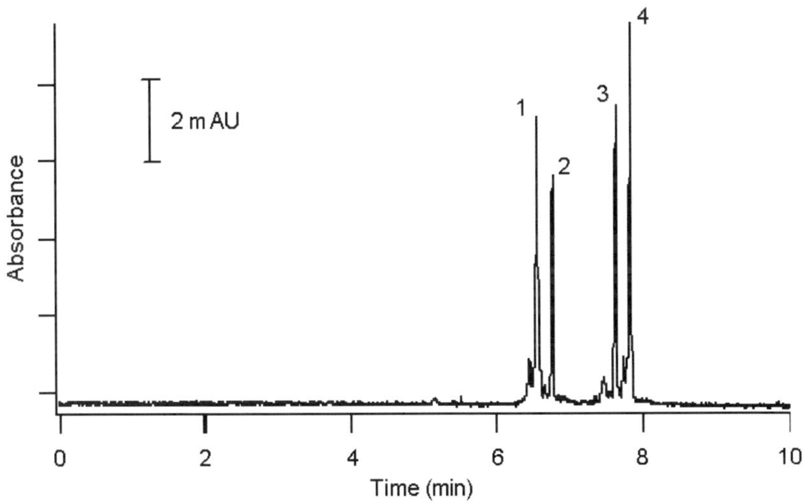

Fig. 2. Separation of four basic proteins: (1) α-chymotrypsinogen A, (2) ribonuclease A, (3) cytochrome C, (4) lysozyme. CE conditions: 47-cm capillary (40-cm to detector), UV detection at 214 nm, –10 kV applied; 25 m*M* phosphate buffer at pH 4.0. Coating procedure: 5 min rinse (20 psi) with 0.1 m*M* DDAB in buffer followed by a 1 min rinse (20 psi) with the separation buffer to flush out excess surfactant; between runs DDAB rinse time is shortened to 2 min. Reprinted in part with permission from **ref. *15***. Copyright 2000 American Chemical Society.

4. Solutions containing DLPC were stored in Nalgene bottles and used within 6 d. Migration time reproducibility and efficiency of protein separations were compromised after this period of time.

3.2.3. Separations With DLPC-Coated Capillaries

1. Protein samples were injected hydrodynamically for 3 s at 0.5 psi.
2. Voltage is applied with a field strength of 425 V/cm (+425 V/cm for cationic protein mixtures, –425 V/cm for anionic protein mixtures).
3. Direct UV detection was at 214 nm. Detector rise time was set to 0.1 s and the data acquisition rate was set to 10 Hz. An example electropherogram for cationic proteins is shown in **Fig. 3A** and anionic proteins is shown in **Fig. 3B**.
4. Between runs, the capillary was rinsed with 0.1 m*M* DLPC in 20 m*M* Tris-HCl, pH 7.4, buffer containing 20 m*M* $CaCl_2$ (5 min, 20 psi) followed by 20 m*M* Tris-HCl, pH 7.4, (1 min, 20 psi).

3.3. Monitoring EOF

The EOF is a good indicator of the characteristics of the coatings described earlier. Therefore, explicit procedures are given for monitoring of high EOF (appropriate for uncoated capillaries and DDAB coated capillaries) and suppressed EOF (appropriate for DPLC-based coatings).

3.3.1. High EOF (>1.5 × 10^{-4} cm^2/Vs) Monitoring

1. Equilibrate a capillary with buffer (5 min rinse at 20 psi).
2. Hydrodynamically inject 1.0 m*M* mesityl oxide (0.5 psi for 3.0 s).
3. A constant voltage (~300 V/cm) is applied, and the signal monitored at 254 nm.
4. The EOF is then calculated using the expression:

$$\mu_{EOF} = \frac{L_d L_t}{t_{EOF} V}$$

where L_d is the capillary length from the inlet to the detector, L_t is the total length of the capillary, t_{EOF} is the migration time of the mesityl oxide peak, and V is the applied voltage. By convention, V is positive when the cathode (negative electrode) is positioned at the detection end of the capillary, and negative when the anode (positive electrode) is at the detector.

3.3.2. Low EOF (<1.5 × 10^{-4} cm^2/Vs) Monitoring (**20**; see **Note 13**)

1. Equilibrate a capillary with buffer (5 min rinse at 20 psi).
2. A first 1.0-m*M* mesityl oxide plug (A in **step 1** of **Fig. 4**) was injected using low pressure (0.5 psi) for 3.0 s.
3. The sample vial is replaced by a buffer vial and low pressure (0.5 psi) is applied for 2 min to push the first mesityl oxide plug into the capillary.

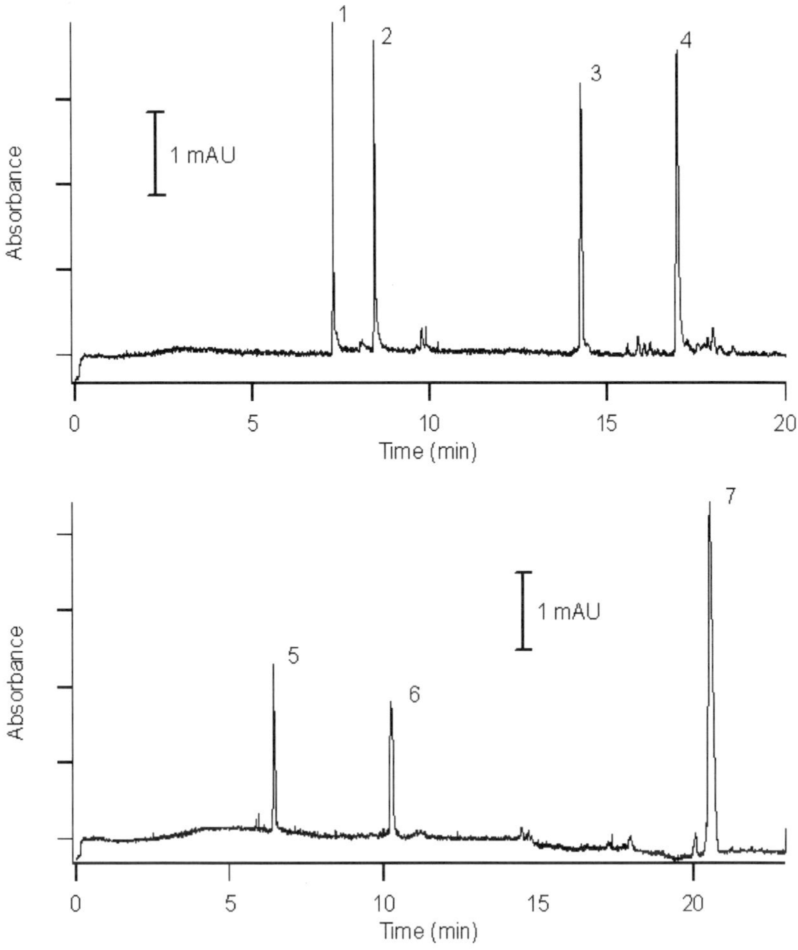

Fig. 3. (**Top**) Separation of cationic proteins at pH 7.4. Peaks, (1) lysozyme, (2) cytochrome c, (3) ribonuclease A, (4) α-chymotrypsinogen A. (**Bottom**) Separation of anionic proteins at pH 7.4. Peaks, (5) insulin chain A, (6) trypsin inhibitor, (7) α-lactalbumin. Experimental conditions: 50-cm capillary (40 cm to detector); temperature, 25°C; separation buffer, 20 m*M* Tris-HCl at pH 7.4; sample, 0.1 mg/mL protein mixture in water; applied voltage, (**Top**) +20 kV, (**Bottom**) –20 kV; direct UV detection at 214 nm. Coating procedure: 20 min rinse with 0.1 m*M* DLPC in 20 m*M* Tris-HCl, pH 7.4, buffer containing 20 m*M* $CaCl_2$ followed by a 1-min rinse with separation buffer to remove excess DLPC; between runs, DLPC rinse time was shortened to 5 min. Reprinted in part with permission from **ref. *19***. Copyright 2002 American Chemical Society.

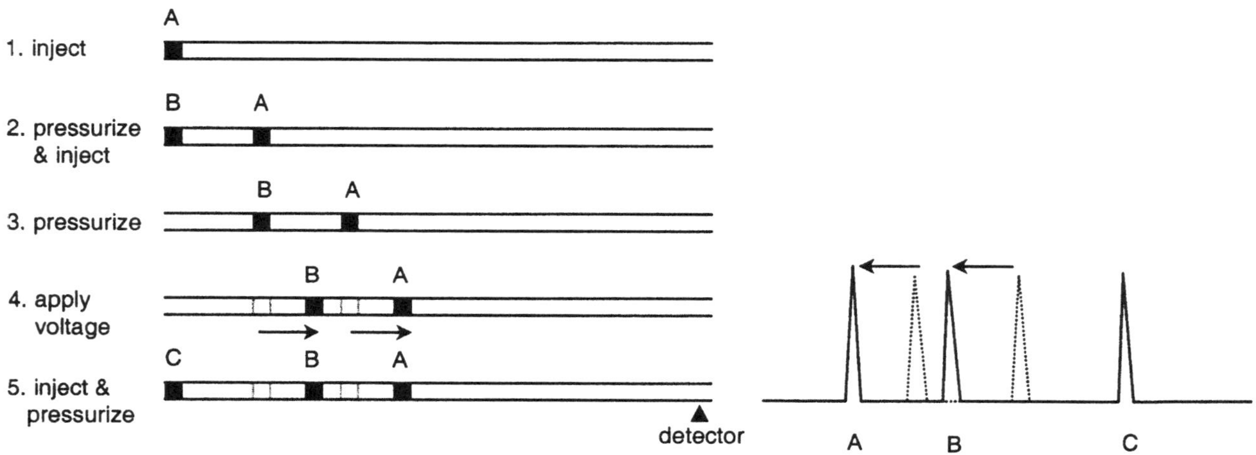

Fig. 4. EOF measurement by sequential injection. Shadowed bands and peaks represent the imaginary position if zero EOF is present (*see* **Subheading 3.3.2.** for details).

4. The vial containing the 1.0-m*M* mesityl oxide is placed at the capillary inlet and a second mesityl oxide plug is introduced by an identical low-pressure injection (0.5 psi for 3.0 s), as depicted in **step 2** of **Fig. 4**.
5. The sample vial is replaced by the buffer vial and 0.5 psi pressure is applied for 2 min to push both mesityl oxide plugs into the capillary. This point is depicted in **step 3** of **Fig. 4**. The spacing between A and B provides a measure of how far the pressure push has moved the bands into the capillary.
6. A constant voltage (200 V/cm) is applied for 3 min (t_{volt} in seconds) causing the two plugs to move within the capillary, as shown in **step 4** of **Fig. 4**. The shadowed bands in **step 4** show the position of bands A and B if no EOF were present.
7. The vial containing the 1.0-m*M* mesityl oxide is placed at the capillary inlet, and a third marker was then injected (0.5 psi for 3 s), as in **step 5** of **Fig. 4**.
8. The buffer vial is repositioned at the capillary inlet and all three bands are pushed to the detector using 0.5 psi pressure. Detection was at 254 nm. The resultant response is shown in **Fig. 4**, in which bands A and B have been shifted from their hypothetical position if no EOF were present (shadowed peaks).
9. The EOF mobility is then calculated using *(20)*:

$$\mu = \frac{\left[t_C - \left(t_B - t_A\right) - t_B\right] \nu}{t_{volt}} \div \frac{V}{L_t}$$

where t_A, t_B, and t_C are the migration times for peaks A, B, and C (in seconds) and t_{volt} is the length of time that the voltage is applied (in seconds). V is the applied voltage and L_t is the total length of the capillary. The low pressure mobilization velocity ϖ, is expressed by *(20)*:

$$\nu = \frac{L_d}{t_{p3} + t_{inj}/2 - t_{delay}}$$

where t_{inj} is the injection time and t_{delay} is the time delay of pressure application from the beginning of data collection. In our experience, t_{inj} and t_{delay} (< 1 s) were found to be insignificant compared to t_C (>10 min, with uncertainty of 3–6 s) and so were approximated as zero.

4. Notes

1. Solutions of strong base attack glass and, therefore, should be stored in polyethylene containers such as Nalgene vials and bottles (*see* also **Note 4**).
2. Preparation of phosphate buffers from orthophosphoric acid and NaOH is not critical in this work. More typical recipes using potassium salts could be used. However, our labs use the H_3PO_4/NaOH preparation to avoid the presence of potassium in phosphate buffers as that causes precipitation of SDS in other applications. H_3PO_4/NH_4OH can be also used as it is more compatible with matrix-assisted laser desorption/ionization mass spectrometry (MALDI–MS), compared

to Na^+ or K^+. Nevertheless, the peak efficiency may be slightly different when different buffer ions or concentrations are used.

3. When DDAB is prepared in low pH buffers, it dissolves quite readily. However, for DDAB to dissolve in neutral pH buffers, stirring for up to 30 min is often required.
4. Nalgeneware, rather than glassware, should be used for the preparation and storage of surfactant solutions. Lower efficiency was observed when glass vials were used. It is speculated that surfactant is lost via adsorption onto the glass surfaces (*see* also **Note 1**).
5. Mesityl oxide is volatile. This is not of significant concern in the procedures described herein, where mesityl oxide is solely used as an EOF marker. However, care must be taken if mesityl oxide is to be used in quantitative measurements. Solutions should be stored in sealed containers in order to maintain the concentration. When used in open sample vials, solutions should be replaced daily.
6. When performing electrokinetic fraction collection in CE, a small amount (e.g., 5 μL) of conducting solution must be placed in the collection vial to maintain electrical contact between the electrode and the capillary outlet. The electrophoretic buffer is normally used, however, the metal cations (Na^+ or K^+) from the buffers can suppress the signal in MALDI–MS. The use of a dilute HCl solution can be used alternatively to minimize the presence of Na^+ and K^+, and, in turn, enhance the MALDI–MS signal ***(18)***. In addition to H^+, other nonmetallic cations such as NH_4^+ and $Tris^+$ are also found to be compatible with MALDI–MS.
7. The stability of the coating is inversely related to the capillary diameter (i.e., the coating is less stable in a 75-μm capillary than the 50-μm capillary described herein). The greater volume to surface-area of larger diameter capillaries results in greater reequilibration of the surfactant in the bilayer into the bulk solution.
8. If a capillary is stored overnight and used the next day, the capillary should be rinsed with 50 m*M* sodium dodecylsulfate (SDS) in water (5 min, 20 psi) to remove the old coating and any adsorbed species. The capillary can then be recoated by following the steps for capillary pretreatment.
9. Excessive rinsing of the capillary will gradually wash the bilayer off the capillary and thus reduce the integrity of the coating.
10. The peak efficiency observed from a brand-new capillary was typically lower in the first few runs. After this break-in period, the peak efficiency improves and stabilizes to the maximum value. The voltage application appeared to shorten this initial break-in period.
11. The presence of calcium in the DLPC solution is mandatory. Calcium is a strong fusogenic agent that helps formation of the bilayer at the capillary wall. Without calcium, bilayer formation is very slow (>1 h) and surface coverage is incomplete (protein adsorption occurs as indicated by low peak efficiency).
12. Sonication was found to promote dissolution of DLPC in buffer. Without sonication, the DLPC solution never clears and poor coating formation at the capillary wall is observed.

13. This procedure requires a constant instrument pressure. This is available on Beckman instruments, but the applied pressure from the Agilent instrument varies by about 5%. A slight modification of the Agilent instrument enhances the consistency and precision of low EOF determinations *(21)*. The pressure line connected to the inlet vial is divided into two separate lines using a T-junction. One tube is connected to the inlet vial as usual. The other line goes outside the instrument to an on/off valve which is further connected to a variable leak. If the on/off valve is closed, the instrument pressure control works as normal. When the valve is open, there is a small leak of air through the valve. The instrument responds to the pressure drop caused by this leak with more frequent pressure adjustments, which results in better than 1% constancy in the applied pressure.

Acknowledgments

This work was supported by the Natural Sciences and Engineering Research Council of Canada and by the Universities of Alberta and Western Ontario. NEB gratefully acknowledges Postdoctoral Fellowship support from NSERC and Eli Lilly.

References

1. Mazzeo, J. R. and Krull, I. S. (1994) Modification of capillaries and buffers for enhanced separations in capillary zone electrophoresis and capillary isoelectric focusing of biopolymers, in *Handbook of Capillary Electrophoresis* (Landers, J. P., ed.), CRC, Boca Raton, FL, pp. 495–512.
2. Rodriguez, I. and Li, S. F. Y. (1999) Surface deactivation in protein and peptide analysis by capillary electrophoresis. *Anal. Chim. Acta* **383,** 1–26.
3. Horvath, J. and Dolnik, V. (2001) Polymer wall coatings for capillary electrophoresis. *Electrophoresis* **22,** 644–655.
4. Towns, J. K. and Regnier, F. E. (1990) Polyethyleneimine-bonded phases in the separation of proteins by capillary electrophoresis. *J. Chromatogr.* **516,** 69–78.
5. Figeys, D. and Aebersold, R. (1997) Capillary electrophoresis of peptides and proteins at neutral pH in capillaries covalently coated with polyethyleneimine. *J. Chromatogr. B* **695,** 163–168.
6. Huang, X. Y., Doneski, L. J., and Wirth, M. J. (1998) Surface-confined living radical polymerization for coatings in capillary electrophoresis. *Anal. Chem.* **70,** 4023–4029.
7. Hjerten, S. (1985). High-performance electrophoresis—elimination of electroendoosmosis and solute adsorption. *J. Chromatogr.* **347,** 191–198.
8. Righetti, P. G., Gelfi, C., Verzola, B., and Castelletti, L. (2001) The state of the art of dynamic coatings. *Electrophoresis* **22,** 603–611.
9. Yeung, K. K.-C. and Lucy, C. A. (1997) Suppression of electroosmotic flow and prevention of wall adsorption in capillary zone electrophoresis using zwitterionic surfactants. *Anal. Chem.* **69,** 3435–3441.

10. Baryla, N. E. and Lucy, C. A. (2000) Simultaneous separation of cationic and anionic proteins using zwitterionic surfactants in capillary electrophoresis. *Anal. Chem.* **72,** 2280–2284.
11. Melanson, J. E., Baryla, N. E., and Lucy, C. A. (2001) Dynamic capillary coatings for electroosmotic flow control in capillary electrophoresis. *Trends Anal. Chem.* **20,** 365–374.
12. Yao, Y. J., Khoo, K. S., Chung, M. C. M., and Li, S. F. Y. (1994) Determination of isoelectric points of acidic and basic-proteins by capillary electrophoresis. *J. Chromatogr. A* **680,** 431–435.
13. Erim, F. B., Cifuentes, A., Poppe, H., and Kraak, J. C. (1995) Performance of a physically adsorbed high-molecular-mass polyethyleneimine layer as a coating for the separation of basic-proteins and peptides by capillary electrophoresis. *J. Chromatogr. A* **708,** 356–361.
14. Cordova, E., Gao, J., and Whitesides, G. M. (1997) Noncovalent polycationic coatings for capillaries in capillary electrophoresis of proteins. *Anal. Chem.* **69,** 1370–1379.
15. Melanson, J. E., Baryla, N. E., and Lucy, C. A. (2000) Double-chained surfactants for semipermanent wall coatings in capillary electrophoresis. *Anal. Chem.* **72,** 4110–4114.
16. Baryla, N. E., Melanson, J. E., McDermott, M. T., and Lucy, C. A. (2001) Atomic force microscopy imaging of cationic surfactant wall coatings in capillary electrophoresis. *Anal. Chem.* **73,** 4558–4565.
17. Yassine, M. M. and Lucy, C. A. Factors affecting the stability of semi-permanent surfactant coatings in capillary electrophoresis. *Anal. Chem.*, in press.
18. Yeung, K. K.-C., Kiceniuk, A. G., and Li, L. (2001) Capillary electrophoresis using a surfactant-treated capillary coupled with offline matrix-assisted laser desorption ionization mass spectrometry for high efficiency and sensitivity detection of proteins. *J. Chromatogr. A* **931,** 153–162.
19. Cunliffe, J. M., Baryla, N. E., and Lucy, C. A. (2002) Phospholipid bilayer coatings for the separation of proteins in capillary electrophoresis. *Anal. Chem.* **74,** 776–783.
20. Williams, B. A. and Vigh, G. (1996) Fast, accurate mobility determination method for capillary electrophoresis. *Anal. Chem.* **68,** 1174–1180.
21. Porras, S. P. (2002) Non-aqueous solvents in capillary zone electrophoresis: The effect of methanol and acetonitrile on physicochemical properties of analytes, Ph.D. thesis, Dept. Chem., Univ. Helsinki, Helsinki, Finland.

2

Capillary Coating for Protein Separation Based on Si-O and Si-C Covalent Bond Formation for Capillary Electrophoresis With Laser-Induced Fluorescence Detection

Hossein Ahmadzadeh, Norman J. Dovichi, and Sergey Krylov

Summary

Protein adsorption to capillary walls is one of the major complications in protein analyses with capillary electrophoresis (CE). Coating the capillary with different materials is used to reduce the adsorption. This chapter overviews different approaches used for capillary coating and concentrates on those utilizing Si-O and Si-C covalent bonds. The apparatus and methods are presented for capillary coating using Si-O and Si-C chemistry. Furthermore, procedures are described for monitoring the quality of coating.

Key Words

Capillary electrophoresis; coating; covalent bond; laser-induced fluorescence; protein.

1. Introduction

The theory of capillary electrophoresis (CE) predicts that the efficiencies of separation for large biological polymers, such as nucleic acids and proteins, should be on the order of 10^6 theoretical plates because of their low diffusion coefficients. In practice, the results are close to the theoretical values only for nucleic acids, whereas for the proteins, the efficiencies are much below those estimated. The major reason for low quality of separation of proteins is their adsorption to the capillary wall, which results in peak broadening. This problem has hindered method development, and, therefore, the full potential of CE in protein separation has yet to be realized ***(1–3)***. One of the most efficient

From: *Methods in Molecular Biology, vol. 276: Capillary Electrophoresis of Proteins and Peptides*
Edited by: M. A. Strege and A. L. Lagu © Humana Press Inc., Totowa, NJ

ways to reduce the protein–wall interaction is to coat the inner wall of the capillary with a layer of an appropriate polymer. Such a polymer should first eliminate the negative charge (inherent to the silica surface) on the capillary wall, and second, it should create a hydrophilic layer (that reduces hydrophobic adsorption of proteins) on the surface.

Capillaries are typically coated with a polymer in two steps. Initially, a bifunctional reagent is covalently linked through the first functional group to silica on the surface of the capillary wall to form a sublayer with another functional group exposed and available for attaching the second layer. Then, a monomer is bound to the exposed functional group of the bifunctional reagent and polymerized as a top layer that is covalently linked to the sublayer. **Fig. 1** schematically depicts the general strategy of two-layer capillary coating.

Hjerten was the first to apply silane chemistry to coating capillaries for CE *(4)*. He used a Si-O bond to create a sublayer and polymerized acrylamide (AA) as a top layer. The disadvantage of such a coating is its susceptibility to alkaline hydrolysis. Both the Si-O bond and polyacrylamide are hydrolyzed at pH >8.0. Such hydrolysis results in the destruction of the coating or the formation of a layer of polyacrylate that is capable of strongly adsorbing proteins and regenerating the electroosmotic flow (EOF). Novotny used the Grignard method to change the Si-O bond to a Si-C bond, which is much more resistant to alkaline hydrolysis *(5)*. This method results in a physically more stable coating than the silane-coupling chemistry. However, because it still uses polyacrylamide, which is hydrolyzed at pH >8.0, this coating is inapplicable to protein separations at basic pHs.

Righetti et al. introduced and optimized the performance of several novel polymers of AA substitutes. The monomers studied were dimethylacrylamide (DMA), N-acryloylaminoethoxyethanol (AAEE), and acryloylaminopropanol (AAP; ***6–19***). AAEE and AAP showed much less protein adsorption on the capillary wall than AA, whereas DMA was more adsorptive for proteins than AA. These properties are ascribed to the much higher hydrophilicity of AAP and AAEE compared with DMA. However, despite the use of the hydrophilic polymers, Righetti's method still involves a sublayer based on a Si-O bond that is vulnerable to alkaline hydrolysis. The chemical formulas of the monomers introduced by Righetti are shown in **Fig. 2**. The reaction steps in making Si-O and a Si-C bonds followed by polymerizing a monomer as a top layer are depicted in **Figs. 3** and **4**.

In **ref. *20*** Ahmadzadeh and Dovichi reported coating of capillaries with AA, DMA, AAEE, and AAP based on either Si-O or Si-C sublayers. The procedures involved either the coupling of a silane bifunctional reagent (3-methacryloxypropyltrimethoxysilane) with the silanol groups (Si-O bond) or silanol chlorination followed by the Grignard-coupling of vinyl magnesium bromide

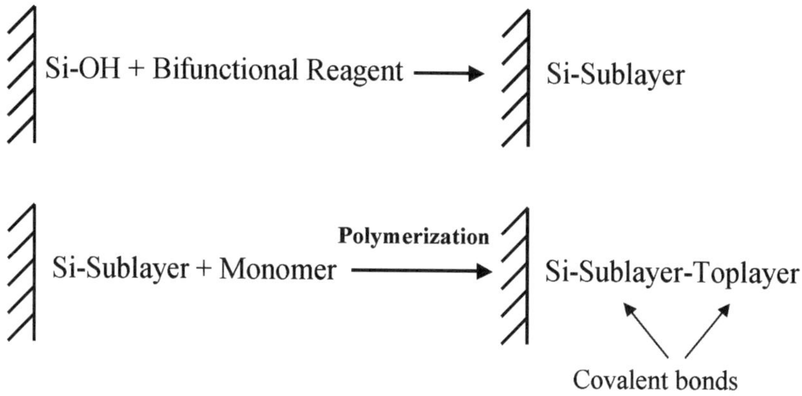

Fig. 1. Schematic diagram showing the strategy for permanent coating.

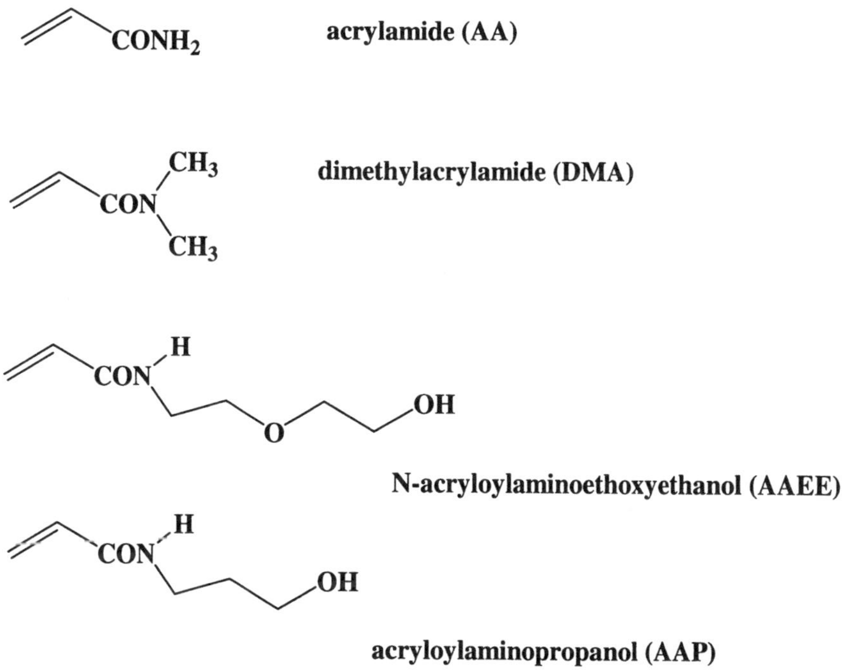

Fig. 2. Chemical formulas of the monomers introduced in **ref. *20***.

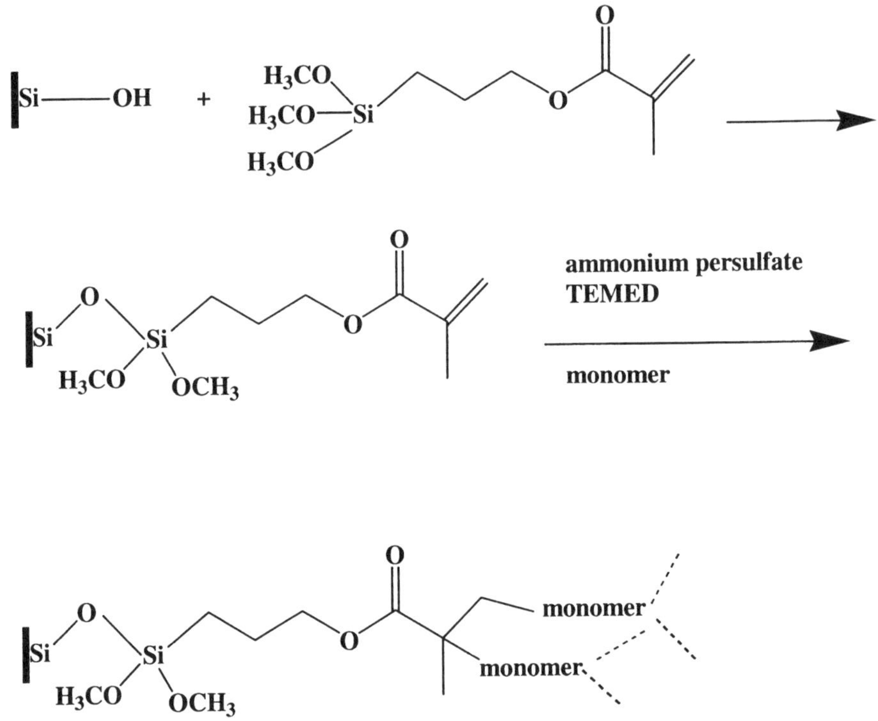

Fig. 3. Chemical equations and reaction steps for coating the capillaries based on Si-O bond formation and polymerization of a monomer of choice as a top layer.

(Si-C bond). The Grignard-based sublayer has been employed further to compare the performance of all four monomers. The procedures proved to be very practical for reducing protein adsorption on capillary walls in CE.

As aforementioned, the coating of capillaries can be based either on silanization or on the Grignard reaction (*see* **Figs. 3** and **4**). In the former method, 0.1 *M* NaOH is flushed through the capillary (4–5 m long, 50 μm id) for 1 h followed by flushing with water for another hour. Finally, a 4% solution of 3-methacryloxypropyltrimethoxysilane in a 1:1 mixture of glacial acetic acid and water should be prepared, and the capillary has to be flushed with this solution for 20 min. The silanization reaction goes to the completion within 1 h. Next, the capillary should be flushed with water for 10 min. Ammonium persulfate (4 μL of freshly prepared, 10%) and tetramethylethylenediamine (TEMED) (1 μL) should be added to 1 mL of a 3% solution of a monomer. This undegased solution should immediately be flushed through the capillary. After 1 h, the polymerization reaction is complete and the gel, that is not

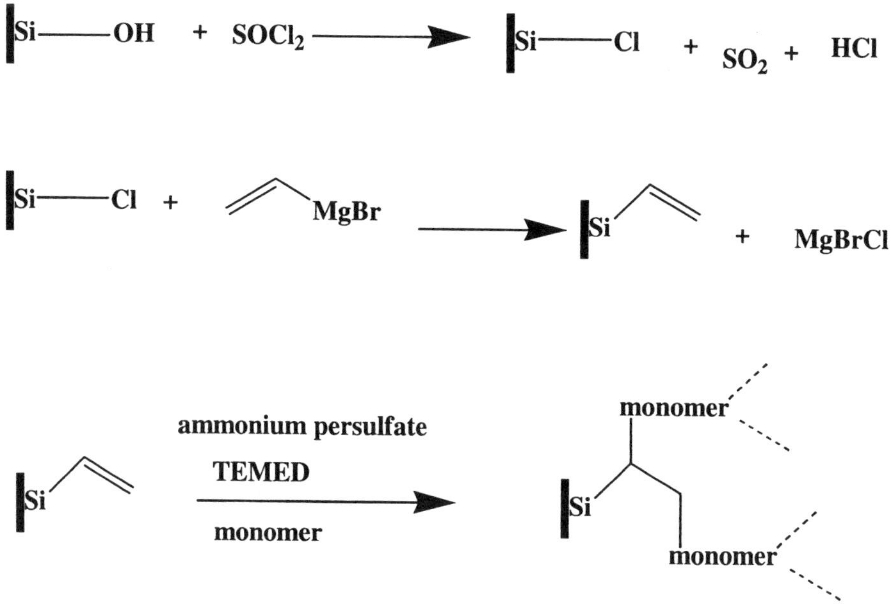

Fig. 4. Chemical equations and reaction steps for coating the capillaries based on Si-C bond formation and polymerization of a monomer of choice as a top layer.

immobilized on the walls, is flushed out and the capillary is filled with water. The capillary has to be stored in water, and before doing a CE run the water inside the capillary ought to be replaced manually with running buffer using a syringe with a proper fitting.

The instrumentation setup for capillary coating is shown in **Fig. 5**. The outlet of a low dead-volume fitting is connected to a 3-cm-long piece of 1/16-in outer-diameter (od) Teflon tubing. The capillary is threaded through the Teflon tube, the connector, and thc tee. The capillary tip should be located near the middle of the Reacti-vial inside a disposable vial. For some solvents (like THF) and some reagents (like vinyl magnesium bromide), this disposable vial has to be made of glass. When the capillary is inside the disposable vial, which is, in turn, in the Reacti-vial, the top nut has to be tightened while the capillary is in place. It is recommended to flush the solutions through the capillary using nitrogen gas at 20 psi dehydrated with a moisture trap. The coating procedure has four steps that are described later in this chapter.

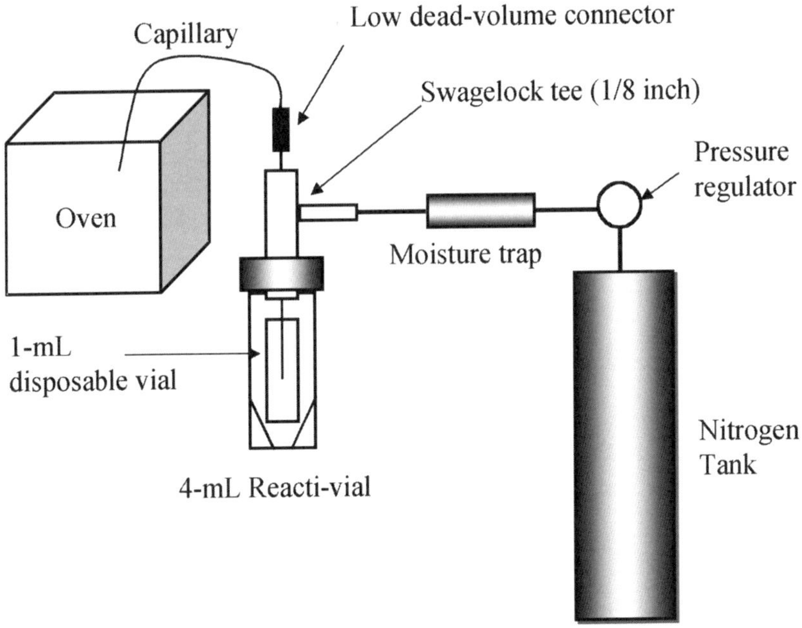

Fig. 5. Apparatus for delivering the reagents into the capillaries for the coating processes. Nitrogen gas pressure is used to force different reagents through the capillary. The reagents are held in a disposable vial placed inside the Reacti-vial. The fittings are used for the air-tight connection of the capillary to the reagent vials.

2. Materials

2.1. Capillary Coating Based on Si-O Bond Formation

1. Doubly distilled deionized water (*see* **Note 1**).
2. Concentrated acetic acid: add 0.5 mL of concentrated acetic acid to 0.5 mL of water (solution A).
3. 0.1 *M* NaOH.
4. Silane reagent, 3-methacryloxypropyltrimethoxysilane: add 40 μL of the silane reagent to 1 mL of solution A (solution B).
5. A 3% solution of a monomer in water.
6. TEMED.
7. Ammonium persulfate (APS) 10% in water, freshly prepared.

2.2. Capillary Coating Based on Si-C Bond Formation

1. Doubly distilled deionized water (*see* **Note 1**).
2. Type 4-A molecular sieves (*see* **Note 2**).
3. Thionyl chloride (*see* **Note 3**).
4. 1 *M* Vinyl magnesium bromide (*see* **Note 4**).

5. Anhydrous THF (*see* **Note 4**).
6. A 3% solution of a monomer in water, freshly prepared.
7. TEMED.
8. APS 10% in water.

2.3. Fluorescent Labeling of Proteins and Peptides

1. 2.5 m*M* Sodium tetraborate (10 m*M* borate), pH 9.4 (*see* **Note 1**).
2. 1 n*M* Protein solution (*see* **Note 5**).
3. Dry 5-furoyl quinoline-3-carboxaldehyde (FQ) 100 nmoles (*see* **Note 6**).
4. 25 m*M* KCN solution in either water or 10 mM borate (*see* **Note 7**).

2.4. In-House CE With Laser-Induced Fluorescence Detector

1. An in-house constructed CE instrument (*see* **Note 8**).
2. A laser-induced fluorescence (LIF) detector based on a sheath-flow cuvet (*see* **Note 8**).
3. Fused-silica capillaries, 50 μm id, 150 μm od (Polymicro Technologies, Phoenix, AZ).
4. A 0–30 kV dc power supply (CZE 1000, Spellman, Plainview, NY).
5. A 488-nm argon-ion laser to be operated at 12 mW (Model 2211-55 SL, Uniphase, San Jose, CA or Model Innova 90-4, Coherent, Mountain View, CA).
6. A ×6.3 microscope objective (Melles Griot, Nepean, ON, Canada) and a ×60, 0.7 NA microscope objective (Mo-0060LWD, Universe Kokagu, Oyster Bay, NY).
7. An interference filter centered at 615 nm with a 45-nm transmittance window (Omega Optical, Brattleboro, VT).
8. A photomultiplier tube (PMT) (R1477, Hamamatsu, Middlesex, NJ).
9. A 16-bit data acquisition board (NB-MI0 16 XH-18, National Instruments, Austin, TX).
10. A personal computer.

2.5. Capillary Coating Apparatus

Fig. 5 shows a simple apparatus that could be constructed to deliver reagents into the capillary. Reagents should be prepared in 1-mL disposable vials. Each reagent vial has to be placed inside a 4-mL Reacti-vial. A hole should be drilled through the lid of the Reacti-vial to accommodate a one-eighth inch stainless steel Swagelock tee. This tee is held in place with the aid of copious amounts of Teflon tape. One side of the tee is connected with Teflon tubing to a nitrogen cylinder. A guard column filled with molecular sieves (type 4-A), as a moisture trap, should be connected to the nitrogen tank to remove the water impurity from nitrogen gas before it is being used to purge the capillaries. The top side of the Swagelock tee has to be connected to a one-sixteenth inch adaptor. This adaptor is connected to a low dead-volume fitting for the capillary. For the temperature-controlled steps, the capillary is placed inside an oven.

3. Methods

3.1. Capillary Coating Procedures

3.1.1. Method 1: Coating Based on Si-O Bond Formation

3.1.1.1. Capillary Pretreatment

1. Cut a 4–5-m long piece of fused-silica capillary, 50 μm id.
2. Condition the capillary by flushing with 0.1 *M* NaOH for 3 h using 20 psi nitrogen pressure.
3. Flush the capillary with water for 1 h using 20 psi nitrogen pressure.

3.1.1.2. Silane Reaction for Si-O Bond Formation

1. Flush the capillary with silane solution B (*see* **Subheading 2.1.4.**) for 1 h using 20 psi nitrogen pressure.
2. Flush the capillary with water for 10 min using 20 psi nitrogen pressure.

3.1.1.3. Polymerization Step

1. Cut the capillary into 1-m long sections.
2. Coat each section with a monomer (*see* **Note 9**). For this, prepare a 3% solution of a monomer by diluting the stock solution in water. Add 1 μL of TEMED and 4 μL of 10% ammonium persulfate to 1 mL of the monomer, and then immediately flush the polymer through the capillary at 60 psi for 5 min.
3. After 1 h of polymerization, flush the gel that is not immobilized on the capillary wall out of each capillary and fill the capillaries with water.
4. Store these coated capillaries in water, and flush the running buffer through them prior to use (*see* **Note 10**).

3.1.2. Method 2: Coating Procedure Based on Si-C Bond Formation

3.1.2.1. Capillary Pretreatment

1. Cut a 4–5-m long piece of fused-silica capillary, 50 μm id.
2. Condition the capillary by flushing 0.1 *M* NaOH for 3 h at a 20 psi nitrogen pressure.
3. Flush the capillary with water for 1 h using 20 psi nitrogen pressure.
4. Flush the capillary with methanol for another hour using 20 psi nitrogen pressure.
5. Dry the capillary in an oven at 140°C for 8 h or overnight by flushing nitrogen gas through it at a pressure of 5 psi.

3.1.2.2. Chlorination of Silanol Groups (*see* **Note 11**)

1. The following day, while the capillary is inside the oven and nitrogen is flowing through it, reduce the oven temperature to 65°C and then flush thionyl chloride under 20 psi nitrogen pressure for 30 min. Check the outflow to identify any plugged capillaries, and test for acidity using pH paper (*see* **Note 3**).

2. After 30 min, seal one end of the capillary with a GC septum while the outlet end remains connected to the vacuum line for an additional 15 min to remove the excess of thionyl chloride. This step should be done while the capillary is inside the oven. Then, seal the outlet end of the capillary using a GC septum, and heat the capillary in the oven for 8 h or overnight (*see* **Note 12**).

3.1.2.3. Grignard Reaction (Si-C Bond Formation)

1. Prepare a fresh solution of 0.25 *M* vinyl magnesium bromide in dry THF (solution C) under nitrogen atmosphere (*see* **Note 13**).
2. Submerge a freshly cut inlet end of the capillary into solution C and uncap the outlet end of the capillary, cut a few centimeters from the outlet end and place it in a tube containing methanol. Apply nitrogen pressure (20 psi) to the inlet end of the capillary to rinse it with solution 1 for 30 min (*see* **Note 14**).
3. Cap the capillary at both ends using a GC septum, and heat it for 6–8 h or overnight at 70°C.
4. Uncap the capillary and cut 5–10 cm off both ends.
5. Rinse the capillary with anhydrous THF for 30 min followed by water for another 30 min (*see* **Notes 15** and **16**).

3.1.2.4. Polymerization Step

1. Cut the capillary into 1-m long sections.
2. Coat each section with a monomer (*see* **Note 9**) as top-layer using the following procedures. Prepare a 3% solution of the monomer by diluting the stock solution in water. Add 1 μL of TEMED and 4 μL of 10% APS to 1 mL of the monomer, and then immediately flush the polymer through the capillary at 60 psi for 5 min.
3. After 1 h of polymerization, flush the gel out of each capillary using nitrogen and replace the polymer with water.
4. Store these coated capillaries in water, and flush a run buffer through them prior to use (*see* **Note 10**).

*3.1.3. Method 3: FQ Labeling of Proteins and Peptides (see **Note 17**)*

1. Add 7.5 μL of 10^{-7} *M* protein solution to 100 nmol of dry FQ, and then add 2.5 μL of 5 m*M* KCN in water ***(21)***.
2. Stop the reaction after 30 min by adding 740 μL of the run buffer (*see* **Note 18**).

*3.1.4. Method 4: EOF Mobility Measurement (see **Note 19**)*

To evaluate the quality of coating, measure the EOF for each capillary by the current monitoring method ***(20,22)*** using the following procedures.

1. Fill the capillary with 10 m*M* borate at pH 9.4, apply high voltage and monitor the current.
2. After the current stabilizes, change the run buffer to 8 m*M* borate. The current will decrease and reach a constant value after a certain period of time (Δt). Measure this time (*see* **Note 20**).

3. Calculate the velocity of EOF (ν_{EOF}) and the mobility of EOF (μ_{EOF}) using the following formulas:

$$\nu_{EOF} = L/\Delta t \quad (1)$$

$$\mu_{EOF} = \nu_{EOF}/E = L^2/V\Delta t \quad (2)$$

where *L* is the capillary length and *V* is the applied voltage.

4. Notes

1. Prepare all aqueous solutions with Milli-Q deionized water and filter them using a 0.2-μm filter. Then degas the buffers to prevent the formation of air bubbles during electrophoresis.
2. The molecular sieve is a moisture trap for N_2 gas that is used to deliver the reagents into the capillary. If not trapped, the traces of water will cause the precipitation of vinyl magnesium bromide and, hence, plug the capillary.
3. Do not return the excess of thionyl chloride to the original bottle. The excess should be decomposed by the addition of water to the solution under a fume hood.
4. THF is a solvent for vinyl magnesium bromide. It has to be absolutely dry for the reaction to be successful. Traces of water will cause precipitation inside the capillary and, hence, failure of the process. Keep THF in a desiccator.
5. Dissolve the proteins in water if stacking is desired and dissolve it in the run buffer otherwise.
6. Prepare a stock solution of 10 m*M* FQ in methanol; aliquot 10 μL of the solution into 500-μL microcentrifuge tubes and remove the solvent under vacuum using a Speed Vac (Savant Instruments Inc., Farmingdale, NY). The dried FQ (100 nmoles) aliquots should be stored at –20°C. These precautions are necessary because it has been observed that FQ slowly degrades while in solution even if the solution is stored at –20°C.
7. Potassium cyanide is highly poisonous. It reacts rapidly with acids to form lethal HCN gas. Stock solutions should be made in a basic buffer and an experimenter should be aware of any change to acidic pH during the experiment. The waste containing KCN should be neutralized by adding a 1% solution of NaOH followed by slowly adding bleach.
8. An in-house constructed CE-LIF instrument with a detector based on a sheath-flow cuvet is described in detail elsewhere ***(21–24)***. Here we are briefly outlining the construction of the instrument. Unless otherwise stated, fused-silica capillaries were 40 cm long, 50 μm id, and 141 μm od. The electric field was applied to an inlet end of the capillary from a 0–30 kV dc power supply. The excitation was provided by the 488-nm line of an argon-ion laser operated at 12 mW. The laser beam was focused approx 30 μm from the outlet tip of the capillary with a ×6.3 microscope objective. Fluorescence was collected by a ×60, 0.7 NA long-working-distance microscope objective, filtered with a spatial filter, and a 615 DF 45

band-pass filter to remove stray and scattered light. Fluorescence was imaged onto a photomultiplier biased at 1 kV. The photocurrent was passed through a current-to-voltage converter and a low-pass filter (RC = 47 ms) and then digitized with a 16-bit data acquisition board connected to a personal computer.

9. This step is valid for each monomer, AA, DMA, AAEE, or AAP.
10. If the gel is left inside the capillary for a period longer than 3 h in this step, a gel-filled capillary that is suitable for capillary gel electrophoresis is obtained. If the reaction time is too long (much more than 3 h) then the capillaries can be irreversibly plugged.
11. Before the chlorinating step, check the color of thionyl chloride. Replace the solution with a fresh one if it is faint.
12. In some cases this step can be repeated to obtain a more rugged coating.
13. To prepare this solution, add 0.75 mL of dry THF and 0.25 mL of 1 *M* vinyl magnesium bromide in THF into a tube filled with nitrogen gas.
14. If the capillary is plugged and no flow is observed, place the capillary back inside the oven for a short period of time. If it is still plugged, check for blockage using a microscope and cut the capillary to appropriate lengths to remove the blocked parts of the capillary and save the rest.
15. The Grignard step could be repeated to get more rugged capillaries.
16. Initially, about 10% of the capillaries could be plugged during the Grignard reaction step. We observed that the failure rate dropped to zero in winter, when the relative humidity in the laboratory decreased and as experienced was gained in handling the reagents. Novice experimenters in humid environments may expect difficulty with this sublayer coating step.
17. The labeling protocol is a modification of the procedure first introduced by Novotny in **ref. *18***.
18. This FQ-labeled protein, with a final concentration of 10^{-9} *M*, could be used to evaluate the coated capillaries. In earlier studies, we showed that a 30-min reaction time was optimal for obtaining the highest intensity of the fluorescence signal ***(20)***.
19. EOF measurements can be used to evaluate the quality of coating; the lower the EOF mobility the better the quality. For Si-O based coatings, the reduction of EOF should be in the order of 100 times as compared to uncoated capillaries. For Si-C based coatings, the EOF should be 1000 times less than that for uncoated capillaries.
20. We have proved that this current monitoring method to measure EOF is as accurate as the neutral marker method, especially for coated capillaries (*see* **ref. *20***).

References

1. Wirth, M. J., Fair Bank, R. W., and Fatumbi, H. O. (1997) Mixed self-assembled monolayers in chemical separations. *Science* **275,** 44–47.
2. Jorgenson, J. W. and Lukacs, K. D. (1981) Zone electrophoresis in open-tubular glass capillaries. *Anal.Chem.* **53,** 1298–1302.

3. Schoneich, C., Huhmer, A. F. R., Rabel, S. R., et al. (1995) Separation and analysis of peptides and proteins. *Anal. Chem.* **67,** 155R–181R.
4. Hjertten, S. (1985) High-performance electrophoresis. Elimination of electroendosmosis and solute adsorption *J. Chrom.* **347,** 191–198.
5. Cobb, K. A., Dolnik, V., and Novotny, M. (1990) Electrophoretic separations of proteins in capillaries with hydrolytically-stable surface structures. *Anal. Chem.* **62,** 2478–2483.
6. Chiari, M., Micheletti, C., Nesi, M., Fazio, M., and Righetti, P. G. (1994) Towards new formulations for polyacrylamide matrices: N-acryloylaminoethoxyethanol, a novel monomer combining high hydrophilicity with extreme hydrolytic stability. *Electrophoresis* **15,** 177–186.
7. Chiari, M., Nesi, M., and Righetti, P. G. (1994) Capillary zone electrophoresis of DNA fragments in a novel polymer network: poly(N-acryloylaminoethoxyethanol). *Electrophoresis* **15,** 616–622.
8. Gelfi, C., Perego, M., Libbra, F., and Righetti, P. G. (1996) Comparison of behavior of N-substituted acrylamides and celluloses on double-stranded DNA separations by capillary electrophoresis at 25°C and 60°C. *Electrophoresis* **17,** 1342–1347.
9. Simo-Alfonso, E., Gelfi, C., Sebastiano, R., Citterio, A., and Righetti, P. G. (1996) Novel acrylamido monomers with higher hydrophilicity and improved hydrolytic stability: II. Properties of N-acryloylaminopropanol. *Electrophoresis* **17,** 732–737.
10. Blanco, S., Clifton, M. J., Joly, J. L., and Righetti, P. G. (1996) Protein separation by electrophoresis in a nonsieving amphoteric medium. *Electrophoresis* **17,** 1126–1133.
11. Gelfi, C., Perego, M., and Righetti, P. G. (1996) Capillary electrophoresis of oligonucleotides in sieving liquid polymers in isoelectric buffers. *Electrophoresis* **17,** 1407–1475.
12. Chiari, M., Nesi, M., Ottolina, G., and Righetti, P. G. (1994) Separation of charged and neutral isotopic molecules by micellar electrokinetic chromatography in coated capillaries. *J. Chromatogr. A* **680,** 571–577.
13. Righetti, P. G. (1996) *Capillary Electrophoresis in Analytical Biotechnology.* CRC, Boca Raton, FL.
14. Miertus, S., Righetti P. G., and Chiari, M. (1994) Molecular modeling of acrylamide derivatives: the case of N-acryloylaminoethoxyethanol versus acrylamide and trisacryl. *Electrophoresis* **15,** 1104–1111.
15. Simo-Alfonso, E., Gelfi, C., Sebastiano, R., Citterio, A., and Righetti, P. G. (1996) Novel acrylamido monomers with higher hydrophilicity and improved hydrolytic stability: I. Synthetic route and product characterization. *Electrophoresis* **17,** 723–737.
16. Zhukov, M.Y., Ermakov, S.V., and Righetti, P.G. (1997) Simplified mathematical model of irreversible sample adsorption in capillary zone electrophoresis. *J. Chromatogr. A* **766,** 171–185.
17. Ermakov, S.V, Zhukov, M.Y., Capelli, L., and Righetti, P.G. (1995) Wall adsorption in capillary electrophoresis. Experimental study and computer simulation. *J. Chromatogr. A* **699,** 297–313.

18. Beale, S. C., Hsieh, Y. Z., Wiesler, D., and Novotny, M. (1990) Application of 3-(2-furoyl) quinoline-2-carbaldehyde as a fluorogenic reagent for the analysis of primary amines by liquid chromatography with laser-induced fluorescence detection. *J. Chromatogr.* **499,** 579–587.
19. Chiari, M., Nesi, M., Sandoval, J. E., and Pesek, J. J. (1995) Capillary electrophoretic separation of proteins using stable, hydrophilic poly (acryloylaminoethoxyethanol)-coated columns. *J. Chromatogr. A* **717,** 1–13.
20. Ahmadzadeh, H. and Dovichi, N. J. (2000) Instrumentation, capillary coating and labeling chemistry for capillary electrophoresis with laser-induced fluorescence detection. Ph.D. thesis, Univ. Alberta, Canada vol. 1, 137–188.
21. Pinto, D., Arriaga, E. A., Craig, D., et al. (1997) Picomolar assay of native proteins by capillary electrophoresis pre-column labeling sub-micellar separation and laser-induced fluorescence detection. *Anal. Chem.* **69,** 3015–3021.
22. Huang, X., Gordon, M. J., and Zare, R. N. (1988) Current-monitoring method for measuring the electroosmotic flow rate in capillary zone electrophoresis. *Anal. Chem.* **60,** 1837–1838.
23. Wu, S. and Dovichi, N. J. (1989) High-sensitivity fluorescence detector for fluorescein isothiocyanate derivatives of amino acids separated by capillary electrophoresis. *J. Chromatogr.* **480,** 141–155.
24. Zhang, J. (1994) Ph.D. thesis, Univ. Alberta, Canada.
25. Cheng, Y. F. and Dovichi, N. J. (1988) Sub-attomole amino acid analysis by capillary zone electrophoresis and laser-induced fluorescence. *Science* **242,** 562–564.
26. Cheng, Y. F., Wu, S., Chen, D. Y., and Dovichi, N. J. (1990) Interaction of capillary zone electrophoresis with a sheath-flow cuvette detector. *Anal. Chem.* **62,** 496–503.

3

On-Column Labeling Reaction for Analysis of Protein Contents of a Single Cell Using Capillary Electrophoresis With Laser-Induced Fluorescence Detection

Hossein Ahmadzadeh and Sergey Krylov

Summary

This chapter presents methods for capillary electrophoresis (CE) fingerprinting of proteins in a cell extract and in single cells. A custom-made CE instrument with laser-induced fluorescence (LIF) detection, used for the analyses, is described. Detailed procedures are given for: (1) on column labeling of proteins with a fluorogenic reagent, 5-furoyl quinoline-3-carboxaldehyde, (2) CE separation of labeled proteins, (3) preparation of a protein extract from cultured cells, and (4) manipulations associated with analyses of proteins in single cells. More than 20 relevant publications are cited in this chapter to assist the reader with adopting the presented methods.

Key Words

Capillary electrophoresis; labeling; laser-induced fluorescence; proteins; single cell.

1. Introduction

1.1. Laser-Induced Fluorescence (LIF) Detection

Optical detection systems based on light absorbance and fluorescence are commonly used in capillary electrophoresis (CE). Typically, on-column detection is employed in which light is focused on the capillary. Light absorbance detection is simpler and less expensive; it is used in most commercial CE instruments. The sensitivity of light absorbance detection in CE, however, is relatively low because of short optical path lengths determined by the inner diameters of conventional capillaries (typically less than 100 μm) *(1)*. Fluorescence detection is more sensitive than light absorbance detection. Fluorescently labeled amino acids have been reported to be 15 times more sensitive when

From: *Methods in Molecular Biology, vol. 276: Capillary Electrophoresis of Proteins and Peptides*
Edited by: M. A. Strege and A. L. Lagu © Humana Press Inc., Totowa, NJ

using fluorescence detection as compared to light absorbance detection *(2)*. However, with conventional lamp light sources for fluorescence excitation, the gain in sensitivity is not very significant. In contrast, a substantial improvement in sensitivity can be achieved if lasers are used as light sources *(3,4)*. Using lasers allows pumping much higher light energy into a sample and thus increasing the yield of fluorescence. It should be noted that if the laser beam intensity is too high, the analyte may decompose photochemically in a process called photobleaching, which makes further increase in light intensity senseless. Photobleaching is one of the fundamental limitations of LIF detection *(1)*.

In 1985, Zare et al. reported the first application of on-column LIF detection in CE *(5)*. The researchers used a 325-nm line of a 5-mW helium-cadmium laser to excite fluorescence of dansyl-amino acids. Fluorescence was collected with optical fibers, spectrally filtered, and quantified with a photomultiplier tube (PMT). The reliable detection of femtomole amounts of the analyte was reported. The major problem of on-column LIF detection is a relatively large amount of light scattered by the capillary walls. Scattered light contributes to the background, which, in turn, limits the sensitivity of an on-column LIF detection system *(6)*. To decrease the background, fluorescence should be measured in a flow chamber with good optical quality to eliminate or reduce light scattering. One of the solutions is in the use of square capillaries that are currently commercially available. Because flat sides of square capillaries produce much less scattered light than cylindrical walls of traditional capillaries, there is still significant light scattering on the walls of square capillaries. An alternative option is to use a postcolumn fluorescence chamber in which the detection is carried out at a large distance from the chamber walls. In this chapter, we describe LIF detection in such a chamber, called a sheath flow cuvet, and its application to the analysis of proteins and peptides.

1.2. LIF Detection in a Sheath-Flow Cuvet

Figure 1 shows a schematic diagram of a sheath-flow cuvet with a capillary inserted in the cuvet. The cuvet is a 250-μm-square chamber that holds the capillary in its center. The sheath flow enters from the top of the cuvet and exits from the bottom of the cuvet to the waste reservoir. The velocity of the sheath flow is chosen to be higher than that of the analyte's migrating in the capillary so that the analyte stream is focused hydrodynamically in the center of the sheath-flow cuvet upon its exiting the capillary. The velocity of the sheath flow should not exceed a certain level at which the regime of sheath flow switches from laminar to turbulent. In the absence of turbulence, there is no mixing of the analyte stream with the sheath fluid except for that caused by diffusion. The contribution of diffusion to the sample dilution is, however, negligible owing to a short time (less than 1 s) required for detection.

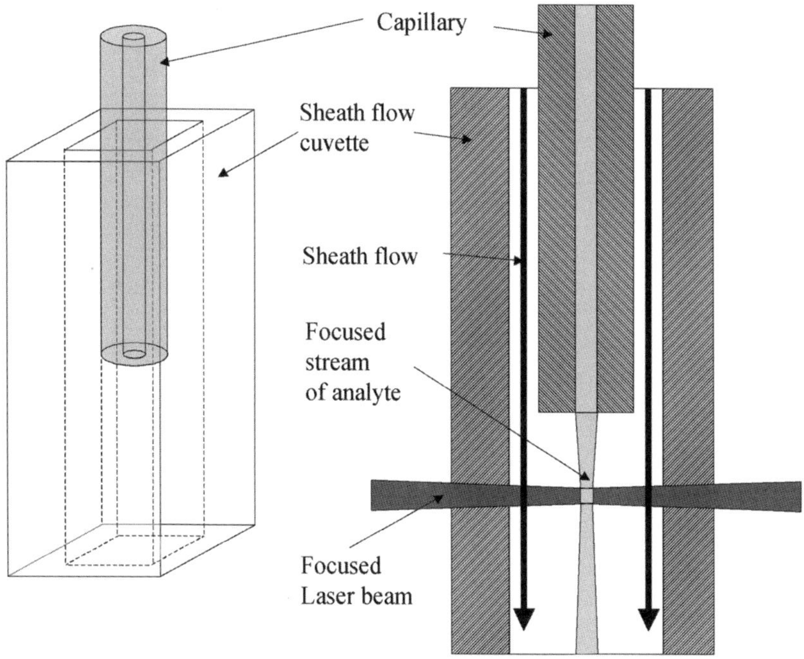

Fig. 1. Schematic diagram of a sheath-flow cuvet with a capillary being inserted in the cuvet (left) and the cross section of the cuvet and capillary showing the sheath-flow direction and hydrodynamic focusing of the analyte exiting the capillary (right).

LIF detection in a sheath-flow cuvet has several advantages over on-column LIF detection. Light scattering in a sheath-flow cuvet is significantly reduced because of the flat windows of the cuvet *(7)*. A large distance from the analyte stream to the walls of the cuvet allows easy separation of light scattered by the walls from fluorescence using a diaphragm. Moreover, if the sheath fluid is the same as the run buffer, then the refractive indexes of two media are the same and no scattering occurs on the interface between the sheath flow and the analyte stream. A little amount of scattered excitation light that gets into the detection optics is filtered with an interference optical filter. Finally, no contamination occurs inside the cuvet window because the analyte stream does not have a contact with the optical windows of the cuvet when the sheath fluid is flowing.

In the 1980s, Dovichi et al. showed that LIF detection in a sheath-flow cuvet allowed the analysis of extremely low levels of fluorescent analytes. For aqueous rhodamine-6G, the concentration detection limit was reported to be 8.9×10^{-14} *M* at two times of the background noise *(4)*. For fluorescently tagged arginine, the mass detection limit of 1100 molecules was achieved using LIF detection in a sheath-flow cuvet *(3)*. Dovichi and Zarrin also used the sheath-flow cuvet as a light scattering detector for particle size analysis *(8–10)*. The concentration detection limit was about 1000 particles per mL. Particles as small as 45 nm in radius were detected using the right-angle collection geometry. Since then, the sheath flow-based LIF detector has been used routinely in Dr. Dovichi's laboratory for high-sensitivity detection *(6,11–15)*.

High-sensitivity LIF detection systems found many applications in ultrasensitive biological analyses. For example, in DNA sequencing, the total amount of sequencing sample introduced into the capillary is about 1 attomol. The amount of each fragment averages perhaps 1 zeptomole *(16)*.

1.3. Fluorescence Labeling Chemistry

Despite being the most sensitive detection technique, LIF suffers from the limitation that most biologically important molecules are not naturally fluorescent. Those that are fluorescent often require UV excitation. For example, aromatic amino acids and proteins containing aromatic amino acids require excitation in the region of 260–300 nm. There are no inexpensive and reliable UV lasers in this spectral region. That is why the labeling chemistry for fluorescent modification on nonfluorescent molecules plays an important role in high-sensitivity LIF detection. There are a vast number of publications on the subject of changing nonfluorescent molecules to fluorescent molecules by chemical labeling reactions. The Molecular Probes catalog is one of the best resources, containing more than 5000 references on labeling and detection of a large number of analytes. The analytes of interest are amino acids, peptides, proteins, nucleotides, enzymes, and antibodies *(17)*.

When selecting a labeling reagent, it is much more appropriate to choose first the spectral properties that match wavelengths of commercially available lasers *(18)*. There are several inexpensive commercial lasers that provide a number of excitation wavelengths. For example, a KrF laser provides excitation wavelength in the UV region, at 248 nm. A He–Cd laser generates at 325 and 442 nm. A He–Ne laser emits three wavelengths: 543.5, 594, and 633 nm. Finally, an Ar-ion laser generates several lines, most intensive of which are 488 and 514.5 nm. It is recommended that the labeling reagent chosen to derivatize a molecule have an excitation wavelength that matches one of the listed lines of commonly used lasers. Diode lasers that emit in the range of 650 to 700 nm have recently been added to the list of reliable and relatively inexpensive lasers. Accordingly, there is currently a great effort in the development of new dyes that absorb strongly in the red and emit in the near infrared regions.

Labeling reagents with strong absorbance in the range of 440 to 600 nm are still most useful and common. Molecular Probes offers a complete list of this kind of fluorescent labeling reagents *(17)*. There are labeling reagents for almost all of the functional groups, but most of the labeling reagents that are commercially available are amine-labeling reagents. There are few labeling reagents for other functional groups. For example, 5-bromomethyl fluorescein (5-BrMF) has been used to label carboxylic acids ***(1)***. Here we discuss the use of the amine-labeling reagent, 5-furoyl quinoline-3-carboxaldehyde (FQ), for on-column labeling of proteins and peptides.

Protein labeling chemistry suffers from one intrinsic problem. There are typically many primary amines (lysine side chains) present in a protein molecule. This problem imposes the possibility of multiple labeling of the protein. For example, if a protein contains N primary amino groups including lysine side chains and N-terminal amino group, then a total of 2^{N-1} possible multiple labeling products can be formed ***(19–21)***. Such products result in band broadening and make CE separation of labeled proteins a rather formidable task. One way to alleviate the multiple-labeling problem is to protect all primary amino groups with an amine-protecting agent, such as phenyl isothiocyanate. The result of such a reaction is the phenyl thiocarbamyl derivative ***(22)***. Then, the protein has to be reacted with an acid. Acid cleavage produces a free N-terminal amino group that could be labeled with an amine-labeling reagent. This chemistry replaces a positively charged ε-amino group of the lysine side chain with a positively charged phenyl carbamyl group. The overall charge of the protein remains constant, and solubility is not affected significantly.

The labeling techniques can be categorized into precolumn, postcolumn, and on-column labeling methods. Precolumn labeling is the most common technique used for CE. Almost all off-line precolumn labeling approaches use a large volume of analytes and reagents. In practice, however, a small fraction of the labeled molecules is injected into the capillary and the rest of the labeled analyte is wasted. A postcolumn labeling reaction is compatible with a sheath flow designs. An on-column "T" has been constructed to introduce labeling reagent for a postcolumn detector during the separation ***(23–25)***. The major drawback of a postcolumn labeling reaction is the slow reaction rate, which limits the detection limits.

On-column labeling drastically reduces the volume of required analytes and reagents. In this method, the fluorescent labeling reagents and the proteins are injected into the capillary and allowed to react to completion under proper conditions. Then, the capillary is immersed inside the run buffer, high voltage is applied and the labeled proteins are separated. The amount of proteins and fluorescent labeling reagents injected into the capillary are typically in the nanoliter range. Here we are presenting the technique for on-column labeling of proteins with the FQ amine-labeling reagent.

On-column labeling is especially advantageous when the protein content of single cells is analyzed. The amount of proteins in a typical eukaryotic cell is in the range of 10^{-15} mol. It is very hard to process this minute amount of protein in the precolumn format because of sample dilution, evaporation, and sample loss on the walls of a reaction reservoir. The on-column format is, in contrast, ideally suited for single cells. The capillary plays the role of a reaction reservoir of a size comparable with the size of the single cell. Sample evaporation from the capillary is negligible while sample dilution and sample loss on the capillary walls can be minimized and well controlled.

2. Materials

2.1. Fluorescent Labeling and Separation of Proteins and Peptides

1. 2.5 m*M* Sodium tetraborate (10 m*M* borate), pH 9.4 (*see* **Note 1**).
2. 11 m*M* Sodium pentasulfate (SPS), in 50 m*M* phosphate buffer at pH 6.8.
3. 1 n*M* Protein solution (*see* **Note 2**).
4. Dry FQ, 100 nmoles (*see* **Note 3**).
5. 25 m*M* KCN, in either water or 10 m*M* borate (*see* **Note 4**).
6. Prepare phosphate-buffered saline (PBS) solution by mixing 8 g of NaCl, 0.2 g of KH_2PO_4, and 0.2 g of KCl and diluting the mixture to 1.00 L in distilled water, pH 6.8.

2.2. In-House CE Instrument With LIF Detector

1. An in-house constructed CE-LIF instrument (*see* **Note 5**).
2. A detector based on a sheath-flow cuvet (*see* **Note 5**).
3. Fused-silica capillaries, 50 μm id, 150 μm od (Polymicro Technologies, Phoenix, AZ).
4. A 0–30 kV dc power supply (CZE 1000, Spellman, Plainview, NY).
5. A 488-nm Ar-ion laser to be operated at 12 mW (Model 2211-55 SL, Uniphase, San Jose, CA or Model Innova 90-4, Coherent, Mountain View, CA).
6. A ×6.3 objective (Melles Griot, Nepean, ON, Canada) and a ×60, 0.7 NA microscope objective (Mo-0060LWD, Universe Kokagu, Oyster Bay, NY).
7. An interference filter centered at 615 nm with a 45 nm transmittance window (Omega Optical, Brattleboro, VT).
8. A photomultiplier tube (PMT) (R1477, Hamamatsu, Middlesex, NJ).
9. A 16-bit data acquisition board (NB-MI0 16 XH-18, National Instruments, Austin, TX).
10. A computer.
11. An inverted microscope (Model Olympus IX70, Carsen Group, Toronto, Ontario).
12. An in-house built cell injector (*see* **Note 6**).

3. Methods

3.1. Method 1: On-Column Labeling Reaction of Proteins and Peptides and CE Separation of the Labeled Products

1. Hydrodynamically inject the mixture of protein and KCN at 11 kPa pressure for 3 s.
2. Rinse the capillary tip with run buffer twice to minimize contamination.

3. Hydrodynamically inject the FQ solution at 11 kPa pressure for 1 s.
4. Immerse the capillary tip for 30 s in a vial containing run buffer preheated to 65°C.
5. Immerse the capillary tip in a vial containing run buffer at room temperature.
6. Perform the CE separation by applying an electric field of 400 V/cm (*see* **Note 7**).

3.2. Method 2: Preparation of Protein Extract From the Cells

1. Wash about a million cells with PBS buffer 5 times.
2. Resuspend the cells in 100 μL of water.
3. Sonicate the cells for 20 min.
4. Centrifuge the suspension at 600*g* for 10 min.
5. Mix 3 μL of the supernatant with 2 μL of a 10 m*M* solution of KCN.

3.3. Method 3: On-Column Labeling Reaction of the Proteins Extracted From the Cells and CE Separation of the Labeled Products

1. Hydrodynamically inject the mixture of protein extract and KCN (prepared in **Subheading 3.2.**) at 11 kPa pressure for 3 s (the injected volume is 250 pL).
2. Rinse the capillary tip with run buffer to minimize contamination.
3. Hydrodynamically inject a 10 m*M* FQ solution at 11 kPa pressure for 1 s.
4. Immerse the capillary tip in a vial containing run buffer preheated to 65°C and incubate for 3 min.
5. Perform the CE separation by applying an electric field of 400 V/cm (*see* **Note 7**).

3.4. Method 4: On-Column Labeling Reaction and CE Separation of the Protein Content of Single Cells

1. Place 50 μL of cell suspension (10^4–10^5 cells per 1 mL of PBS containing 2.5 m*M* NaCN) on a microscope glass slide and let the cells to settle down.
2. Observe the cells through an inverted microscope.
3. Place a capillary in a vertical position over a cell of choice and inject the cell into the capillary by a pulse of partial vacuum (11 kPa) applied to the distal end of the capillary.
4. Inject a plug of 10 m*M* FQ solution by a 1 s pulse of partial vacuum applied to the distal end of the capillary. The capillary tip should be placed in a vial containing the SPS running buffer, and the vial should be placed in a 65°C ultrasound bath for 30 s to lyse the cell. After lysis and labeling reaction are complete, perform the separation at an electric filed of 400 V/cm.

4. Notes

1. Prepare all buffers with Milli-Q deionized water and filter using a 0.2-μm filter.
2. Dissolve the protein in water if stacking is desired, otherwise dissolve it in buffer.
3. Prepare a stock solution of 10 m*M* FQ in methanol; aliquot 10 μL into 500 μL micro centrifuge tubes and remove the solvent under vacuum using a Speed Vac (Savant Instruments Inc., Farmingdale, NY). The dried FQ aliquots, which are 100 nmol each, should be stored at –20°C. These precautions are necessary since it is observed that FQ degraded slowly in solution even if the solution was stored at –20°C.

4. Potassium cyanide is highly poisonous. It reacts rapidly with acids to form lethal HCN gas. Stock solutions should be made in a basic buffer and the experimenter should be aware of any change to acidic pH during the experiment. Neutralization of the waste containing KCN should be made by addition of 1% NaOH solution followed by slow addition of bleach.
5. Experiments were carried out using an in-house constructed CE-LIF instrument with a detector based on a sheath-flow cuvet ***(1,3,11,12,26)***. Unless otherwise stated, fused-silica capillaries were 50 μm (ID), 150 μm od. Electrophoresis was carried out with a 0–30 kV dc power supply. The excitation of fluorescence was provided by the 488-nm line of an Ar-ion laser operated at 12 mW. The laser beam was focused approx 30 μm from the tip of the capillary with a ×6.3 objective. Fluorescence was collected by a ×60, 0.7 NA microscope objective. An interference filter centered at 615 nm with a 45-nm transmittance window was used to remove stray light scattered light. Fluorescence was imaged onto a PMT biased at 1000 V. The photocurrent was passed through a current-to-voltage converter and a low-pass filter (RC = 47 ms) and then digitized with a 16-bit data acquisition board connected to a Macintosh computer.
6. A cell injector is a multifunctional device that facilitates: (1) cell injection, (2) CE separation of cellular component, as well as (3) cleaning the capillary. The details on its construction are described elsewhere ***(27)***.
7. Positive polarity for uncoated capillary and negative polarity for a coated capillary.

References

1. Ahmadzadeh, H. and Dovichi, N. J. (2000) Instrumentation, capillary coating and labeling chemistry for capillary electrophoresis with laser-induced fluorescence detection. Ph.D. thesis, Univ. Alberta, Canada, Chs. 1 and 6.
2. Albin, M., Weinberger, R., Sapp E., and Moring, S. (1991) Fluorescence detection in capillary electrophoresis: evaluation of derivatizing reagents and techniques. *Anal. Chem.* **63,** 417–422.
3. Wu, S. and Dovichi, N. J. (1989) High-sensitivity fluorescence detector for fluoresceinisothiocyanate derivatives of amino acids separated by capillary electrophoresis. *J. Chromatogr.* **480,** 141–155.
4. Dovichi, N. J., Martin, J. C., Jett, J. H., Trkula, M., and Keller R. A. (1984) Laser-induced fluorescence of flowing samples as an approach to single-molecule detection in liquids. *Anal. Chem.* **56,** 348–354.
5. McGuffin, V. L. and Zare, R.N. (1985) Laser fluorescence detection in microcolumn liquid chromatography: application to derivatized carboxylic acids. *Appl. Spectrosc.* **39,** 847–853.
6. Swerdlow, H., Zhang, J. Z., Chen, D. Y., et al. (1991) Three DNA sequencing methods using capillary gel electrophoresis and laser-induced fluorescence detection. *Anal. Chem.* **63,** 2835–2841.
7. Lynos, J. W. and Faulkner, L.R. (1982) Optimization of flow cells for fluorescence detection in liquid chromatography. *Anal. Chem.* **54,** 1960–1964.

8. Zarrin, F. and Dovichi, N. J. (1985) Particle counting by laser light scatter for capillary hydrodynamic chromatography. *Anal. Chem.* **57,** 1826–1829.
9. Zarrin, F., Risfelt, J. A., and Dovichi, N. J. (1987) Light scatter detection within the sheath-flow cuvette for size determination of multicomponenet submicrometer particle suspensions. *Anal. Chem.* **59,** 850–854.
10. Zarrin, F., Bornhop, D. J., and Dovichi N. J. (1987) Laser Doppler velocimetry for particle size determination by light scatter within the sheath-flow cuvette. *Anal. Chem.* **59,** 854–860.
11. Cheng, Y. F. and Dovichi, N. J. (1988) Subattomole amino acid analysis by capillary zone electrophoresis and laser-induced fluorescence. *Science* **242,** 562–564.
12. Cheng, Y. F., Wu, S., Chen, D. Y., and Dovichi, N. J. (1990) Interaction of capillary zone electrophoresis with a sheath-flow cuvette detector. *Anal. Chem.* **62,** 496–503.
13. Wu, S. and Dovichi, N. J. (1992) Capillary zone electrophoresis separation and laser-induced fluorescence detection of zeptomole quantities of fluoresceinthiohydantoin derivatives of amino acids. *Talanta* **39,** 173–178.
14. Chen, D. Y., Swerdlow, H. P., Harke, H. R., Zhang, J. Z., and Dovichi N. J. (1991) Low-cost, high-sensitivity laser-induced fluorescence detection for DNA sequencing by capillary gel electrophoresis. *J. Chromatogr.* **559,** 237–246.
15. Zhang, J. Z., Chen, D. Y., Wu, S., Harke, H. R., and Dovichi, N. J. (1991) High-sensitivity laser-induced fluorescence detection for capillary electrophoresis. *Clin. Chem.* **37,** 1492–1496.
16. Camilleri, P. (1993) *Capillary Electrophoresis Theory and Practice*. CRC, Boca Raton, FL, pp. 25–64.
17. Haugland, R. P. (1996) *Handbook of Fluorescent Probes and Research Chemicals*. Molecular Probes, Eugene, OR.
18. Novotny, M. (1988) Recent advances in microcolumn liquid chromatography. *Anal. Chem.* **60,** 500A–510A.
19. Liu, J., Hsieh, Y. Z., Wiesler, D., and Novotny; M. (1991) Design of 3-(4-carboxybenzoyl)-2-quinolinecarboxaldehyde as a reagent for ultrasensitive determination of primary amines by capillary electrophoresis using fluorescence detection. *Anal. Chem.* **63,** 408–412.
20. Camilleri, P. (1993) *Capillary Electrophoresis Theory and Practice*. CRC, Boca Raton, FL, pp. 311–370.
21. Pinto D., Arriaga E. A., Craig D., et al. (1997) Picomolar assay of native proteins by capillary electrophoresis precolumn labeling submicellar separation and laser-induced fluorescence detection. *Anal. Chem.* **69,** 3015–3021.
22. Zhao J. Y., Waldron K. C., Miller J. Z., Harke H. R., and Dovichi N. J. (1992) Attachment of a single fluorescent label to peptides for determination by capillary zone electrophoresis. *J. Chromatogr.* **608,** 239–242.
23. Tsuda T., Kobayashi Y., Hori A., Matsumoto T., and Suzuki O. (1988) Post-column detection for capillary zone electrophoresis. *J. Chromatogr.* **456,** 375–381.
24. Roze D. J. and Jorgenson J. W. (1988) Post-capillary fluorescence detection in capillary zone electrophoresis using o-Phthaldialdehyde. *J. Chromatogr.* **447,** 117–131.

25. Pentoney S. L., Huang X., Burgi D. S., and Zare R. N. (1988) On-line connector microcolumns: application to the on-column o-Ohthaldialdehyde derivatization of amino acids separated by capillary zone electrophoresis. *Anal. Chem.* **60,** 2625–2629.
26. Zhang, J. (1994) Ph.D. thesis, Univ. Alberta, Canada.
27. Krylov, S. N., Starke, D. A., Arriaga, E. A., et al. (2000) Instrumentation for chemical cytometry. *Anal. Chem.* **72,** 872–877.

4

Covalent and Noncovalent Labeling Schemes for Near-Infrared Dyes in Capillary Electrophoresis Protein Applications

John Sowell, Jozef Salon, Lucjan Strekowski, and Gabor Patonay

Summary

Capillary electrophoresis (CE) is experiencing increased use in the field of separation science. Part of its growing popularity of capillary electrophoresis can be attributed to the high efficiency of the separations achievable with the technique, making it an attractive tool for bioanalytical applications. Laser-induced fluorescence (LIF) is a common detection method for CE. One of the problems frequently experienced when using visible LIF detection is matrix autofluorescence which has the effect of degrading the overall sensitivity of the technique. However, the use of near-infrared (NIR) laser induced fluorescence nearly eliminates matrix autofluorescence, as very few molecules have intrinsic fluorescence in this region. This chapter describes the use of covalent and noncovalent labeling schemes for tagging biomolecules with near infrared dyes. To fully appreciate the advantages that the NIR LIF technique can supply, we also review applications that employ detection schemes other than NIR LIF. Specific applications to be discussed include drug–protein studies by CE, as well as capillary electrophoretic immunoassays.

Key Words

Capillary electrophoresis; capillary electrophoretic immunoassay; covalent; drug binding constant; dye displacement; human serum albumin; labeling; laser-induced fluorescence; near infrared; noncovalent.

1. Introduction

One of the most common detection schemes used in protein analysis is laser-induced fluorescence (LIF). Fluorescence detection schemes offer high sensitivity and selectivity because the fluorophore is attached to the species of interest and, therefore, functions as a reporter molecule. The majority of fluorescence

From: *Methods in Molecular Biology, vol. 276: Capillary Electrophoresis of Proteins and Peptides*
Edited by: M. A. Strege and A. L. Lagu © Humana Press Inc., Totowa, NJ

detection schemes use visible fluorophores, with the fluorescein–Ar-ion laser couple being one of the most popular. However, a problem often encountered with visible LIF is autofluorescence from the sample matrix. Autofluorescence arises from the intrinsic fluorescence that many biological compounds have in the visible region of the spectrum. The presence of matrix autofluorescence degrades both sensitivity and selectivity, often preventing low concentration applications that are actually the most important in bioanalytical and biomedical applications.

Very few biological molecules possess intrinsic fluorescence in the near-infrared (NIR) region of the spectrum (650–1100 nm). Consequently, because of the absence of interfering photocounts, NIR LIF detection can potentially improve sensitivity and selectivity. The improvements in detection limits available with NIR LIF detection arise from the lack of intrinsic fluorescence of matrix compounds in this region of the spectrum. Because of this lack of interference, detection becomes detector limited as opposed to background limited, which is the case in the lower wavelength region. The NIR region is, therefore, well suited for protein analysis and bioanalytical applications where autofluorescence of biological compounds is a concern at shorter wavelengths. Additional improvements in sensitivity arise from decreased light scatter. Light scatter is dependent on the wavelength of detection by $1/\lambda^4$. As a result, a typical detection wavelength of 820 nm offers a sixfold reduction in scatter over detection at 500 nm.

The development of bioanalytical NIR fluorescence applications during the past decade can be attributed to advances in solid-state technology. Because of its high intensity and narrow bandwidth, the laser is the most common excitation source used in fluorescence detection schemes. Visible lasers, such as the Ar-ion, are often expensive, bulky, and have limited operational lifetimes. Diode lasers are the sources most often used with NIR LIF detection and do not have any of these disadvantages. They are rugged, inexpensive, compact, and have long operational lifetimes, although they require additional optics. The typical signal transducer for visible fluorescence detection is the photomultiplier tube (PMT). Although red-sensitive PMTs are available, they are often quite expensive and have limited lifetimes. Avalanche photodiodes (APD) make a much more attractive choice, offering high quantum yields in the NIR. Additionally, APDs are inexpensive, rugged, compact, and have long operational lifetimes. The complementary nature of diode lasers and APDs combined with the desirable photophysical properties of NIR dyes make for an attractive technique for protein applications. To fully appreciate NIR LIF advantages and utility, we will first review general capillary electrophoresis (CE) theory and applications.

1.1. Introduction to CE

CE is generally an important analytical method owing to the high speed and efficient separation of such analytes as proteins, peptides, sugars, and many others *(1–3)*. The speed of the separation depends on high electric fields and, in some cases, short column lengths. The current trend is toward increasing the speed of the separation by miniaturization of the columns in length and internal diameter, which reduces injection volumes of the analytes and results in minimal zone variance. Detection by the small injection volumes for CE coupled to LIF has become very applicable because of the high sensitive detection and availability of a wide scope of fluorescent tags and lasers. Detection limits for CE-LIF have been reported from attomole to zeptomole and even to single molecules *(4,5)*.

The work of an ultraviolet light-emitting diode (LED)-induced fluorescence was first carried out by Hillebrand et al. *(6)*, who separated and detected within femtomole range a mixture of amino acids labeled with fluorescamine. A UV-LED with emission maximum at 370 nm ($\Delta\lambda$ = 12 nm) was used as an excitation source for fluorescamine-derivatized analytes. A scheme of the experimental setup is shown in **Fig. 1**.

In the past few years, some researchers have published articles showing the utility of LED detection in CE by using labeling dyes that absorb in the visible or near UV region. Some of the disadvantages linked to LED in CE are that the diverging beam of an LED cannot be focused to a micrometer spot size as could a laser beam in LIF. Another problem is the signal-to-noise ratio (S/N) in the measurement. The limits of detection (LODs) achieved in the experiment were in the femtomole range. However, the LODs observed on commercial CE equipment with LIF detection usually fall in the attomole range *(7)*, which is three orders of magnitude lower. Despite this, LED-induced fluorescence detection in CE is useful for routine analyses of proteins, peptides, and amino acids, considering the relatively low optical power of the source, as compared to lasers. However, it is clear that because of the optical considerations discussed above, low detection limits are limited by the method.

In 1998, Vicki et al. *(8)* described the use of nonaqueous capillary electrophoresis (NACE) with LIF to improve detection sensitivity. The nonaqueous solution is helpful in lowering detection limits owing to the minimization of quenching effects. Quenching can occur by different mechanisms. Collisional quenching occurs when the excited-state fluorophore is deactivated upon contact with another molecule in solution (such as O_2, halogen, amine, or acrylamide), which is called the quencher ***(9)***. There are two main types of quenching, dynamic and static. Dynamic quenching may occur when an excited molecule undergoes a collision with a quencher, producing a nonradiative

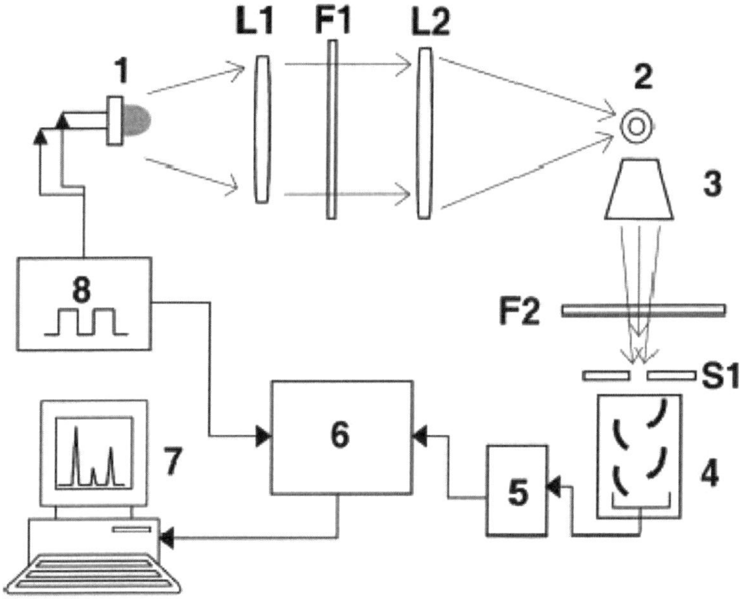

Fig. 1. Scheme of the LED-based fluorescence detection. (1) LED; L1, L2 quartz lenses; F1 a low- and F2 a high-pass filter; (2) capillary detection window; (3) microscope objective; (4) photomultiplier (PMT); (5) amplifier; (6) boxcar; (7) computer; (8) generator. Used with permission from **ref.** ***6***.

energy transfer. This type of quenching depends on the temperature and viscosity of the medium. By increasing the temperature and decreasing the viscosity, a collision appearance is reinforced. Long-range quenching is also a nonradiative energy transfer, in which dipole interactions between the fluorophore and quencher occur ***(10)***. In order to avoid quenching of a sample, which results in poor detection limits, degassing techniques such as the purge and freeze-pump-thaw methods, or additives in the buffer such as micelles and organic solvents ***(11,12)*** are often used. The detection limit results for 8-aminonapnthalene-1,3,6-trisulfonic acid (ANTS)-derivatized maltotriose by using the nonaqueous buffers (*N*-methylformamide [NMF] and formamide) and degassed techniques were used.

The addition of organic solvents affects the viscosity and polarity of the medium. Commonly, a more viscous and less polar medium can minimize dynamic and long-range quenching. In **Fig. 2**, the viscosity and polarity values of formamide and NMF are compared against water. At higher viscosity, the mobilities of the fluorophore and the possible quenchers are decreased which reduces the probability of a collision that could result in dynamic

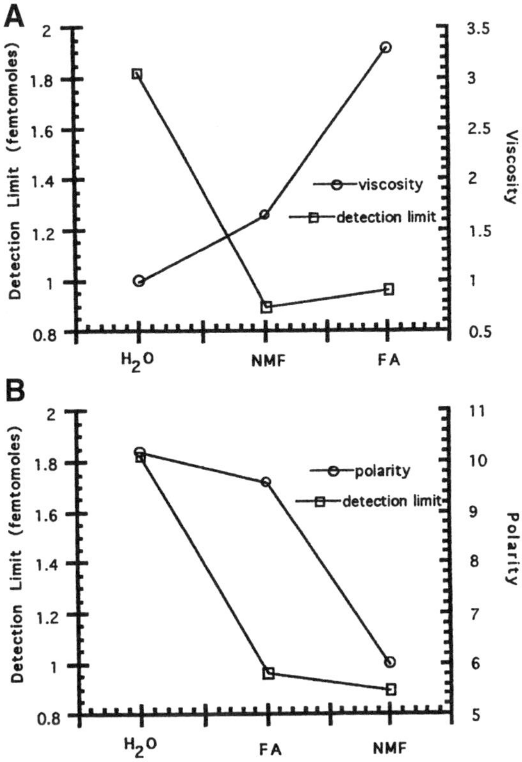

Fig. 2. Viscosity and polarity trends for the detection limit of ANTS-derivatized maltotriose. Used with permission from **ref.** ***8***.

quenching. Long-range quenching is also minimized because of a decrease in polarity of the nonaqueous media. When fluorophores are attached to large biomolecules, quenching effects are less important because the biomolecules shield the fluorophores from collisional deactivations.

Recently, researchers have been interested in the NIR as a viable alternative to visible fluorescence in many bioanalytical applications. Work with dyes that fluoresce in the NIR region (~670–1000 nm) possess many advantages over visible fluorescence, owing to significant reductions in scattering effects (6× greater at 520 vs 820 nm), mostly absent background fluorescence, and low detector noise.

One of the major advancements in NIR fluorescence detection has been incorporation of semiconductor diode lasers for LIF detection. The main advantages of using a semiconductor laser as an excitation source are its low cost, long usable life, stable output, and relatively high powers in the near-IR.

The entire diode laser assembly is small (several cm) and can be run using a simple dc battery. An example of the diode-based near-IR LIF detector is shown in **Fig. 3** *(13)*.

Work has been done involving the separation and detection of six amino acids labeled with the NIR dye NN382 (compound **9** in **Fig. 6**) (obtained from Li-Cor), which contains an isothiocyanate group for covalent attachment of the dye to a primary amine *(13)*. Four polar sulfonate groups not only increase aqueous solubility but also enhance the photochemical stability of the dye *(14)*.

Ummadi and Weimer have successfully developed a quantitative and sensitive method to monitor the metabolism of amino acids during bacterial growth of *Brevibacterium linens* BL2 in a chemically defined medium (CDM) under the CE–LIF conditions needed to achieve an optimal and sensitive separation using precolumn amine derivatization with CBQCA *(14)*. The experiment involves micellar electrokinetic capillary chromatography (MEKC) using a borate buffer with sodium dodecyl sulfate (SDS) and tetrahydrofuran (THF) additives. MEKC is a technique that involves the use of charged micelles to separate charged and uncharged molecules by means of a pseudomicellar phase that is created through hydrophobic interactions between solute molecules and detergent. The CE/MEKC method *(17)* was used to identify 18 CBQCA-derivatized amino acids (*see* **Fig. 4**). Seventeen of them were quantitated at attomolar concentrations (*see* **Table 1**).

Johansson and et al. *(18)* have described a technique for real-time imaging through optical fiber array-assisted LIF of capillary electrophoretic enantiomer separations. An optical fiber array was constructed for collection and transport of the fluorescence along the capillary to the charge-coupled devive (CCD) camera to detect the achiral separation of dansyl-DL-amino acids (DNS-Aas) in a capillary during an electrophoretic run.

The separation conditions of enantiomers using natural β-cyclodextrins added to a phosphate buffer containing a small amount of organic modifier were similar to conditions developed by Ward et al. *(18)*. In order to get a faster enantiomer separation, the method was modified by using a shorter capillary, buffer pH was lowered, and the β-cyclodextrin concentration was raised. The experiment was carried out in darkness to prevent interference from ambient light, thereby increasing the exposure time and subsequently lowering the limit of detection.

Biomolecule labeling can be difficult if the biomolecule of interest does not have a reactive moiety for attachment of a fluorophore, or if the reactive moiety has an important role in the biomolecular activity. In these cases, noncovalent labeling may be a viable alternative. The main disadvantage of noncovalent labeling is the reversibility of the label-biomolecule interaction.

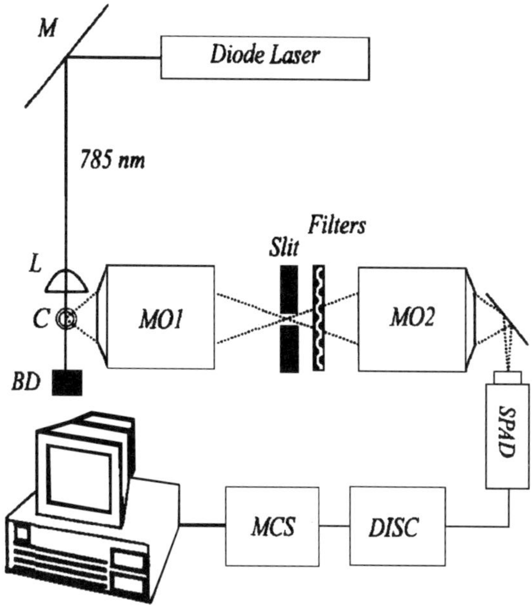

Fig. 3. Block diagram of the diode-based near-IR LIF detection system. M, mirror; L, laser focusing lens; C, capillary; BD, beam dump; MO1, collection microscope objective; MO2, focusing microscope objective; SPAD, single photon avalanche diode; DISC, discriminator; MCS, multichannel scaler. Used with permission from **ref.** ***13***.

1.2. Noncovalent Labeling Scheme for NIR Dyes

1.2.1. Design and Synthesis of NIR Dyes for Labeling of Proteins

Several classes of organic molecules and metal complexes exhibit absorption and fluorescence in the NIR region of the electromagnetic spectrum (>650 nm) ***(20–25)***. Of these, cyanine dyes have been developed as practical bioanalytical reagents, because of their relative ease of synthesis, high extinction coefficients (up to 300,000 $M^{-1}cm^{-1}$), and acceptable fluorescent quantum yields (up to 50%) and Stokes' shifts (up to 50 nm). The general structure of cyanine dyes is given in **Fig. 5**. The molecules contain two heterocyclic moieties A and B (in most cases, nitrogen heterocycles) joined by an odd number of methine (CH) groups in which (m+1) π electrons are distributed over m atoms ***(20)***. This produces a delocalized cation across the polymethine chain where the heterocyclic end units act both as an electron donor and an electron acceptor. Depending on the nature of the heterocyclic groups and the length of the polymethine chain, this unique electronic feature gives this class of cyanines a wider absorption span than any other known class of dyes. In general, the extension of the

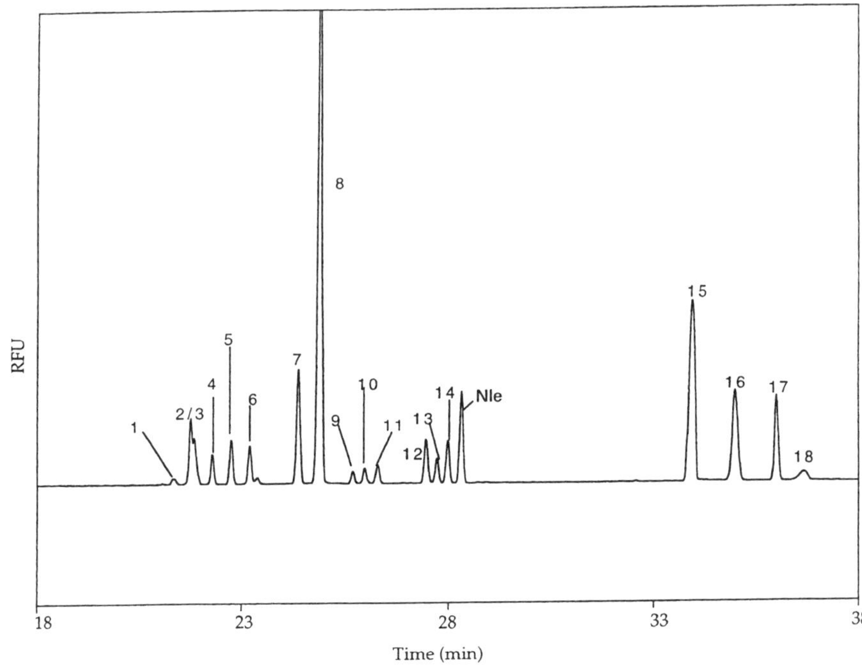

Fig. 4. Optimum conditions for separation of an amino acid standard solution using 6.25 m*M* sodium borate, 150 m*M* SDS, 10 m*M* THF (pH 9.66) buffer and run conditions of 24 kV, 25°C. 1 Ser, 2 Gln, 3 Met, 4 Asn, 5 Thr, 6 Tyr, 7 Ala, 8 Gly/Val, 9 His, 10 Pro, 11 Cys, 12 Ile, 13 Phe, 14 Leu, Nle, 15 Glu, 16 unknown, 17 Arg, 18 Asp. Used with permission from **ref. *14***.

polymethine chain by a vinyl group (two CH units) produces a bathochromic shift of about 100 nm, and starting with certain pentamethine derivatives ($n = 2$ in **Fig. 5**) the absorption and fluorescence are observed in the NIR region ***(23)***. Unfortunately, the length extension of the polymethine chain also results in a dramatic decrease of the dye stability in solution, especially in the presence of molecular oxygen under normal light conditions. For the most part, the instability is a result of self-stacking of planar molecules in solution, especially in aqueous solution, to form hypsochromic oligomers called H-oligomers ***(20)***. These aggregates show blue-shifted absorption bands. It is now understood that H-oligomer formation leads to a triplet dye species that allows endogenous triplet oxygen to undergo intersystem crossing via energy resonance transfer from its triplet ground state ($^3O_2$) to the destructive singlet species ($^1O_2$). A reaction of $^1O_2$ then destroys the chromophore.

Table 1
CBQCA Derivatization of Amino Acids in CDM

Amino acid	Time for complete derivatization (h)	Stability time (h)	Rate of derivatization (μM/h)	Lowest detectable concentration (aM)[a]
1. Ser	12	34	79	0.8
2. Gln	12	34	57	0.6
3. Met	6	24	22	0.1
4. Asn	12	34	63	0.6
5. Thr	14	34	60	0.7
6. Tyr	12	28	23	0.2
7. Ala	12	34	94	0.9
8. Gly/Val	12	24	182	1.1
9. His	6	34	54	0.3
10. Pro	12	34	145	1.5
11. Cys	20	34	8	0.1
12. Ile	12	34	13	0.1
13. Phe	12	34	10	0.1
14. Leu	12	36	13	0.1
Nle	6	34	24	ND
15. Glu	12	28	227	2.3
16. Unknown	12	34	—	ND
17. Arg	12	34	24	0.3
18. Asp	12	34	63	0.6

[a]This is the lowest detection limit of $S/N \geq 5$.
Reproduced with permission from **ref. *14***.

There are two practical ways of decreasing H-aggregation of cyanines in solution, thereby increasing their stability. One approach is to introduce bulky or flexible substituents at the chromophore so that the aggregation is inhibited by steric factors. For electrostatic reasons, the aggregation is also strongly inhibited for dyes substituted with a number of anionic groups, normally sulfonate groups (SO_3^-) ***(26)***. These two approaches are further exemplified by the structures of highly stable dye labels **7** and **9** (*see* **Fig. 6**). Dye **7** shows strong noncovalent binding with proteins, whereas the isothiocyanate group (N=C=S) at **9** (NN382) allows for a covalent attachment of the NIR chromophore to a protein by the reaction of the isothiocyanate functionality with a primary amine group on the protein. In order to minimize aggregation of the dyes **7** and **9**, their end-heterocyclic units contain a bulky $C(CH_3)_2$ moiety and are substi-

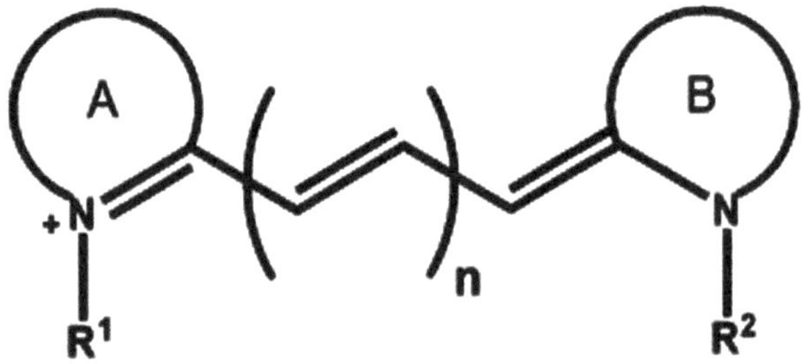

Fig. 5. A general structure of cyanine dyes.

tuted with a flexible, anionic sulfonatobutyl group ($[CH_2]_4SO_3^-$). The aggregation is further discouraged by the presence of a nonplanar trimethylene bridge ($CH_2CH_2CH_2$) at the polymethine chain. As an added benefit, this structural feature increases rigidity of the chromophore/fluorophore resulting in an increased efficiency of fluorescence in comparison to the dyes without the trimethylene bridge. Dye **7** has a net charge of –1, whereas **9** is trianionic in solution.

In the synthesis of **7** (*see* **Fig. 6**), an indolenine **1** is quaternized by the reaction with 1,4-butane sultone and the resultant inner salt **3** is then condensed with *bis*(Schiff base) **6** derived from cyclohexanone (**5**) *(27)*. The direct precursor **8** to isothiocyanato-functionalized dye **9** is obtained in a similar way starting with a sodium sulfonate analog **2**. Dye **9** is obtained by treatment of **8** with sodium 4-isothiocyanatophenoxide. An alternative procedure for **9** may involve the synthesis of a 4-aminophenoxy-substituted dye derived from **8** followed by transformation of the free amino group into the isothiocyanate group by treatment with thiophosgene *(27)*.

The discussed displacement reaction of chlorine atom from **8** and similar dyes involves an $S_{RN}1$ mechanism and, accordingly, it is efficient with nucleophiles that are good single-electron donors such as phenols or benzenethiols *(28,29)*. This displacement reaction proceeds smoothly in solvents that support the single-electron-transfer (SET) process, such as *N,N*-dimethylformamide (DMF) or dimethyl sulfoxide (DMSO), and is completely inhibited in aqueous solution. In particular, the lack of reactivity of the chloro-substituted dye **7** in aqueous solution allows for this agent to be used as a noncovalent label for

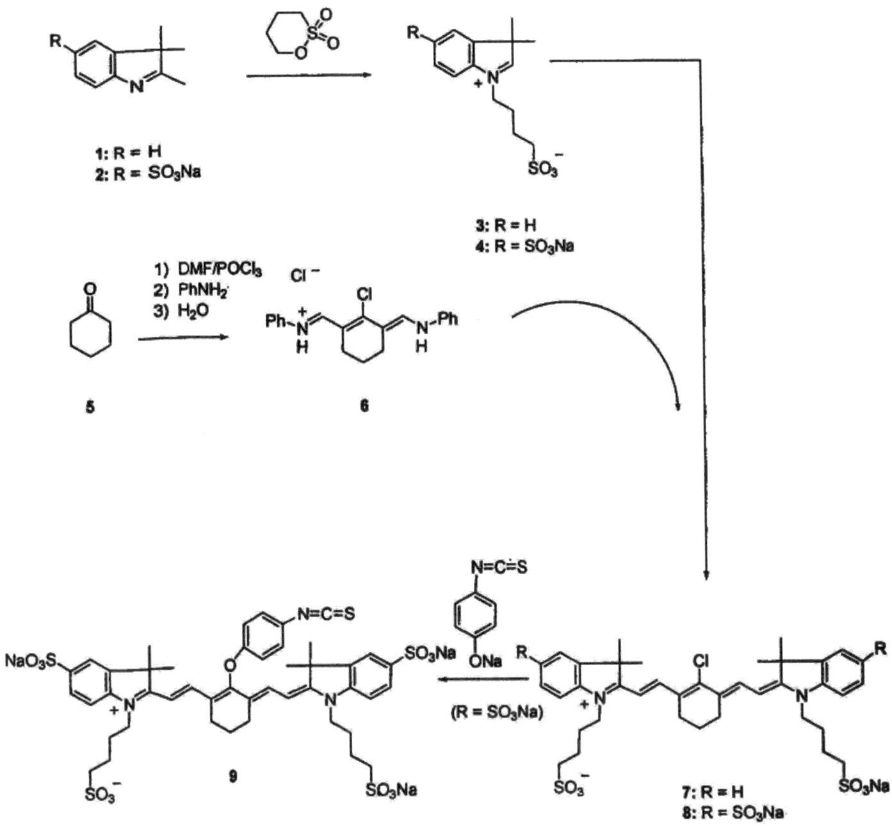

Fig. 6. Synthesis of dyes **7** and **9**.

proteins. The $S_{RN}1$ process can also be used for dechlorination of chloro-substituted dyes by treatment with a phosphine (PR_3) or an aliphatic thiol (RSH) *(30)*.

The current interest in covalent labeling of proteins with an NIR chromophore/fluorophore has been stimulating the synthesis of many functionalized dyes that can selectively be attached to a primary amino group *(31–35)* or a thiol group *(36,37)* of the protein. The isothiocyanate function and the esters derived from *N*-hydroxysuccinimide (NHS) are selective toward a primary amine. Other dyes *(36,37)* are substituted with reactive groups that are said to undergo a selective reaction with a thiol. So far, few of these functionalized dyes have been rigorously tested in bioanalytical applications.

1.2.2. Noncovalent Labeling in CE With NIR LIF Detection to Study Ligand Binding to Human Serum Albumin (HSA) by NIR Dye Displacement

Drug–protein binding is an important process that ultimately determines the fate of a drug once introduced to the body. Drug–protein interactions influence various pharmacological properties such as toxicity, distribution, activity, and excretion. These interactions may be highly specific as is the case with thyroxine and thyroxine binding globulin *(38,39)*. Alternately, the binding may be more general in nature, such as the interaction of many pharmaceutical agents with human serum albumin.

In addition to affecting the pharmacological properties mentioned above, competition among different drugs for the same binding site on a protein may lead to drug–drug interactions. Effects of drug–drug interactions can include increased toxicity and increased/decreased efficacy. Competition may also occur with drugs and endogenous compounds, an example being the displacement of drugs bound to human serum albumin by fatty acids and bilirubin *(40–43)*. Drug stereochemistry is also an important property in protein–drug interactions. Because drug–protein binding may be stereoselective, drug chirality may affect the above mentioned pharmacological properties.

In the past, the most common methods used to study drug–protein interactions were dialysis and ultrafiltration *(38,40,43)*. These methods have several disadvantages. Time is a drawback with dialysis in that days are required for analysis. If the system is unstable over this period of time, then problems arise. Additionally, corrections for free and bound analyte concentrations need to be made in order to correct for analyte adsorption onto the dialysis membrane. Although ultrafiltration is a much faster method, problems still exist with analyte measurement, as well as temperature control.

Because of these problems, much work has been done to find better methods for the study of drug–protein interactions. Many of the newer approaches are CE based. Because of their increased throughput and ease of measurement these methods are more convenient.

Generally, the traditional CE-based methods for binding constant studies necessitate the use of multiple runs (at varying concentrations of either drug or protein) for the determination of a binding constant. Consequently, they tend to be time intensive and consume large amounts of reagents. In some applications, such as the screening of combinatorial libraries, these characteristics are unsuitable. A sensitive, high-throughput method that requires very little analyte is required. For these reasons, a CE-based NIR dye displacement technique for drug–protein binding constant determinations was developed. Drug binding to subdomain IIIA, or site II, of human serum albumin was investigated. The method is sensitive, uses very little reagent, allows for the determination of a binding constant in a single run, and is amenable to automation.

Most LIF detection schemes use a fluorophore that is covalently attached to the analyte. However, derivatization reactions are often time consuming and may require rigorous control of pH in order to achieve satisfactory labeling efficiency. For example, dyes substituted with an isothiocyanate functionality that is reactive toward primary amines are often used for covalent tagging of proteins and peptides. However, amine groups on proteins will only be deprotonated at elevated pH levels that are often not suitable when working with biological samples. A trade-off exists between labeling efficiency and maintaining the structural integrity of the analyte. Another disadvantage with covalent labeling is that the labeled analyte molecules often have different numbers of dye molecules attached to them. This heterogeneous dye–analyte stoichiometry leads to band broadening ***(44)***. Furthermore, purification steps are required when doing covalent labeling to remove excess unreacted dye. On the other hand, noncovalent labeling schemes are significantly faster, generally requiring only a couple of minutes, as opposed to hours or even days for covalent labeling. Control of pH may not be necessary. Additionally, if the stoichiometry of the labeling reaction is known, purification steps may not be necessary.

An example of noncovalent labeling of a protein with a NIR fluorophore is presented using a NIR dye (*see* **Fig. 6**, dye **7**) and human serum albumin (HSA). The experiments include analysis of the stoichiometry of the dye–albumin interactions and determination of a binding constant. Additionally, a method allowing for the determination of drug–albumin binding constants is discussed, as follows.

1.3. Covalent Labeling for Competitive CE-Based Immunoassay for Albumin With NIR LIF Detection

1.3.1. Overview

Immunoassays are widely used analytical tools which take advantage of the highly specific interaction between an antibody and antigen. Immunoassays are used in a wide variety of clinical, medical, environmental, and biochemical applications. The efficacy of the technique relies on two properties of antibodies: high specificity for the target ligand and the strength of the antigen–antibody binding. The high specificity of antibodies allows for the determination of trace amounts of antigen, often in the presence of multiple interfering compounds present at much higher concentrations. The high binding constant values allow for accurate quantitation of antigen. The combination of these two characteristics results in a very powerful analytical tool.

One of the advantages of the immunoassay technique is the flexibility afforded by a wide range of available formats. Usually, immunoassays are carried out in either competitive or noncompetitive formats. Competitive assays are done by labeling either the antibody or antigen depending on what the analyte is. If the analyte is an antibody, antigen is coated on the surface of the microtiter plate. Labeled antibody is then added. Finally, unlabeled antibody from the sample is introduced. The labeled and unlabeled antibody binds competitively for a limited number of sites, i.e., the antigen coated on the surface of the plate. The amount of analyte present is determined through a change in the signal (*see* **Fig. 7**). Alternatively, if the analyte is an antigen, then labeled antibodies are fixed to a solid support. Labeled antigen is then added. Finally, unlabeled antigen from the sample is added (*see* **Fig. 8**). The amount of antigen present in the sample is determined by the change in signal. In general, signal strength decreases with increasing analyte concentration.

Noncompetitive assays operate on a different principle, a limiting amount of one reagent binding to an excess of another reagent. Most often, the solid support is saturated with antigen. Excess antigen is removed through a washing step. The sample antibody is then added, followed by another washing step. Finally, a second antibody (which has a label) specific for the primary antibody is added (*see* **Fig. 9**). Unlike competitive assays, signal intensity increases with increasing analyte concentrations. There exists a multitude of variations on these two basic formats.

1.3.2. Immunoassay Detection Systems

Although many analytes studied by immunoassays are detectable by UV/Vis absorption, this method of detection is not sensitive enough for most applications. As a result, immunoassays must incorporate some form of labeling in order to achieve the desired sensitivity. There are several requirements that the labeling method should meet. The derivitation procedure should be fast and simple. Additionally, the incorporated label should not interfere with antigen–antibody recognition. The most popular labeling methods currently used with immunoassays are radioactive, enzyme, and fluorescent labels.

1.3.3. Radioactive Labels

Radioactive labeling was one of the first labeling techniques used for immunoassays. A radioactive element is incorporated into the structure of the antigen or antibody and is used as a tracer. The energy emitted by the radioactive decay is used to generate a signal. Typical radioactive tracers include ^{125}I, ^{3}H, ^{57}Co, and ^{14}C. There are several drawbacks when using radioactive labels. Safety, limited shelf life, and disposal of radioactive compounds are obvious

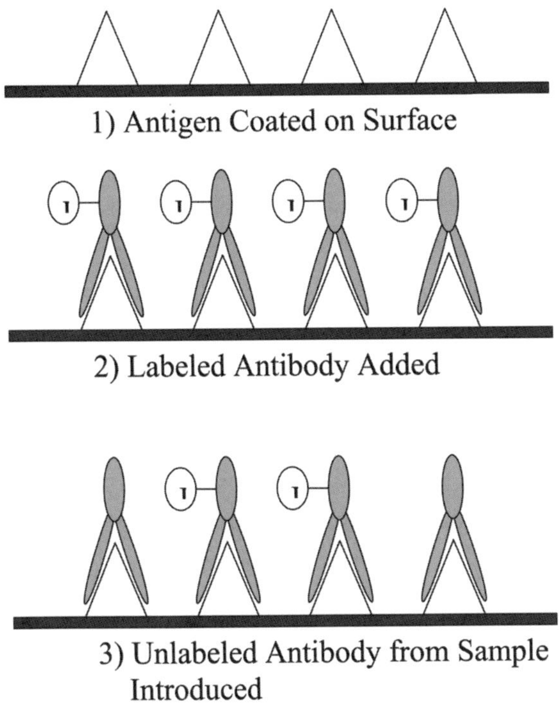

Fig. 7. Schematic of competitive immunoassay where antibody is the analyte of interest.

issues. Radioactive decay may be harmful to the labeled compound as well. Radioactive labeling is less suitable for CE applications, although it is still used in other immunochemistry applications.

1.3.4. Enzyme Labels

Enzyme labels are probably the most widely used labeling scheme in immunoassay today. The principle involves the capacity of the enzyme to act as a catalyst. First, an enzyme is attached to the desired compound. Second, all of the appropriate immunochemistry is carried out, i.e., antigen addition, and so on. Finally, a substrate is added. The enzyme label present converts the substrate into a detectable product. The product is generally colored, fluorescent, or chemiluminescent. A single molecule of enzyme may convert a large number of the substrate molecules, thereby increasing the signal strength and consequently, the sensitivity of the assay.

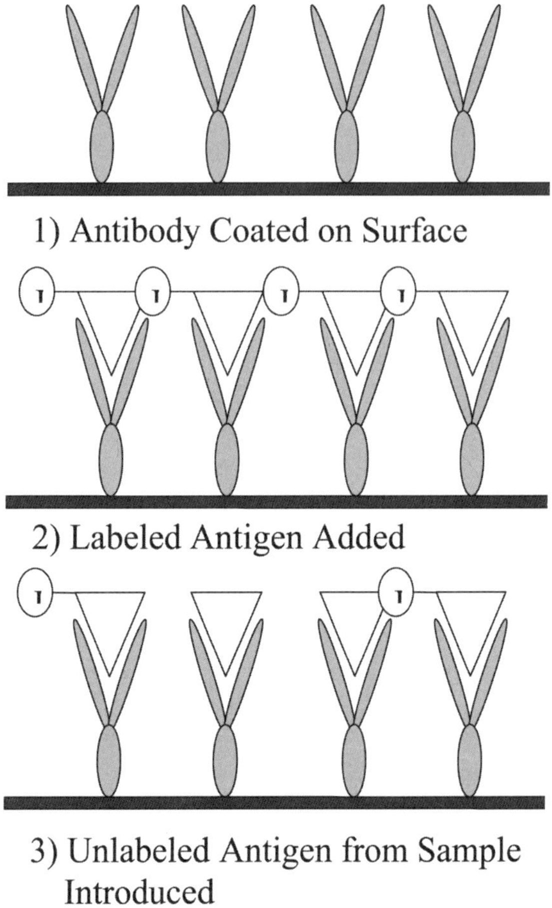

Fig. 8. Schematic of competitive immunoassay where antigen is the analyte of interest.

The two most common enzymes used as labels are alkaline phosphatase and horseradish peroxidase. Horseradish peroxidase is oxidized when reacted with hydrogen peroxide. Upon oxidation, it undergoes a reaction with another substrate, forming a colored or fluorescent product, depending on the substrate used. Alkaline phosphatase catalyzes the hydrolysis of phosphate esters from primary alcohols, phenols, and amines. The hydrolysis products are then detected.

Whereas enzyme linked immunoassays overcome many of the problems associated with radiolabeling, they do have disadvantages. Attachment of an enzyme, typically in the range of 50–100 KDa, may interfere with antibody–antigen interactions. Additionally, the conditions during the signal generation stage need to be carefully controlled, as enzyme immunoassays are

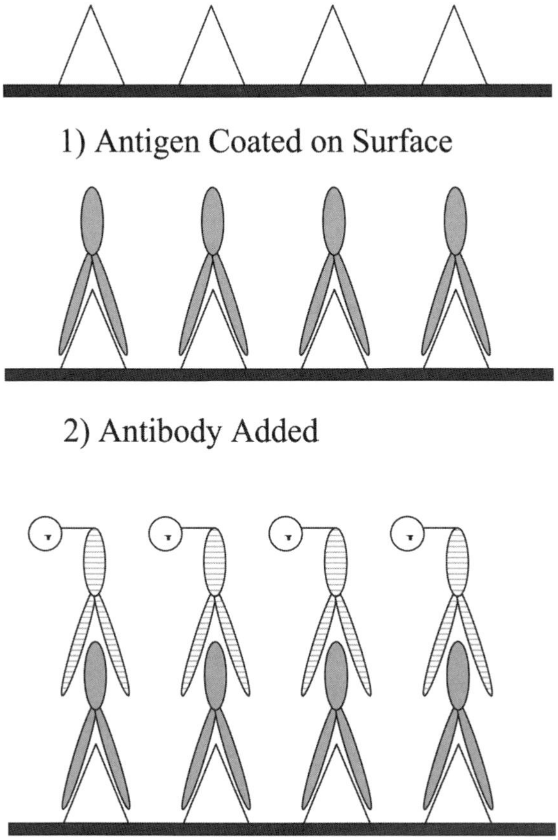

Fig. 9. Schematic of noncompetitive immunoassay.

sensitive to changes in temperature and pH. In addition, the enzyme reaction normally requires more time than is available for CE applications.

1.3.5. Fluorescent Labels

Although still not used as widely as enzyme labels, fluorescent labels are fast becoming the signal generation method of choice for immunoassays. Whereas some enzyme labels can produce fluorescent products, most produce nonfluorescent substrate products, necessitating the use of absorbance detection. Because of the high background noise associated with absorbance detection, the method is inherently less sensitive than fluorescent detection. Because of their smaller size, fluorescent labels are also less likely to interfere with antigen–antibody interactions than enzyme labels. The ideal fluorescent label should have the following characteristics: a high molar absorptivity and quantum yield, a large Stoke's shift in order to minimize light scattering effects, quick and efficient coupling reaction, good solubility under physiological conditions, low noncovalent affinity for biomolecules, photostability and small size. Variations of the basic fluorescence immunoassay may be used to overcome some of the disadvantages of fluorescent detection. Some of these variations include fluorescence polarization, time-resolved fluorescence, and fluorescence energy transfer.

To date, nearly all immunochemical techniques using fluorescent labels have used visible fluorophores. Owing to the often complex nature of the matrix in immunoassays, background autofluorescence is a problem. The high background noise results in loss of sensitivity. On the other hand, the photophysical characteristics of NIR dyes make them well suited for immunochemical applications. Whether working with environmental or biological samples, autofluorescence in the NIR region is almost nonexistent. As a result, background noise is eliminated and sensitivity is improved.

1.3.6. CE-Based Immunoassays

CE-based immunoassays with LIF detection are seeing increased use in clinical, biopharmaceutical, and environmental chemistry. The advantages over conventional immunoassays are high-speed, high-resolution separations in a minimal detection volume, and the ability to simultaneously determine multiple analytes. Immunoassays are analytical techniques which work on the specific reaction between an antibody (Ab) and an antigen (Ag). Antibodies are glycoproteins that consist of two distinct regions, the Fab region and the Fc region. The Fab region is responsible for Ag recognition and binding. One of the advantages of immunoassays is that the technique can be performed in various formats. Commonly, immunoassays are carried out in either the competitive or noncompetitive format, or even on-line.

Competitive assays are based on the fact that the use of at least one reagent is limited ***(45)***. The labelled reagent (Ab^* or Ag^*) together with analyte (Ab or Ag) bind to the limited reagent (Ab or Ag), which may result in examples 1 and 2. Both Kennedy and Chen have done work

$$Ab + Ab^* + Ag(limited) = Ab\text{-}Ag + Ab^*\text{-}Ag + Ab + Ab^* \quad (1)$$

$$Ag + Ag^* + Ab(limited) = Ab\text{-}Ag + Ab\text{-}Ag^* + Ag + Ag^* \quad (2)$$

with competitive CEIA ***(46,47)***. Quantitation is done through changes in peak areas.

$$Ab + Ag^* (excess) = Ab\text{-}Ag^* + Ag^* (excess) \quad (3)$$

$$Ag + Ab^* (excess) = Ab^*\text{-}Ag + Ab^* (excess) \quad (4)$$

Noncompetitive CEIA involves the separation of Ab–Ag complexes from free, labeled Ab or Ag. Thus, there are two possible formats as shown in **Eq. 3** and 4. Quantitation is done through changes in signal strength, i.e., changes in peak area. Schulz has worked a good deal with insulin using noncompetitive capillary electrophoretic immunoassays (CEIA) ***(46)***.

In most CEIA experiments, the immunoreaction is performed off-line. On-line analysis provides for greater automation of the technique. For example, Tao and Kennedy ***(47)*** have developed an on-line competitive immunoassay to monitor insulin concentration in a flowing stream.

CE has proven to be a powerful technique for the separation of large biological molecules such as proteins ***(48)***. When detection schemes such as LIF are used, CE can achieve detection limits in the low pmol range, comparable to or better than most conventional immunoassays. Single molecule detection is even possible when enzyme amplification is used in conjunction with LIF ***(49)***. Owing to its superior separation efficiency and high detection sensitivity, CE has the ability to rapidly separate free Ab and Ag from bound Ab and Ag. Capillary electrophoresis-based immunoassays combine immunological recognition, on-line quantitation, microscale analysis, and automation, offering distinct advantages over conventional, solid-phase immunoassays ***(4)***. In addition to the speed of analysis, CEIA have several other advantages: CEIA (1) consumes much less reagent, (2) is much less labor intensive in that tedious washing and rinsing steps are eliminated, (3) allows for the simultaneous detection of multiple analytes, (4) allows for direct visualization of immunocomplex formation, and (5) utilizes the wide range of detection methods that are also available for CE. Finally, precision is improved with CEIA. With conventional, solid-phase immunoassays, it is impossible to control the orientation of the antibody/antigen during the coating process. Some of the antibodies may adhere to the support in such a way as to make their recognition site inaccessible to antigen. As a result, overall precision suffers. However, CEIA avoids this problem since the immunoreaction is carried out in solution.

2. Materials

2.1. Noncovalent Labeling

2.1.1. Chemicals

1. Human serum albumin, boric acid, warfarin, sodium dibasic phosphate, phosphoric acid, ibuprofen, fenoprofen, ketoprofen, quinidine, chlorpromazine, naproxen, sulfisoxazole, imipramine, clofibrate, tryptophan and sodium hydroxide were obtained from Sigma (St. Louis, MO).
2. 200 m*M* Borate buffer, pH 8.5, was prepared by dissolving the appropriate amount of boric acid in nanopure grade water (Barnstead model D4571 ultrapure water system). The pH was adjusted by the addition of 0.5 *N* sodium hydroxide.

3. 100 m*M* Phosphate buffer, pH 7.2 was prepared by dissolving the appropriate amount of sodium phosphate in water. The pH was adjusted by adding phosphoric acid.
4. Protein, dye, and drug solutions were prepared by dissolving the appropriate amount of each reagent into 100 m*M*, pH 7.2 phosphate buffer.
5. All solutions were stored in the dark at 4°C prior to use.

2.1.2. Apparatus

1. All CE experiments were performed using a modified P/ACE 5000 capillary electrophoresis instrument (Beckman Instruments, Inc., Fullerton, CA). The instrument was interfaced with a proprietary microscope and laser assembly. The laser assembly consisted of a diode laser focused directly onto a fiber optic cable. The arrangement gave roughly 4 mW of power at the capillary interface. Detection was accomplished via a Peltier-cooled, three-stage avalanche photodiode. Three band-pass filters (820 ± 10 nm) were used in order to minimize background noise. The APD signal was demodulated by a lock-in amplifier. The signal was filtered before it arrived at a Beckman 406 A/D converter. The signal was then analyzed by a personal computer.
2. Fluorescence measurements were performed on an ISS K2 multiphasic fluorometer (ISS, Champaign, IL).
3. Absorbance measurements were performed on a Perkin-Elmer Lamda 20 (Perkin-Elmer, Norwalk, CT) UV/Vis/NIR spectrophotometer.

2.2. Covalent Labeling

2.2.1. Chemicals

1. Sodium phosphate, boric acid, HSA, and monoclonal HSA antibodies were purchased from Sigma.
2. The NIR dye NN382 was a gift from LI-COR (Lincoln, NE).
3. All water used was nanopure grade (Barnstead Model D4751).
4. Derivatization buffer was prepared by dissolving the appropriate amount of phosphate buffer into water and adjusting the pH with sodium hydroxide.
5. Run buffer was prepared by dissolving the appropriate amount of boric acid into water. The proper pH was achieved through addition of sodium hydroxide.
6. Final run buffers were filtered through 0.45-μm nylon membrane filters and sonicated for 10 min prior to use.

2.2.2. Preparation of NN382-HSA Conjugate

1. HSA was labeled with the NIR dye NN382 to allow for visualization and quantitation.
2. The labeling procedure consisted of adding 1 mg of NN382 and 1 mg of HSA to 1 mL of 100 m*M*, pH 9.5, phosphate buffer (*see* **Note 1**).
3. The mixture was stirred for 1 h, purified (*see* **Subheading 2.2.3.**) and then quenched by addition of 25 mL of phosphate buffer, 100 m*M*, pH 7.2.
4. The labeled HSA was stored at 4°C prior to use.

2.2.3. Purification of NN382-HSA

1. Because of the excess of dye used in the labeling procedure, it was necessary to purify the NN382-HSA preparation in order to remove the free dye (*see* **Note 2**).
2. Purification was done via size exclusion chromatography using a PD10 column (Amersham Pharmacia Biotech, Sweden).
3. The column was equilibrated with 25 mL of 100 m*M*, pH 7.2, phosphate buffer.
4. The NN382-HSA sample, 2.5 mL, was then introduced to the column.
5. The sample was then eluted with 3.5 mL of phosphate buffer.
6. The efficiency of dye removal was tested via electrophoresis. No free dye peak was observed after purification.

2.2.4. Preparation of Monoclonal Albumin Antibody

1. Stock antibody solutions were prepared by addition of 25 μL of 4.4 mg/mL of monoclonal HSA antibody to 25 mL of Tris-HCl buffered saline, pH 8.3.
2. The antibody mixture was stored at 4°C prior to use.

3. Methods

3.1. Noncovalent Labeling

3.1.1. CE

1. All experiments were performed at 23°C.
2. The primary electrophoresis run buffer was 200 m*M* borate acid, adjusted to a pH of 8.5.
3. All buffers were sonicated and filtered prior to use.
4. All separations were performed using fused silica capillaries with a polyimide coating (Polymicro Technology, Phoenix, AZ).
5. Total capillary length was 57 cm, with injection-to-detection length of 50 cm. The id of the capillary was 50 μm.
6. At the beginning of each day, the capillary was rinsed with 1 *N* sodium hydroxide, followed by 15-min rinses with water and run buffer.
7. All samples were introduced by pressure injection (5 s at 0.5 psi).
8. Voltage (15 kV) was applied over a 30-s ramp time (*see* **Note 3**).
9. Following each run, the capillary was rinsed with run buffer for 5 min.

3.1.2. Noncovalent Labeling and Dye Displacement Procedure

1. The noncovalent labeling of HSA with NIR dye was carried out by mixing aliquots of the protein and dye solutions.
2. The mixture was then vortexed for 2 min (*see* **Note 4**) prior to injection. Various dye–protein ratios and concentrations were used in order to investigate the utility of the dye as a noncovalent label for HSA.
3. Dye displacement experiments were performed by adding an aliquot of the appropriate drug solution to an equilibrated dye–protein solution.
4. This solution was then vortexed for 2 min prior to injection.

5. Enantioselective binding studies were performed by adding aliquots of either D- tryptophan or L-tryptophan to an equilibrated mixture of dye–protein solution.
6. This solution was vortexed for 2 min prior to injection.

3.1.3. Utility of NIR Dye as a Noncovalent Label

As previously mentioned, the use of noncovalent labeling offers several advantages over covalent labeling schemes. The suitability of the dye as a noncovalent label for HSA was investigated. As the concentration of protein is increased, with respect to a fixed dye concentration, the labeled protein peak increases while the free dye peak decreases, suggesting that the dye is indeed suitable as a noncovalent label for HSA. (*See* **Fig. 10**)

3.1.3.1. Optimization of Separation Conditions

The electrophoretic conditions were systematically optimized such that free dye and dye-labeled albumin were resolved. It was necessary to choose a run buffer that did not result in Joule heating, because excessive heating over the time scale of the separation could affect dye-protein and drug–protein interactions (*see* **Note 5**). Boric acid met this criterion. The optimum ionic strength of the run buffer was determined by varying the ionic strength of the run buffer from 25 to 250 m*M*. Free dye and dye-labeled protein were consistently resolved in a reasonable amount of time using a concentration of 200 m*M* boric acid. Protein separations done with CE have the potential to be problematic owing to protein adsorption onto the capillary wall ***(50)***. The use of run buffers with pH values greater than the isoelectric point of the protein has been shown to minimize protein adsorption to the capillary wall. Boric acid buffers with pH values from 6.0 to 9.5 were tested, and it was found that a run buffer with pH 8.5 provided the best results (*see* **Note 6**). Buffers on the extreme ends of physiological pH were avoided because albumin undergoes pH-dependent conformational changes under these pH conditions ***(51)***.

3.1.3.2. Calculation of Stoichiometry and Binding of NIR Dye–HSA Complex

The peak splitting was sometimes observed with the protein peak, suggesting that there may be multiple dye–protein species present, with the dye binding to the protein via hydrophobic ***(50)*** and/or coulombic interactions ***(52)***. Dovichi et al. have demonstrated that excessive peak width such as this is caused by the partial resolution of multiple species ***(43)***. Despite the likely presence of multiple dye–protein species, a number of techniques were used to gain some insight into the nature of the dye–protein interaction; specifically, the stoichiometry of the complex and the dye binding constant was determined.

Saturable binding curves, using both fluorescence and absorbance, were constructed. A fluorescence binding curve is shown in **Fig. 11**. It shows the change in fluorescence intensity of a fixed dye concentration, 5×10^{-6} *M*, relative to increasing concentrations of human serum albumin. A binding constant of 5.5×10^5 M^{-1} was calculated. A definitive stoichiometry could not be determined from the binding curve. Consequently, a Job's plot was used to determine the primary form of the dye–HSA complex (*see* **Fig. 12**). The total concentration of the dye and HSA remained constant; however, the mole fraction of the dye was varied. The fluorescence intensity as a function of dye mole fraction was monitored. The maximum of the Job's plot represents the primary stoichiometric form present. The Job's plot shows a 1:1 stoichiometry between dye and protein. This is in agreement with the data obtained with the binding curve study in that the fluorescence intensity levels off close to the area where the dye concentration is approximately equal to the protein concentration. It should be noted that this stoichiometry is only for the predominant form of the dye–albumin complex. It is likely that multiple stoichiometries exist.

3.1.3.3. Dye Displacement Assay

When a ligand or drug is introduced into an equilibrated mixture of dye and protein, there exists two ways in which the dye and protein can interact. Noncompetitive interactions occur when the dye and drug bind at different locations on the protein. **Fig. 13** illustrates the noncompetitive binding of warfarin. Upon addition of warfarin, a third peak appears, presumably because of the formation of a dye–warfarin–protein complex. The appearance of this complex suggests that warfarin and the dye bind at different locations on the protein. Whereas noncompetitive interactions do not allow for the determination of binding constants, they do provide some indirect information concerning the binding site of the dye. Warfarin is known to bind at site I, or subdomain IIA, on HSA. Owing to the fact that the dye and warfarin do not exhibit a competitive type interaction, it is reasonable to assume that the dye does not bind at site I.

The dye displacement technique is based on the principle that drug binding to albumin may be monitored through competitive interactions between the NIR dye and drug introduced (*see* **Note 7**). A drug is introduced into an equilibrated mixture of dye and protein. The drug and dye bind at the same location on the protein and, because of a shortage of binding sites relative to drug and dye population, they compete for binding. Consequently, drug binding may be monitored through changes in either the free dye concentration or the amount of dye bound to protein, both of which are easily measured. The electropherogram in **Fig. 14A** shows dye labeled protein and free dye. The electropherograms in **Fig. 14B** and **C** have the same concentration of dye and protein as in **Fig. 14A**; however, quinidine (B) and ketoprofen (C), have been added.

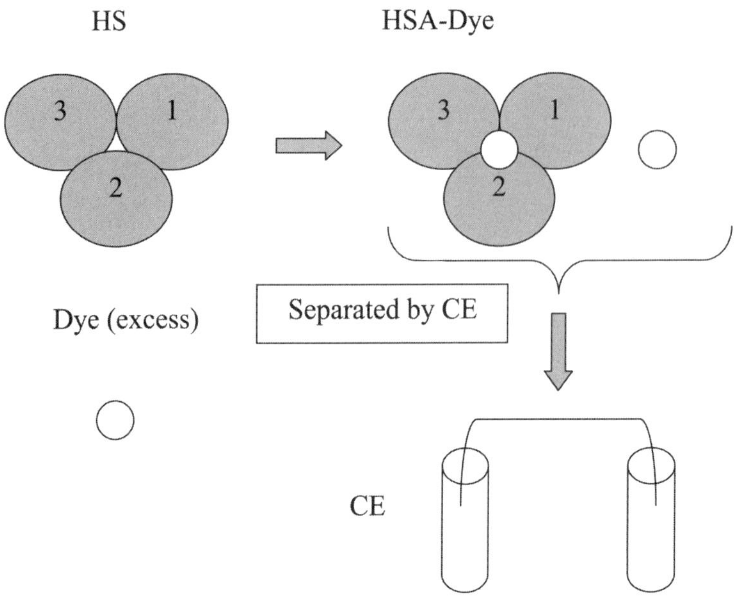

Fig. 10. Schematic for noncovalent labeling of HSA with NIR dye.

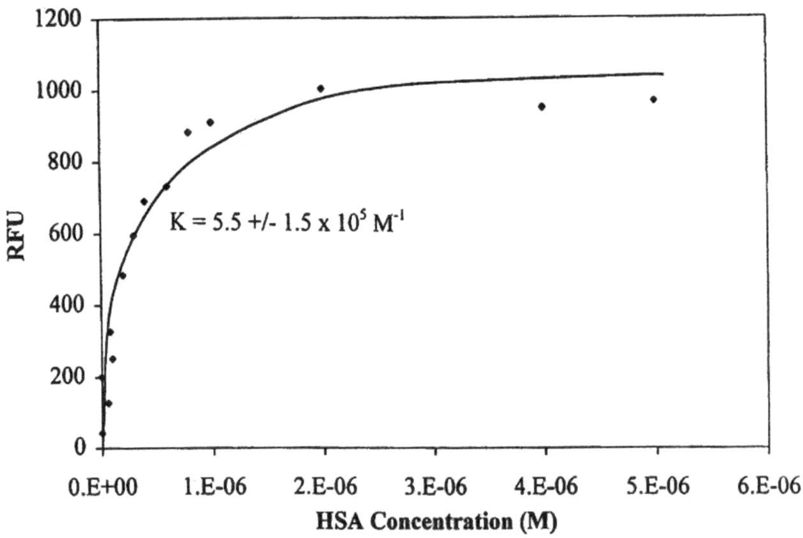

Fig. 11. Fluorescence binding curve for NIR dye and albumin. Used with permission from **ref. *56***.

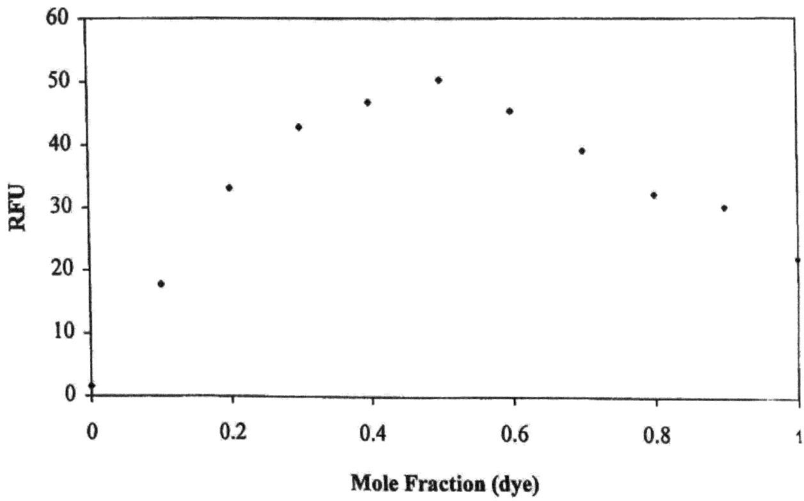

Fig. 12. Job's plot of NIR dye and albumin. Used with permission from **ref. *56***.

Ketoprofen and quinidine are known to bind at site II, or subdomain IIIA, of HSA. It can be seen that as the drug is introduced, the peak area of the labeled protein peak decreases and the peak area of free dye increases. This suggests that upon introduction of drug into the system, the drug displaces a portion of dye from the protein, explaining the changes observed in the peak areas. The degree to which the dye is displaced from the protein upon introduction of drug is directly related to the association constant of the drug. It can be seen in **Fig. 14** that ketoprofen ($K_a = 3.8 \times 10^6 M^{-1}$) displaces more dye than quinidine ($K_a = 7.9 \times 10^4 M^{-1}$). Using this relation, it is possible to construct a graph that allows for the determination of binding constants of drugs that bind to subdomain IIIA of human serum albumin. Four compounds were chosen based on their wide range of binding constants, ketoprofen, quinidine, clofibrate ($K_a = 7.6 \times 10^5 M^{-1}$) and imipramine ($K_a = 1.9 \times 10^5 M^{-1}$). The graph is a plot of % dye bound as a function of binding constant. The binding constants of ibuprofen and naproxen were calculated with thc dye displacement assay. K_a values of $1 \times 10^6 M^{1}$ and $8 \times 10^5 M^{-1}$ were obtained for ibuprofen and naproxen, respectively. The calculated K_a values for these drugs are in agreement with results found in the literature ***(54–56)***. Although the plot is linear over the K_a values, sensitivity is lost when drugs with K_a values greater than or less than an order of magnitude of the dye's binding constants are used. When drugs with these values are used, the ends of the graph level off, and the plot

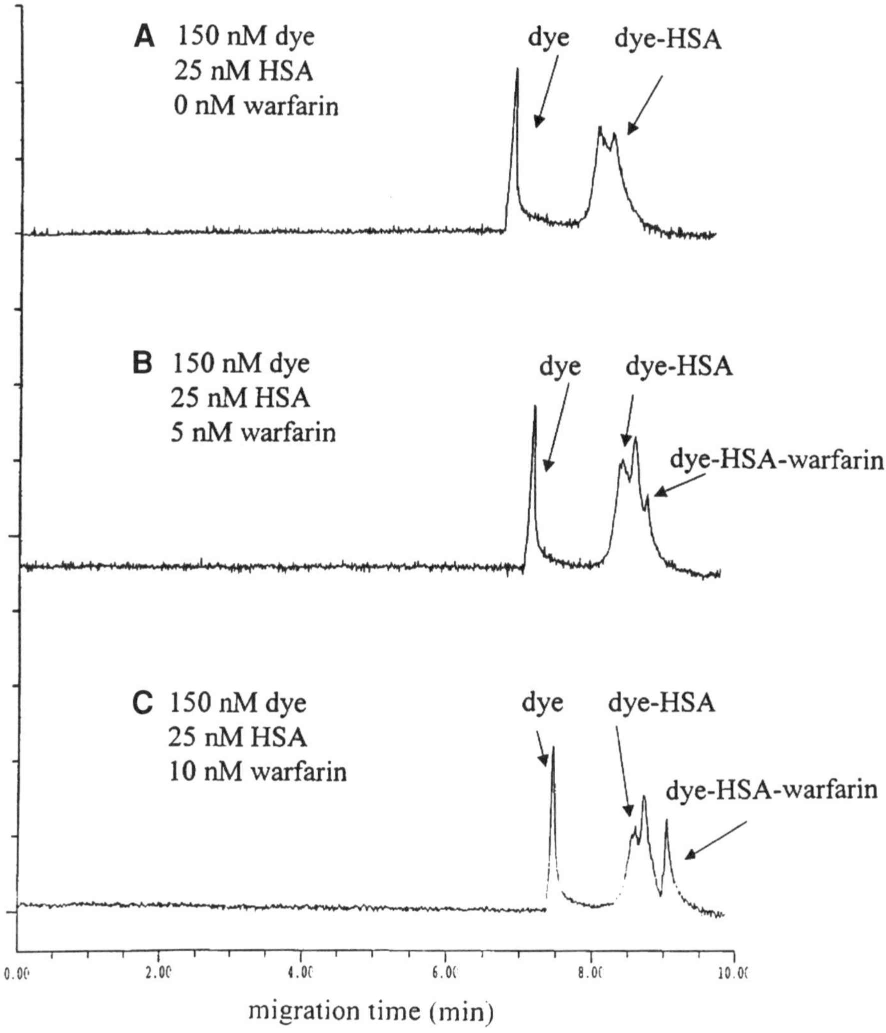

Fig. 13. Electropherograms illustrating noncompetitive binding of warfarin to albumin. Separation conditions: 200 m*M* boric acid, pH 8.5, 15 kV, 30 s ramp time, and 5 s pressure injection (0.5 psi). Used with permission from **ref. *56***.

resembles a binding curve. Consequently, the effective range of the technique is determined by the binding constant of the dye.

3.1.3.4. Enantioselective Binding

It is recognized that enantiomers of drugs may bind to albumin to varying degrees *(38–40)*. Because the biological activity of enantiomeric forms of a drug may be significantly different, it was of interest to determine if the method described could be used to monitor enantioselective binding to albumin. Tryptophan was chosen as a test compound because its enantiomers have binding constants significantly different from albumin. The electropherogram in **Fig. 15A** shows the dye and dye-labeled protein. In **Fig. 15B-C**, D-tryptophan and L-tryptophan were added, respectively. It can be seen that the L-tryptophan displaces more dye than the D-tryptophan, as predicted. In order to better illustrate the enantioselective binding of tryptophan, a large excess of the ligand was used in the experiments.

3.3. Covalent Labeling

3.3.1. CE

1. All experiments were performed on a modified Beckman P/ACE 5000 CE at 23°C.
2. The electrophoresis run buffer consisted of 200 m*M* boric acid, pH 8.5.
3. All separations used fused silica capillaries coated with polyimide (Polymicro Technology, Phoenix, AZ).
4. The capillaries were 57 cm in length with an id of 50 μm (*see* **Note 8**).
5. At the beginning of each day, the capillary was rinsed with 1 *M* sodium hydroxide for 30 min, followed by 15-min rinses with deionized water and run buffer.
6. The samples were introduced by pressure injection (5 s at 0.5 psi). The injected volume was 7 nL.
7. The applied voltage was 15 kV, with a 30-s ramp time (*see* **Note 3**).
8. Following each run, the capillary was rinsed for 2 min with water and run buffer, respectively.

3.3.2. Noncompetitive Immunoassay

A noncompetitive assay was done by mixing 250 μL of 65 n*M* NN382-HSA with 250 μL of 0–65 n*M* HSA antibodies. All dilutions were made using phosphate buffer. pH 7.2. The solutions were vortexed for 5 min on slow speed at room temperature prior to injection (*see* **Note 9**).

The first step in the development of the competitive CEIA for HSA was to determine if the antibody–antigen recognition still existed after the labeling

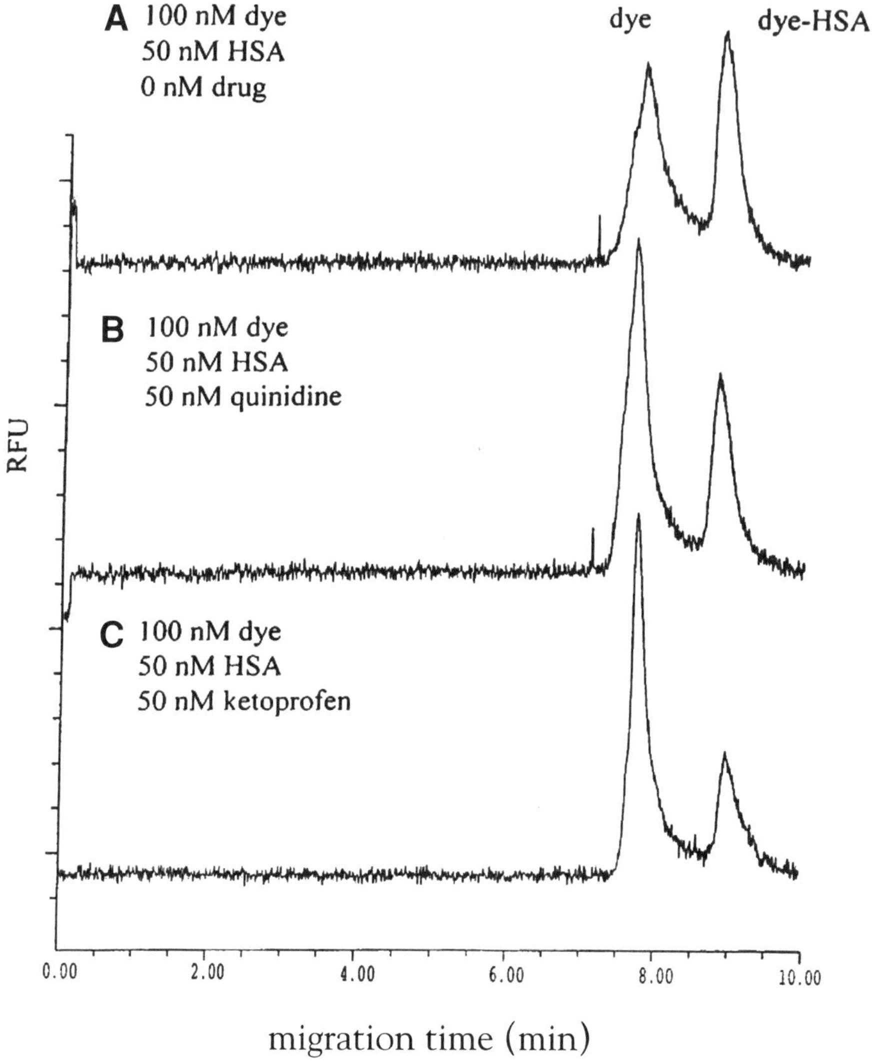

Fig. 14. Electropherograms illustrating competitive binding of various drugs to albumin. Separation conditions: 200 m*M* boric acid, pH 8.5, 15 kV, 30 s ramp time, and 5 s pressure injection (0.5 psi). Used with permission from **ref. *57***.

procedure. A noncompetitive immunoassay format was used to ascertain this. The electropherograms for the noncompetitive assay for HSA are shown in **Fig. 16 A–C**. **Figure 16A** is an electropherogram of NN382-HSA. Upon addition of HSA antibody solution, the peak area of NN382-HSA decreases and a

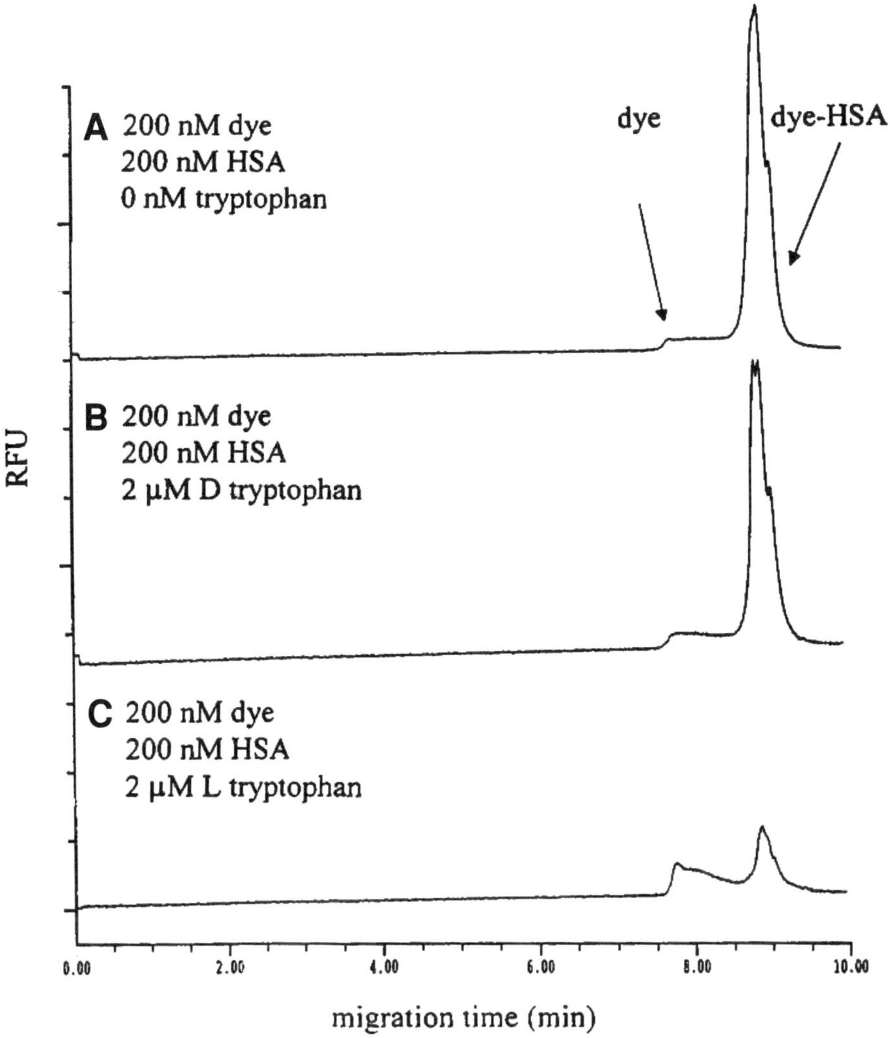

Fig. 15. Electropherograms showing enantioselective binding. Separation conditions: 200 m*M* boric acid, pH 8.5, 15 kV, 30 s ramp time, and 5 s pressure injection (0.5 psi).

new peak appears, presumably because of the formation of the immunocomplex. The trend continues with increasing concentrations of HSA antibody. These results suggest that the dye does not interfere greatly with antibody–antigen recognition.

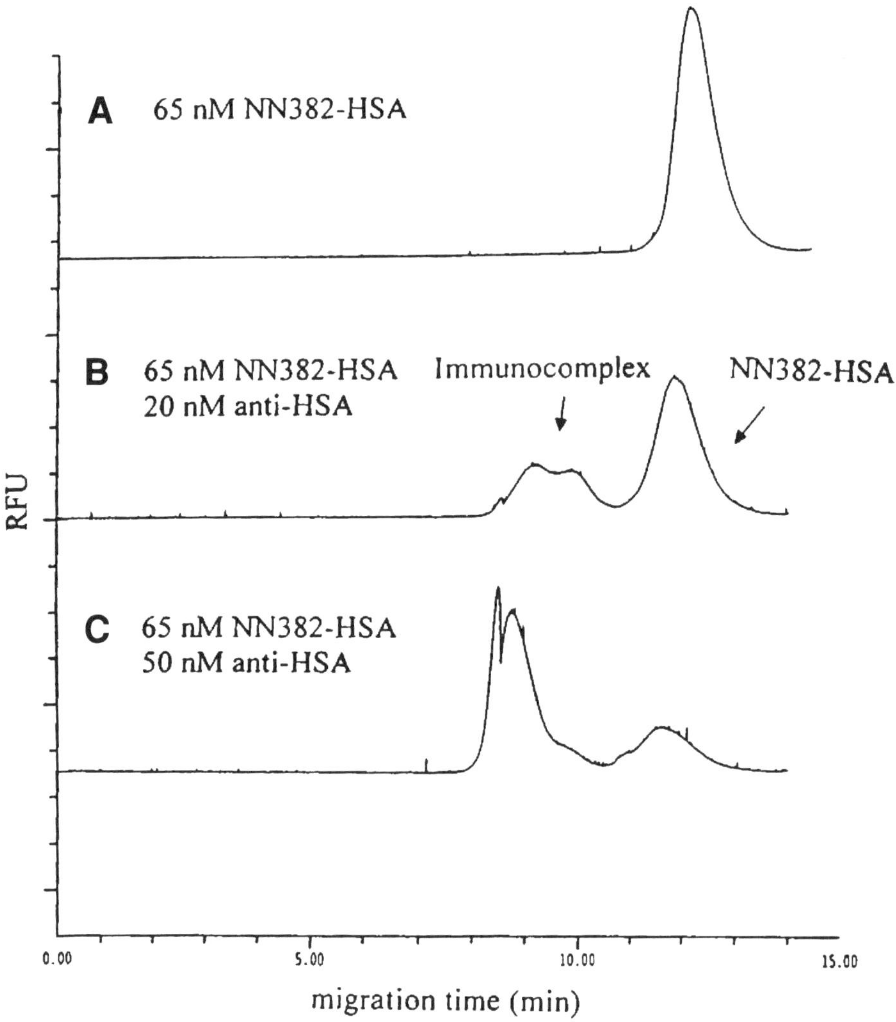

Fig. 16. Electropherograms for noncompetitive immunoassay for albumin. Separation conditions: 200 m*M* boric acid, pH 8.5, 15 kV, 30 s ramp time, and 5 s pressure injection (0.5 psi).

3.3.3. Competitive Immunoassay

Competitive immunoassays were performed by mixing 65 n*M* NN382-HSA, 65 n*M* HSA antibody, and 0–100 μ*M* HSA. All dilutions were made using phosphate buffer, pH 7.2. The solutions were vortexed for 5 min on slow speed at room temperature prior to injection.

Once it had been determined that the integrity of antibody–antigen interaction was intact, the next step was the competitive immunoassay. The detection limit of the assay was determined to be 500 p*M*, nearly a two orders of magnitude improvement over what has been achieved in the past *(14)*. Note that the effective range of the assay may be changed by altering the initial concentrations of NN382-HSA and HSA antibody. The electropherograms for the competitive immunoassay are shown in **Fig. 17A–D**. **Fig. 17A** is an electropherogram of a trace amount of free NN382-HSA and the immunocomplex. Upon addition of unlabeled HSA, the peak area of the NN382-HSA peak increases, whereas the peak area of the immunocomplex decreases. Again, this trend continues with increasing concentrations of unlabeled HSA (*see* **Fig. 17B–D**). The change in peak areas occurs owing to the competitive interaction between labeled and unlabeled HSA. Upon addition of unlabeled HSA, NN382-HSA is displaced from the antibody to some extent, explaining the decrease in the immunocomplex peak and the increase in the free NN382-HSA peak. Using this relation, a curve was constructed in order to determine the concentration of HSA in an unknown sample. The graph was made by plotting percent bound NN382-HSA vs log [HSA] (*see* **Fig. 18**).

3.4. Conclusion

Although covalent labeling is still used in most bioanalytical applications of NIR dyes, noncovalent labeling has many advantages. The labeling procedure is fast and sample clean-up may not be necessary. Additionally, rigorous control of pH is not required, making noncovalent labeling quite attractive when working with biological matrices, with their inherent pH sensitivity. One of the methodologies presented in this chapter utilizes a noncovalent labeling scheme for the development of a dye displacement assay for the determination of drug binding constants to HSA. The method developed allows for the determination of binding constants in a single run. The method is fast and sensitive, uses little reagent, and is amenable to automation.

As seen in the literature, most bioanalytical applications of NIR dyes still use covalent labeling schemes. Covalent labeling schemes eliminate the equilibrium between the dye and the target. They also allow for a much higher degree of specificity. CEIAs using NIR LIF detection is a method that exemplifies the advantages of using NIR dyes for bioanalytical applications. The technique combines the high selectivity of immunoassays, the high separation efficiency of CE and the high sensitivity of NIR LIF detection.

Although there are multiple advantages to CEIAs, they are not without drawbacks. One of the major problems is the labeling scheme. It should be noted that this problem is not unique to NIR dyes; indeed, it applies to all fluorescent

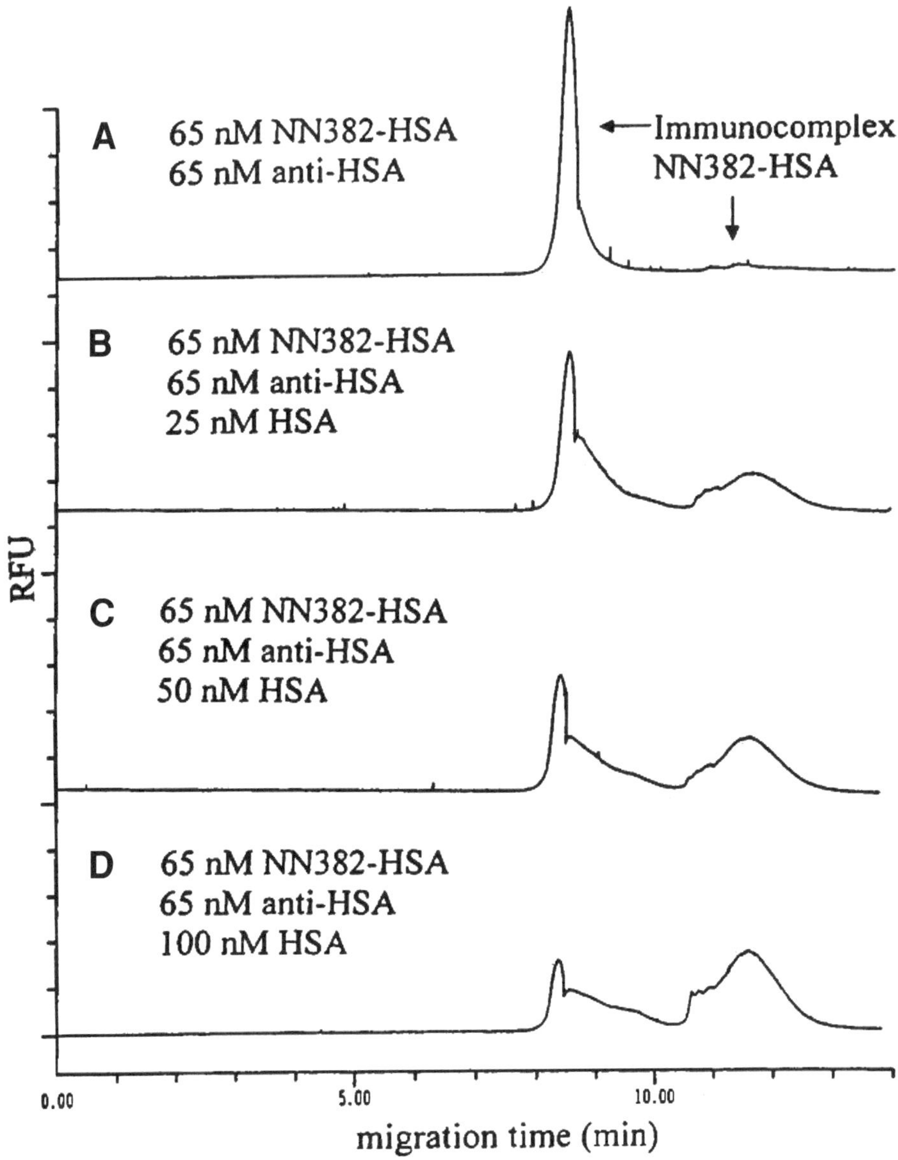

Fig. 17. Electropherograms for competitive immunoassay for albumin. Separation conditions: 200 m*M* boric acid, pH 8.5, 15 kV, 30 s ramp time, and 5 sec pressure injection (0.5 psi).

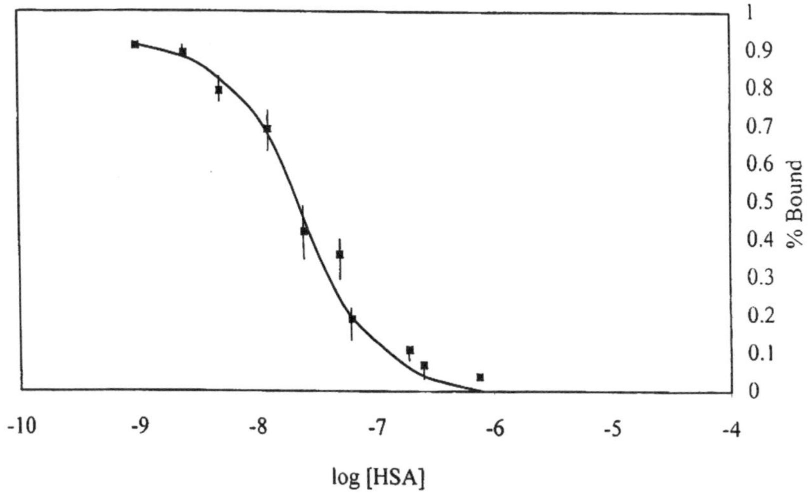

Fig. 18. Calibration curve for determination of albumin concentration.

probes. Most often, fluorescent dyes used for protein labeling contain an isothiocyanate functionality that is reactive toward primary amines. Consequently, elevated pH levels must be used in the labeling procedure (exceeding the value of the pK_a of the amine). Unfortunately, many proteins do not tolerate these elevated pH values so that there is a limit as to how high the pH may be taken during the derivatization reaction. As a result, labeling efficiency suffers and a heterogeneous dye–protein stoichiometry exists. This heterogeneous dye–protein stoichiometry leads to band broadening in chromatographic–type applications. The extent of this problem is directly related to the size of the protein.

An alternative approach would be to use dyes that have functionalities that target less prevalent groups, such as thiols. Because thiols are not nearly as common on proteins as amines, the final stoichiometry of the dye–protein is significantly decreased, thereby eliminating excessive peak broadening. This would allow for the possibility of doing multiple analyte CEIA in a single run. Additional applications could be single, site-specific molecular probes, as well as energy transfer applications.

4. Notes

1. NN382 has an isothiocyanate functionality that is reactive toward primary amines. Elevated pH levels must be used to deprotonate amines. However, many proteins are not stable at elevated pH values. Consequently, a trade-off exists between labeling efficiency and protein stability.

2. Unreacted dye must be removed from the labeled protein solution owing to the fact that the excess dye could potentially noncovalently interact with both protein and antibody.
3. When separating noncovalent complexes via CE, excess heat should be avoided at all costs. Heat has the potential to induce dissociation of the complex. An extended ramp time reduces shock to the sample.
4. In order for the data to be consistent, the solution must be vortexed until it has achieved equilibrium. Otherwise, kinetic effects will be observed.
5. As mentioned in **Note 3**, excess heat should be avoided when separating noncovalent complexes owing to induced dissociation. As a result, buffers that generate little current under the separation conditions are ideal.
6. When working with proteins, buffer pH is potentially very important. If the application requires that the protein's conformation remain intact, physiological conditions should be mimicked as closely as possible. A knowledge of how a protein behaves as a function of pH is necessary.
7. The dye displacement technique for binding constant determination assumes a 1:1 drug:protein binding ratio.
8. Capillary length is a potentially important parameter when doing CEIA. When performing CEIA, the immunocomplex is dissociating, to some extent, over the time-scale of the separation. Dissociation has the effect of decreasing the sensitivity of the measurement. It follows that shorter capillaries will give better sensitivity. However, decreasing the length of the capillary reduces the efficiency of the separation and therefore, degrades resolution. When doing CEIA, or potentially any noncovalent complex separation, a trade-off exists between sensitivity and resolution. It should be noted that the degree of dissociation over the time scale of a separation is dependent on the binding constant of the species involved.
9. The antigen–antibody mixture was vortexed for 5 min to ensure that equilibrium was established. If equilibrium is not established, kinetic effects will be observed.

References

1. Swartz, M. E. (1993) Capillary electrophoretic determination of inorganic ions in a prenatal vitamin formulation. *J. Chromatogr.* **640,** 441–444.
2. Schwartz, H. E., Ulfelder, J. K., Chen, F. A., and Pentoney, S. L. (1994) The utility of laser-induced fluorescence detection in applications of capillary electrophoresis. *J. Capil. Electorophor.* **1,** 36–54.
3. Gegendre, B. L., Williams, D. C., Soper, S. A., Erdmann, R., Ortmann, U., and Enderlein, J. (1996) An all solid-state near-infrared time-correlated single photon counting instrument for dynamic lifetime measurements in DNA sequencing applications. *Rev. Sci. Instrum.* **67,** 3984–3989.
4. Cheng, Y. F. and Dovichi, N. J. (1988) Subattomole amino acid analysis by capillary zone electrophoresis and laser-induced fluorescence. *Science* **242,** 562–564.
5. Lee, Y. H., Maus, R. G., Smith, B. W., and Winefordner, J. D. (1994) Laser-induced fluorescence detection of a single molecule in a capillary. *Anal. Chem.* **66,** 4142–4149.

6. Hillebrand, S., Schoffen, J. R., Mandaji, M., et al. (2002) Performance of an ultraviolet light-emitting diode-induced fluorescence detector in capillary electrophoresis. *Electrophor.* **23,** 2445–2448.
7. Kist, T. B. L., Termignoni, C., and Grieneisen, H.-P. H. (1994) Capillary zone electrophoresis separation of kinins using a novel laser fluorescence detector. *Braz. J. Med. Biol. Res.* **27,** 11–19.
8. Vicki, L., Ward, M., and Khaledi, G. (1998) Nonaqueous capillary electrophoresis with laser induced fluorescence detection. *J. Chromatogr. B* **718,** 15–22.
9. Lackowicz, J. R. (1999) *Principles of Fluorescent Spectroscopy*. Kluwer Academic, New York.
10. Ingle, J. D. and Crouch, S. R. (1998) *Spectrochemical Analysis*. Prentice-Hall, Upper Saddle River, NJ, ch. 12.
11. Davis, G. A. (1972) Dansylglycine as a fluorescent probe for aqueous solutions of cationic detergents. *J. Am. Chem. Soc.* **94,** 5089–5090.
12. Hinze, W. L. (1979) Solution Chemistry of Surfactants. Plenum, New York.
13. Legendre, B. L., Dixie, L., Moberg, D. C., and Soper S. A. (1997) Ultrasensitive near-infrared laser-induced fluorescence detection in capillary electrophoresis using a diode laser and avalanche photodiode. *J. Chromatogr. A* **779,** 185–194.
14. Ummadi, M. and Weimer B. C. (2002) Use of capillary electrophoresis and laser-induced fluorescence for attomole detection of amino acids. *J. Chromatogr. A* **964,** 243–253.
15. Berquist, J., Vona, M. J., Stiller, C.-O., and O'Connor, W. T. (1996) Capillary electrophoresis with laser-induced fluorescence detection: a sensitive method for monitoring extracellular concentrations of amino acids in the periaqueductal grey matter. *J. Neurosci. Methods* **65,** 33–42.
16. Strickland, M., Weimer, B. C., and Broadbent, J. R. (1996) Capillary electrophoresis of Cheddar cheese. *J. Chromatogr. A* **731,** 305–313.
17. Johansson, T., Petersson, M., Johanssons, J., and Nilsson, S. (1999) Real-time imaging through optical fiber array-assisted laser-induced fluorescence of capillary electrophoretic enantiomer separations. *Anal. Chem.* **71,** 4190–4197.
18. Ward, T. J., Nichols, M., Sturdivant, L., and King, C. C. (1995) Use of organic modifiers to enhance chiral selectivity in capillary electrophoresis. *Amino Acids* **8,** 337–344.
19. Tyutyulkov, N., Fabian, J., Mehlhorn, A., Dietz, F., and Tadjer, A. (1991) *Polymethine Dyes: Structure and Properties*. St. Kliment Ohridski University Press, Sofia, Bulgaria.
20. Hamer, F. M. (1964) *The Cyanine Dyes and Related Compounds*. Wiley, New York.
21. Dachne, S., Resch-Genger, U., and Wolfbeis, O. S. (1998) *Near Infrared Dyes for High Technology Applications*, NATO ASI series. Kluwer Academic, Dordrecht, The Netherlands.
22. Peters, A. T. and Freedman, H. S. (1995) *Modern Colorants: Synthesis and Structure*. Blackie Academic and Professional, Glasgow, UK.
23. Fabian, J., Nakazumi, H., and Matsuoka, M. (1992) Near-infrared absorbing dyes. *Chem. Rev.* **92,** 1197–1226.
24. Katritzky, A. R. and Sabongi, G. J. (eds.) (1990) *Infrared Absorbing Dyes*. Plenum, New York.

25. Rao, T. V. S., Huff, J. B., and Bieniarz, C. (1998) Supramolecular control of photophysical properties of cyanine dyes. *Tetrahedron* **54,** 10,627–10,634.
26. Lipowska, M., Patonay, G., and Strekowski, L. (1993) New near-infrared cyanine dyes for labeling of proteins. *Synth. Commun.* **23,** 3087–3094.
27. Strekowski, L., Lipowska, M., and Patonay, G. (1992) Facile derivatizations of heptamethine cyanine dyes. *Synth. Commun.* **22,** 2593–2598.
28. Strekowski, L., Lipowska, M., and Patonay, G. (1992) Substitution reactions of a nucleofugal group in heptamethine cyanine dyes. Synthesis of an isothiocyanato derivative for labeling of proteins with a near-infrared chromophore. *J. Org. Chem.* **57,** 4578–4580.
29. Lipowska, M., Patonay, G., and Strekowski, L. (1995) A novel near-infrared cyanine dye for bioanalytical applications. *Heterocycl. Commun.* **1,** 427–430.
30. Flannagan, J. H., Khan, S. H., Menchen, S., Soper, S. A., and Hammer, R. P. (1997) Functionalized tricarbocyanine dyes as near-infrared fluorescent probes for biomolecules. *Bioconjugate Chem.* **8,** 751–756.
31. Strekowski, L., Mason, C., Lee, H., and Patonay, G. Synthesis of water-soluble near-infrared cyanine dyes functionalized with [(Succuinimido)oxy]carbonyl group. *Heterocycl. Commun.*, in press
32. Strekowski, L., Mason, C., Lee, H., et al. Synthesis of a functionalized cyanine dye for covalent labeling of biomolecules with a pH-sensitive chromophore. *J. Heterocycl. Chem.*, in press.
33. Strekowski, L., Gorecki, T., Mason, J., Lee, H., and Patonay, G. (2001) New heptamethine cyanine reagents for labeling of biomolecules with a near-infrared chromophore. *Heterocycl. Commun.* **7,** 117–122.
34. Strekowski, L., Lipowska, M., Gorecki, T., Mason, C., and Patonay, G. (1996) Functionalization of near-infrared cyanine dyes. *J. Heterocyclic Chem.* **33,** 1685–1688.
35. Ernst, L. A., Gupta, R. K., Mujumdar, R. B., and Waggoner, A. S. (1989) Cyanine dye labeling reagents for sulfhydryl groups. *Cytometry* **10,** 3–10.
36. Gruber, H., Kada, G., Pragl, B., et al. (2000) Preparation of thiol-reactive Cy5 derivatives from commercial Cy5 succinimidyl ester. *Bioconj. Chem.* **11,** 161–166.
37. Retoff, S. and Larsen, P. R. (1989) *Endorcrinology*, Vol. 1. Saunders, Philadelphia, PA.
38. Westphal, U. (1971) *Steroid Protein Interactions*. Springer-Verlag, New York.
39. Kwong, T. C. (1985) Free drug measurements: methodologies and clinical significance. *Clin. Chim. Acta* **151,** 193–216.
40. Svennson, C. K., Woodruff, M. N. Baker J. G., and Lalka, D. (1986) Free drug concentration monitoring in clinical practice. Rational and current status. *Clin. Pharmacokin.* **11,** 450–469.
41. Barre, J., Didey, F., Delion, F., and Tellerment, J. P. (1998) Problems in therapeutic drug monitoring. *Ther. Drug Monit.* **10,** 133–143.
42. Levy, R. H. and Schmidt, D. (1985) Utility of free level monitoring of antiepileptic drugs. *Epilepsia* **26,**199–205.

43. Craig, D. B., and Dovichi, N. J. (1998) Multiple labeling of proteins. *Anal. Chem.* **70,** 2493–2494.
44. Bao, J. J. (1997) Capillary electrophoretic immunoassays. *J. Chromatogr. B* **699,** 463–480.
45. Chen, F. T.-A. and Pentoney, J. (1994) Characterization of digoxigenin—labeled B phycoerythrin by capillary electrophoresis with laser induced fluorescence detection. Application to homogenous digoxin immunoassay. *J. Chromatogr.* **680,** 425–450.
46. Schultz, N. M. and Kennedy, R. T. (1993) Rapid immunoassay using capillary electrophoresis with fluorescence detection. *Anal. Chem.* **65,** 3161–3165.
47. Tao, L. and Kennedy, R. T. (1996) On line competitive immunoassay for insulin based on capillary electrophoresis with laser induced fluorescence detection. *Anal. Chem.* **68,** 3899–3906.
48. Towns, J., Bao, J., and Reigner, F. E. (1992) Synthesis and evaluation of epoxy polymer coatings for the analysis of proteins by capillary zone electrophoresis. *J. Chromatogr.* **599,** 227–237.
49. Xue, Q. and Yeung, E. S. (1995) Differences in chemical reactivity of individual molecules of an enzyme. *Nature* **373,** 681–683.
50. Legendre, B. L. and Soper S. A. (1996) Binding properties of near IR dyes to proteins and the separation of the dye/protein complexes using capillary electrophoresis with laser induced fluorescence detection. *Appl. Spectrosc.* **50,** 1196.
51. Peters, T. (1996) *All About Albumin: Biochemistry, Genetics and Medical Applications*. Academic, San Diego, CA.
52. Sauda, K., Imasaka, T., and Ishibashi, N. (1986) Determination of protein in human serum by high performance liquid chromatography with semiconductor laser fluorometric detection. *Anal. Chem.* **58,** 2649–2653.
53. Hage, D. S., Noctor, T. A., and Wainer, I. W. (1995) Characterization of the protein binding of chiral drugs by high performance affinity chromatography, interactions of R- and S-ibuprofen with human serum albumin. *J. Chromatogr. A* **693,** 23–32.
54. Boga, O. and Borga, B. (1997) Serum protein binding of nonsteroidal anti-inflammatory drugs: a comparative study. *J. Pharmacokinet. Biopharm.* **25,** 63–77.
55. Sakai, T., Maruyama, T., Sako, T., et al. (1999) Stereoselective serum protein binding of ketoprofen in liver diseases. *Enantiomer* **4,** 477–482.
56. Sowell, J., Agnew-Heard, K., Mason, J. Ch., Mama, Ch., Strekowski, L., and Patonay, G. (2001) Use of noncovalent labeling in illustrating ligand binding to human serum albumin via affinity capillary electrophoresis with near-infrared laser induced fluorescence detection. *J Chromatogr. B* **755,** 91–99.
57. Sowell, J., Mason, J. Ch., Strekowski, L., and Patonay, G. (2001) Binding constant determination of drugs toward subdomain IIIA of human serum albumin by near-displacement capillary electrophoresis. *Electrophoresis* **22,** 2512–2517.

5

Capillary Electrophoresis in the Analysis and Monitoring of Biotechnological Processes

Vadim Klyushnichenko

Summary

Capillary electrophoresis (CE) became a versatile technique for analysis of biological macromolecules. We have applied capillary zone electrophoresis (CZE) and SDS-gel CE for the characterization of recombinant proteins during development of major bioprocessing steps, including fermentation, hybridoma cell cultivation, chromatographic purification, and chemical transformation. Rapid SDS-gel CE was developed for the fast analysis of fermentation broth and hybridoma cell culture. The total analysis time was reduced to 4.5 min. We have developed system for fraction collection, which allows analyzing separated proteins by MALDI–TOF-MS. The main advantages of applied techniques were high resolution and selectivity, fast analysis, and high accuracy.

Key Words

Analysis; biotechnological process; capillary electrophoresis (CE); capillary zone electrophoresis (CZE); customized CE; fermentation; high resolution; hybridoma cell culture; IgG, monoclonal antibodies; insulin; MALDI-TOF-MS; $NADP^+$-dependent formate dehydrogenase; optimization; proinsulin; rapid SDS-gel CE; recombinant proteins; SDS-gel CE; SDS-polymer capillary electrophoresis (SDS-polymer CE).

1. Introduction

The growing number of therapeutic recombinant proteins constantly requires new quantitative analytical methods with high variability of techniques, fast development, high reproducibility and simplicity of developed techniques, application of different on-line detectors instead of staining and destaining, high accuracy and sensitivity, the possibility of combination with other ana-

From: *Methods in Molecular Biology, vol. 276: Capillary Electrophoresis of Proteins* and Peptides
Edited by: M. A. Strege and A. L. Lagu © Humana Press Inc., Totowa, NJ

lytical techniques, and relatively low cost of operation. Capillary electrophoresis (CE) is becoming a more and more popular analytical tool for fast and quantitative analysis of recombinant proteins.

The strategies for analysis of proteins and peptides include regular and reversed-charge capillary zone electrophoresis capillary zone electrophoresis (CZE and RC-CZE) ***(1)***, capillary isoelectro focusing (CIEF), micellar electrokinetic chromatography (MEKC), sodium dodecyl sulfate-gel capillary electrophoresis (SDS-gel CE), isotachophoresis (ITP) ***(2)***, and their combination with high-performance liquid chromatography (HPLC), matrix-assisted laser desorption/ionization–time-of-flight mass spectrometry (MALDI–TOF-MS), and electrospray ionization–mass spectrometry (ESI–MS) ***(3,4)***. In particular, the combination of CE and MS has been successfully applied to the analysis of heterogeneity of glycoproteins expressed in mammalian cells ***(5,6)***, enzymatic removal of carbohydrate chains ***(7,8)***, the analysis of modified ***(9)*** proteins or biotechnology derived proteins ***(10,11)***, direct analysis of isoforms ***(12)*** and cytokine fragments ***(13)***, and to peptide mapping ***(14)***.

Compared to chromatographic techniques, CE provides a higher sensitivity to the analysis, which may even be increased by special capillary design, combinational approaches ***(15,16)***, or type of detection ***(17)***. Easy calibration makes this method quantitative and directly applicable to quality assessment of recombinant proteins, for monitoring of complex reactions ***(18–20)***, and in-process testing during cell cultivation, purification, and formulation ***(21–23)***.

In the current work, we summarize our experience with rapid and quantitative analysis of recombinant therapeutic proteins, expressed in large quantities in bacterial or mammalian systems. Subsequent purification of these proteins by chromatographic methods generates high-purity products with low levels of contaminants. Development of methodology for CE separation makes this method a highly efficient, widely applicable technique that can be enhanced by combined analytical techniques.

2. CZE in the Analysis of Step-by-Step Production of Recombinant Human Insulin

CZE is currently one of the most commonly used CE techniques for rapid analysis of proteins. Charged proteins are moving with different velocities in the electrophoretic buffer, based on their electrophoretic mobility, which depend on the protein size and charge at particular pH.

The aim of this work was the development of an analytical HPLC–CE system for each step of the technology of recombinant human insulin production from *Escherichia coli*. Here we show the advantages of CE separations, with the indirect comparison with HPLC.

2.1. Main Steps for Production of Recombinant Human Insulin

This technology consists of five main steps (*see* **Table 1**), accompanied by considerable transformation of protein molecules in terms of size, secondary structure, and charge (*see* **Fig. 1**; *24*). Recombinant proinsulin fusion protein (rPFP, Step 1) produced by recombinant cells was converted to denatured proinsulin (dP) by reaction with BrCN (Step 2) with opened and chaotically closed six -SH groups. After reaction of sulfitolysis, proinsulin-*S*-sulfonate (P-SSO3, Step 3) was reduced to a proinsulin molecule with three correct S–S bridges in the presence of β-mercaptoethanol (Step 4). During the final Step 5, proinsulin was converted to recombinant human insulin (rHI) under the influence of trypsin *(25)*.

There are several chromatographic separations between each technological step, including final RP-HPLC analysis of the final product. Each transformation was analyzed by different types of CE and HPLC in order to compare applicability of these two methods and build a unique analytical system (*see* **Table 1**). Although we have developed a combined HPLC–CE system, electrophoretic analysis was particularly effective in those conditions, where HPLC analysis was not effective because of the low resolution or sensitivity.

2.2. CZE in the Step-by-Step Monitoring

In the first step of the technology, direct analysis of purified rPFP by CZE was not effective, owing to high agglomeration of protein and formation of soluble aggregates (*see* **Fig. 2**). Because of their heterogeneity and high variability of charge to mass ratio, it was almost impossible to identify the peak of monomeric rPFP. It is interesting that we did not see the high variability of aggregates, by size-exclusion HPLC. Only after we treated the sample of rPFP with 2-mercaptoethanol (2-ME) did all forms of the protein convert to monomer, which a forms a monopeak (*see* **Fig. 3**). After the treatment of real cell extract with 2-ME, we were able to identify rPFP peak and proteins contaminating the host cell (*see* **Fig. 4**). Patrick and Lagu demonstrated the same approach in the analysis of rPFP by size-exclusion HPLC *(26)*. In the meantime, MECC separation of rPFP, monomerized by 2-ME, resulted in peak broadening, which excluded the possibility of analysis. Depending on the distribution of molecular weight and concentration of contaminating cell proteins, the SDS-polymer CE may be applicable for this kind of analysis, although the molecular weight of the proteins should differ by at least two times for clear separation.

Another useful feature of CE is its higher sensitivity, compared to HPLC, which allows better quantification of protein impurities or products of degradation at extremely low levels. For example, according to the requirements of

Table 1
Main Steps of the Technology for the Production of Recombinant Human Insulin

No.	Step	Main* and by-products	Molecular mass (kD)	Analytical methods
1	Isolation of fusion protein	**rPFP**	17	SE-HPLC
		Dimeric rPFP	34	CZE
		HMWP	40–70	
2	Proinsulin denaturation	**Denatured proinsulin**	9	SE-HPLC
		rPFP	17	
		HMWP	70	
3	Proinsulin sulfitolysis	**Proinsulin-*S*-sulfonate**	9.5	RP-HPLC
		Incompletely sulfonated proinsulin	9–9.3	CZE
		RPFP	17	
		Fusion protein-*S*-sulfonate	17.5	
4	Proinsulin renaturation	**Proinsulin**	9	SE-HPLC
		Structural analogs and oligomers	9, 18–36	CZE
5	Insulin production	Insulin	5.8	SE-HPLC
		Insulin analogs	5.7	RP-HPLC
		Insulin derivatives	5.7	CZE
		Proinsulin		MECC
		LMWP-HMWP	6–36	

*Main products are in bold.
HMWP, high-molecular-weight proteins; LMWP, low-molecular-weight proteins.

U.S. and British pharmacopoeias, the amount of recombinant proinsulin in purified insulin should be less than 0.1% *(27,28)*. The efficiency of HPLC is limited in this application by several thousand theoretical plates and high noise-to-peak ratio, which makes the analysis of such small amounts of impurities by HPLC virtually impossible. It is also difficult to quantify such impurities by slow immunoassays and polyacrylamide gel electrophoresis (PAGE) with overloaded lanes.

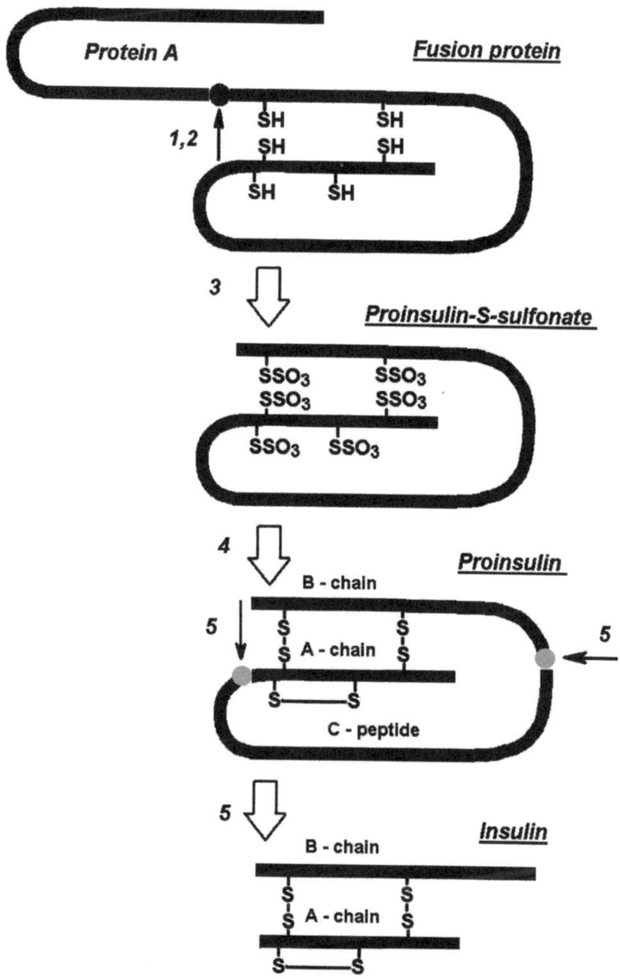

Fig. 1. Transformation of fusion protein into insulin: 1, rPFP; 2 , transformation of rPFP to linear proinsulin and protein A fragment; 3 and 4 , formation of recombinant human proinsulin (rHP); 5, C-peptide restriction out of rHP molecule, accompanied by the formation of recombinant human insulin (rHI). Reprinted with permission from **ref.** ***25***.

After optimization of buffer type and pH, we have established optimal conditions for separation of insulin and proinsulin with high selectivity (*see* **Fig. 5**). Even when the capillary was overloaded with real sample of purified insulin, the peak of proinsulin was detected at the same retention time (*see* **Fig. 6**). Despite the fact that the insulin peak was broad and contaminated with other

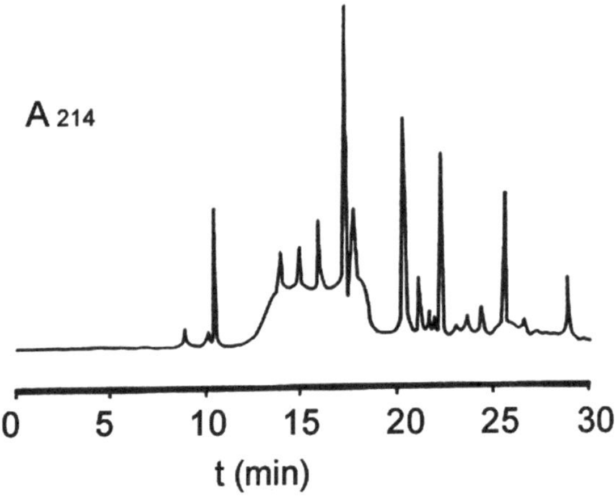

Fig. 2. Separaton of oligomers of rPFP by CZE. Reprinted with permission from **ref.** *25*.

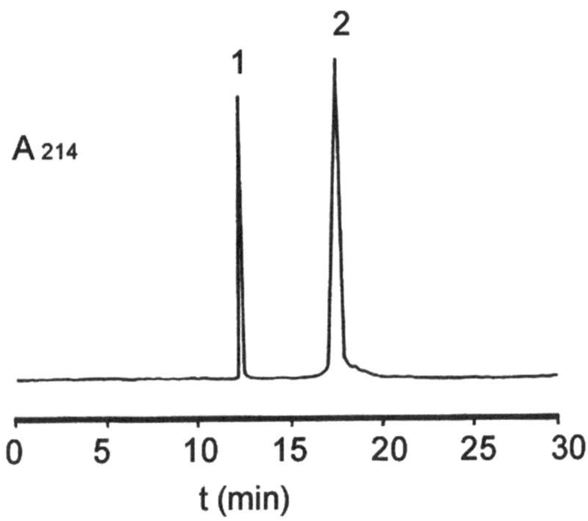

Fig. 3. Analysis of pure rPFP, treated with 2-mercaptoethanol (ME). Peaks were identified by migration time of standards: 1, ME; 2, rPFP. Reprinted with permission from **ref.** *25*.

minor derivatives, the resolution between the insulin and proinsulin peaks was high, as was the selectivity of separation. The level of proinsulin in the mixture was less than 0.1%, although close to the detection limit.

Despite wide commercialization of different types of technologies for production of recombinant human insulin, the further optimization of the process is continuing in many countries, because of the strategic importance of this protein in pharmaceutical development ***(29)***.

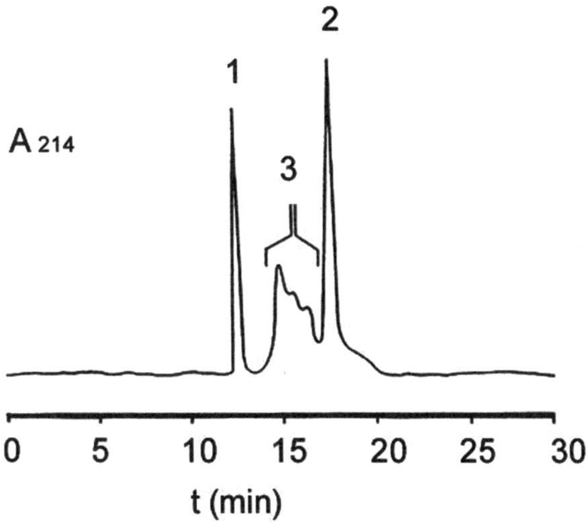

Fig. 4. Analysis of cell extract, containing rPFP. Peaks 1 and 2 are like in **Fig. 3**. Peak 3, contaminating proteins. Reprinted with permission from **ref.** ***25***.

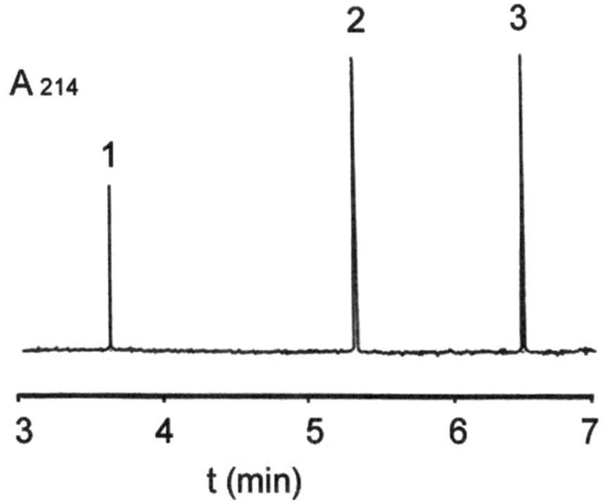

Fig. 5. Model separation of rHP and rHI by CZE under optimal conditions. Peaks were unidentified by migration times of standards: 1, internal marker of electroosmotic flow; 2, rHP; e, rHI; injection by voltage +5kV during 1.1s. Protein concentration: 0.3 mg/mL. Analysis was performed using Applied Biosystems 270A CE system. Conditions: fused silica capillary, supplied by Beckman, id 100 μm, 52-cm length, 40-cm effective length; temperature 28°C; voltage 20 kV; buffer 0.03 *M* Na-Phosphate (pH 11.2); detection UV-214 nm. Reprinted with permission from **ref.** ***25***.

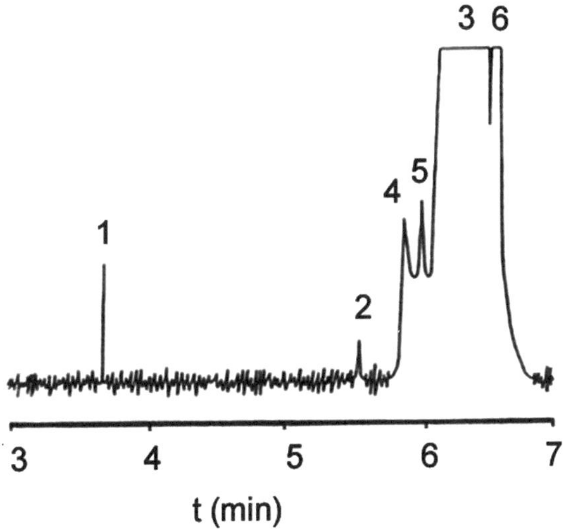

Fig. 6. Determination of minor impurity of rHP in rHI (purified by RP-HPLC), by CZE. 1, internal marker of electroosmotic flow; 2, rHP; 3, rHI; 4, diarginine insulin; 5, arginil-(AO)-insulin; 6, desamido insulin. Conditions same as in **Fig. 5**, except injection was done by voltage +5 kV during 8 s. Reprinted with permission from **ref.** *25*.

2.3. Optimization of CZE

Many parameters can be used for optimizing CZE separation. Most common are type of buffer, pH *(30)*, ionic strength, organic solutions, denaturing agents *(31)*, temperature, length and id of the capillary *(32)*, type and time of injection, protein microheterogeneity, protein–capillary wall interactions *(33)*, and others. Frequently, it is very important to have maximal separation of model compounds, because the resolution and selectivity of the separation of real mixtures is lower owing to contamination with impurities.

We have optimized CZE separation of insulin and proinsulin using a standard approach: adjusting the pH value of the phosphate buffer and thus providing a maximal difference between the charges of the proteins (*see* **Fig. 7**) *(34)*. According to our data, the best selectivity was found at pH 7.0–11.2, whereas the optimal time was 15–25 min. The resolution between peaks was higher at basic pH, and optimal pH of the buffer was determined to be pH 11.2.

In another experiment, we have studied the influence of conductivity and pH of Na-phosphate buffer upon the number of theoretical plates of the proinsulin peak (*see* **Fig. 8**). The parabolic form of 3D plot with main maximum at

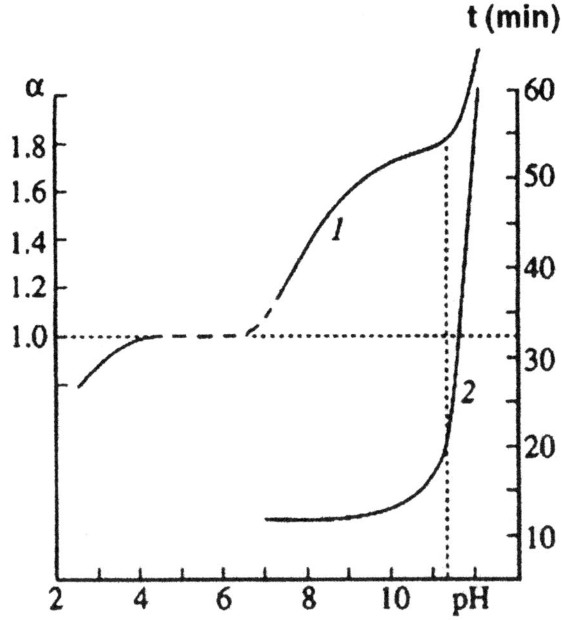

Fig. 7. Effect of buffer pH on: *1*, selectivity of CZE separation of insulin and proinsulin; and *2*, the migration time of insulin. Conditions: fused silica capillary, supplied by Beckman, id 100 μm, 52-cm length, 40-cm effective length; temperature 28°C; voltage 10 kV; buffer 0.01–0.03 *M* Na-phosphate pH 2.5–5.0 and 7.0–12.0); conductivity approx 4 mS/cm; concentration of protein in sample 0.1 mg/mL; injection by voltage at +5 kV during 0.1 s; detection UV-214 nm.

pH 11.2 and conductivity at 4 mS makes clear the conditions for the sharpest peak of proinsulin. If several analyzed proteins have similar nature, like insulin, proinsulin, and desamido insulin, the most effective conditions for their separation should be found from their independent plots.

2.4. Materials

1. Water was purified on Milli-Q system (Millipore, MA).
2. 0.1 *M* Na-borate buffer, pH 9.3.
3. 0.01–0.03 *M* Na-phosphate buffer, pH 2.5–5.0 and 7.0–12.0); conductivity approx 1–10 mS/cm.
4. All buffers were filtered through GVWP filters (Millipore) and degassed for 20 min.

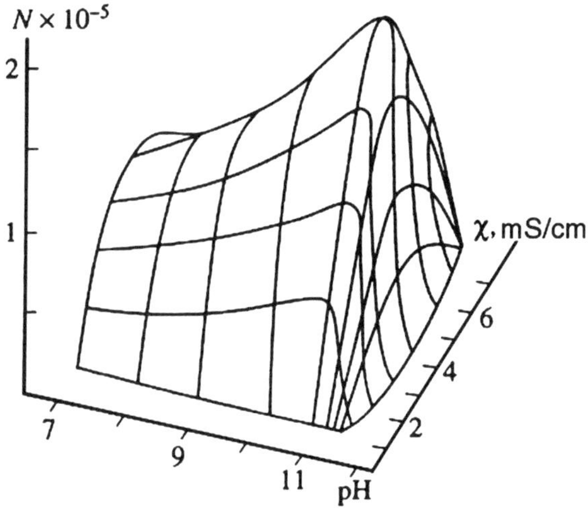

Fig. 8. Effect of conductivity χ and pH of buffer solution on the selectivity of CZE of proinsulin. Experimental conditions: fused silica capillary, supplied by Beckman, id 100 μm, 52-cm length, 40-cm effective length; temperature 20°C; voltage +10 kV; buffer 0.03 *M* Na-phosphates, pH 7.0–12.0; concentration of protein in sample 0.1 mg/mL; injection by voltage at +5 kV during 0.1 s; detection UV-214 nm. *N* is the number of theoretical plates.

2.5. Methods

1. Analysis was performed using the Beckman PACE 2010 system (except where stated otherwise).
2. Fused silica capillary, supplied by Beckman, id 100 μm, 87-cm length, 80-cm effective length.
3. Conditions: temperature 28°C; voltage 10 kV; buffer 0.1 *M* Na-borate, pH 9.3; concentration of protein in sample 1 mg/ml; injection by voltage at +5 kV during 5 s.
4. Detection: on column UV-214 nm.
5. Reductive degradation of protein samples by 2ME was performed with heating at 100°C for 1–3 min.
6. For the separation, we used specimens of insulin, proinsulin, denatured proinsulin, proinsulin-*S*-sulfonate, and fusion rPFP, obtained from different steps of the preparation of recombinant human insulin at Shemyakin and Ovchinnikov Institute of Bioorganic Chemistry, Russian Academy of Sciences. For identification of human insulin we used a standard specimen (Atlanta, GA, cat. no. 83/500, Chemie- und Hadelsgesellshaft, Heidelberg, Germany).
7. For optimization of pH CZE separations we have used a fused silica capillary, supplied by Beckman, id 100 μm, 52-cm length, 40-cm effective length.

8. Conditions: temperature 28°C; voltage 10 kV; buffer 0.01–0.03 *M* Na-phosphate, pH 2.5–5.0 and 7.0–12.0; conductivity approx 1–10 mS/cm; concentration of protein in sample 0.1 mg/mL; injection by voltage at +5 kV during 0.1 s; detection: on column UV-214 nm.

2.6. Notes

1. There are certain limitations for the sample preparation of insulin and insulin-containing proteins. Insulin is a hydrophobic protein with isoelectric point (p*I*) close to 5.2. This protein is practically insoluble in aqueous solutions at neutral pH. The solubility of insulin increases at both acidic (pH < 2.0) or basic (pH >8.0) extreme conditions, where the stability of protein itself may be affected during long treatment. Therefore, the prepared samples should be separated immediately.
2. Proinsulin and its derivatives, as well as rPFP, have broader ranges of solubility. However, they should be treated in the same way as insulin for better comparison of separation results.

3. Analysis of Hybridoma Cell Culture Process by SDS-Polymer CE

SDS-polymer CE is a fast, powerful, and sensitive method for qualitative and quantitative analysis of biopolymers. Comparison of SDS-polymer CE with regular SDS-polyacrylamide gel electrophoresis (SDS-PAGE) and SE-HPLC shows that the capillary format can be faster (it does not require staining and destaining) and automated with on-line detection ***(35)***. It requires only small sample and buffer volumes ***(36,37)***, and allows for detection of small differences in the molecular mass of aggregated proteins ***(38,39)***. There are many parameters to consider in order to develop high-performance analysis. Different types of polymers have been used in the separation medium to optimize SDS-polymer CE ***(40)***. In order to minimize electro-osmotic flow (EOF), the inner surface of the capillary should be modified by linear polyacrylamide through Si-O-Si-C or Si-C linkages ***(41,42)***. Depending on the type of modification, and the stability of the surface, the quality and stability of analysis show significant differences ***(43,44)***. Such parameters as temperature id and length of the capillary, and the applied electric field are also important for the efficiency of protein separation and the life span of the column ***(45)***.

The objective of this work was the monitoring of real biotechnological process by adopting SDS-polymer CE in order to carry out on-line analysis and to obtain data to regulate the process ***(46)***. During cultivation of hybridoma cells, there is a change in the concentration of proteins contained in the medium and the main product (IgG). Great differences between the pure protein peak and the peak of the same protein in a complex medium are observed, the latter usually being more complex and wider. In order to optimize SDS-polymer CE, sample preparation, injection application of real fermentation broth, and some modification of SDS-polymer CE for fast analysis had to be considered.

3.1. Calibration of Capillary

Semilinear calibration curves in the range 14–200 kDa were obtained by running proteins with different molecular mass (*see* **Fig. 9**). Because of the complex nature of proteins, they can differ in terms of charge, hydrophobicity, structure, and composition (glycoproteins), and they may interact with SDS molecules, polymer, and capillary wall, and this may effect their migration in SDS-polymer solution *(47)*.

3.2. Sample Injection

One of the main advantages of CE is the possibility of analyzing small volumes of sample placed into special vials for automated injection. The volume injected from the vial (with the 5–50 μL of the sample) into the capillary is usually 0.1–0.2 pL, for a total capillary volume of 0.5–2.0 μL. We have observed that peak heights after multiple injection of protein are different in the case of injection from a single or from different vials containing 50 μL of the protein solution (*see* **Fig. 10**).

Upon injection from the single vial, the height of the next peak was always up to 25–30% lower. We have found that this occurs because small droplets of polymer buffer are transferred by the end of the capillary along with the polymer solution (*see* **Fig. 11**). The peak height does not decrease if the injection comes from different vials.

3.3. Analysis of the Main Protein Components

The cell media for the hybridoma cultures consists initially of the main protein components: bovine serum albumin (BSA), transferrin, and insulin. The concentration of IgG increases during the cultivation period. In order to determine each component correctly with changing concentration, we have analyzed their electrophoretic properties and stability under identical conditions. BSA used for cell cultivation exhibits one major and three minor peaks (*see* **Fig. 12**).

Transferrin exhibits a double peak with molecular mass 74–80 kDa and insulin appears as a peak close to the peak of low molecular mass substances (*see*s **Fig. 13**). BSA and transferrin were relatively stable. Their peaks did not change during storage under different conditions.

Degradation of pure IgG (obtained at the Institute of Enzyme Technology) to three peaks was observed after storage of protein solution and attributed to proteolytic degradation owing to microbial contamination (*see* **Fig. 14**). The decomposition of the main peak and the appearance of some additional peaks of lower molecular mass were observed (*see* **Fig. 14A–D**).

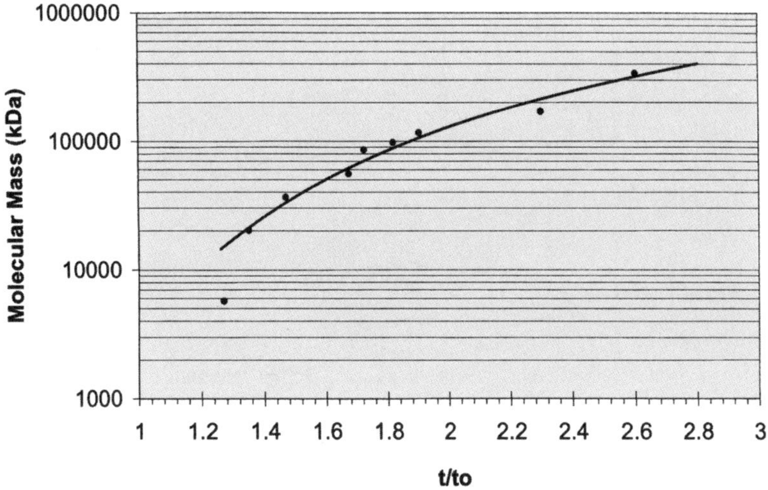

Fig. 9. Calibration curve for SDS-polymer CE. *t*, elution time of protein; *to*, elution time of marker (Orange G).

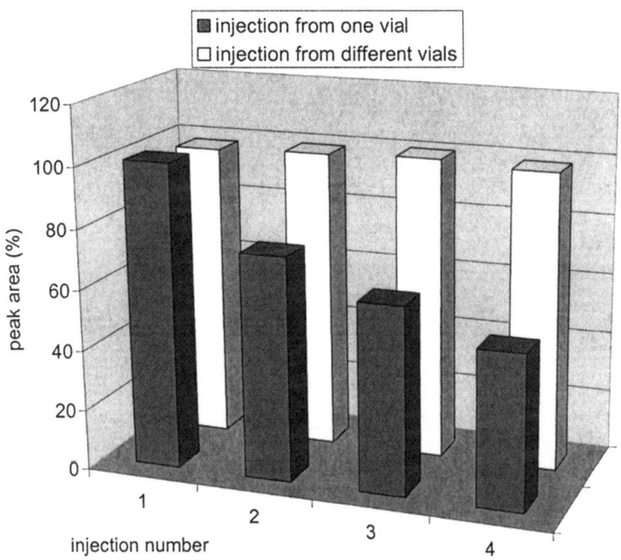

Fig. 10. Multiple injection from single and different vials.

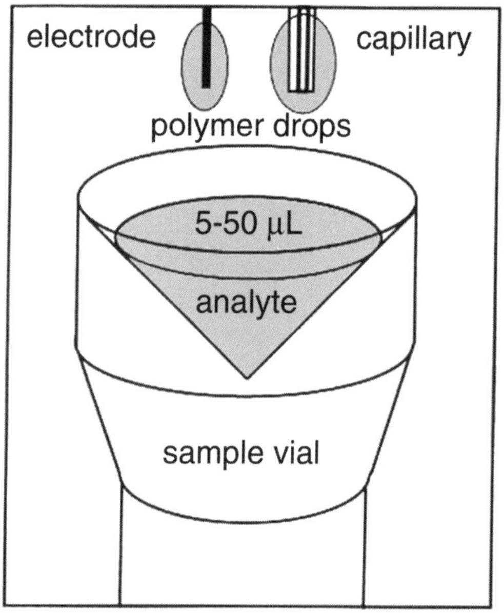

Fig. 11. Transfer of the small droplets of polymer buffer to the vial with sample by capillary and electrode.

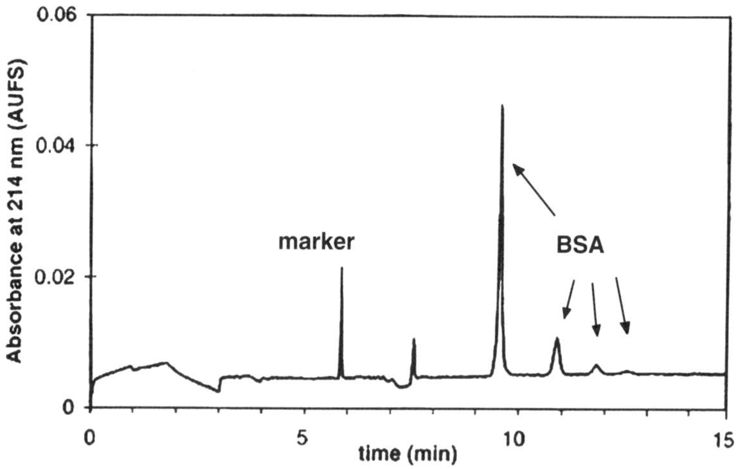

Fig. 12. Analysis of BSA by SDS-polymer CE. Capillary length 27 cm; 20 cm effective length; voltage 8.1 kV; electrokinetic injection for 20 s, 5 kV.

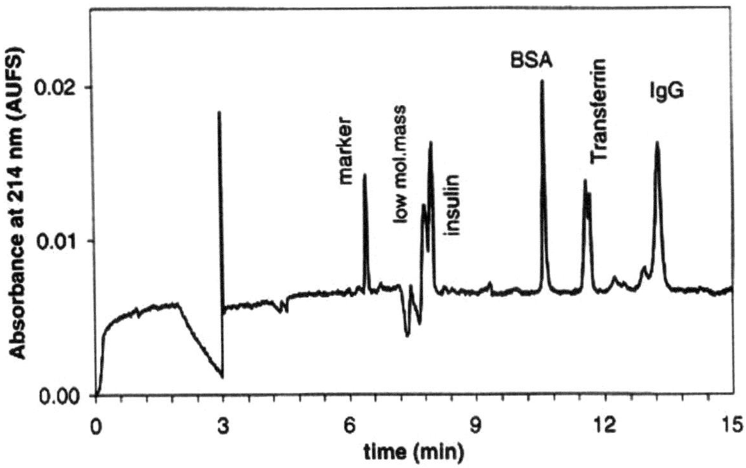

Fig. 13. Separation of pure proteins by SDS-polymer CE following electrokinetic injection for 20 s. Voltage 5 kV. Other conditions as in **Fig. 12**.

After reductive degradation with 2ME IgG exhibits two peaks, H- and L-chains, corresponding to the proteins with 25 and 50 kD (*see* **Fig. 14E**). Thus, it was possible to identify the first peak (14.5 min), as an L-chain and assume that the group of three peaks (19–22 min) was IgG and degraded IgG after the loss of one and two L-chains (*see*s **Fig. 14B–D**). IgG degradation by microbial contamination during storage of protein solution at room temperature for zero, 24, and 48 h was also observed by CZE (*see* **Fig. 15**), and the decrease in the main peak (IgG) area and the appearance of the sharp peak at 18 min were also observed. In the meantime, the SDS-polymer CE provided better resolution and identification of newly appearing compounds (*see* **Figs. 14** and **15**).

3.4. Sample Injection and Peak Area Calibration

A calibration curve of a peak-vs-protein concentration was established for IgG, insulin, BSA, and transferrin to provide a basis for the analysis of real samples from the cell broth (*see* **Fig. 16**). The time of injection was varied from between 10 and 99 s by pressure or electrokinetic mode. For injection by pressure, the most suitable range was 20–60 s. For short pulses of injection, the resolution is higher, but the peak area is smaller and the noise level is higher. On electrokinetic injection the resulting resolution and peak area were higher (*see* **Fig. 13**). The protein concentration used for calibration varied between 0.01–1.00 mg/mL. At protein concentration lower than 0.01 mg/mL the peak-

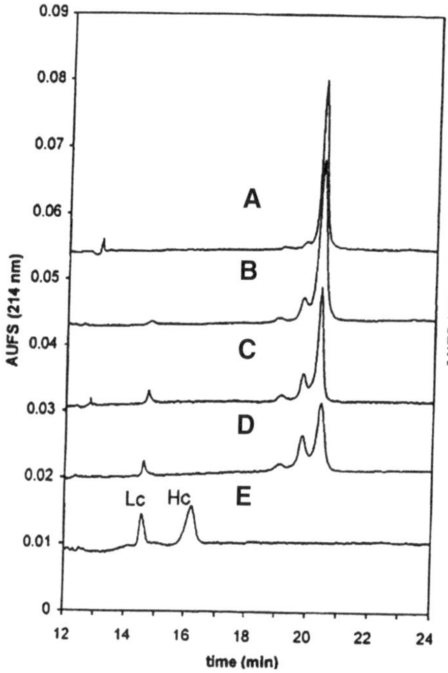

Fig. 14

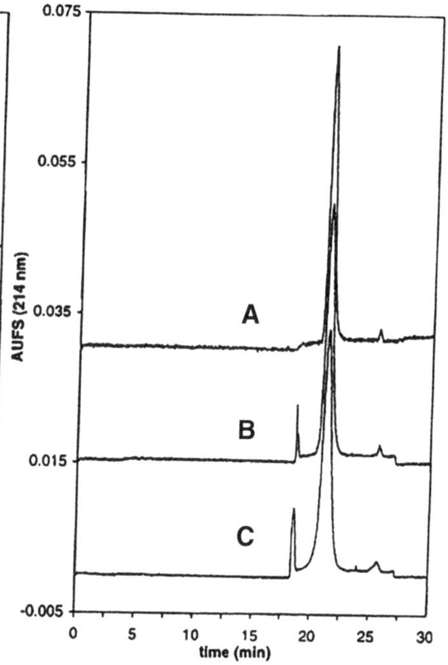

Fig. 15

Fig. 14. Analysis of proteolytic degradation of IgG during storage of the protein solution at room temperature. *A*, 0 h; *B*, 24 h, *C*, 48 h; *D*, 72 h; *E*, IgG after reductive degradation by 2-ME with heating. Conditions of separation: coated capillary 37-cm length; 12 kV; electrokinetic injection 40 s, 5kV. Abbreviations: Lc, light chain; Hc, heavy chain.

Fig. 15. CZE analysis of the proteolytic degradation of IgG. *A*, 0 h; *B*, 24 h; *C*, 48 h. Conditions of separation: 0.1 *M* Na-borate buffer, pH 9.0; bare silica capillary 60 cm in length; voltage 10 kV.

to-noise ratio was found to be lower than 6/1 for separation of pure proteins and 3/1 in the case of fractions from culture broth. Thus, for correct quantitative measurement of proteins in process samples, preconcentration may be beneficial.

3.5. Sample Preparation for CE

The concentration of each protein component of initial cell culture media and at the end of cultivation is usually very small and it is hard to quantify them properly without sample concentration (*see* **Table 2**).

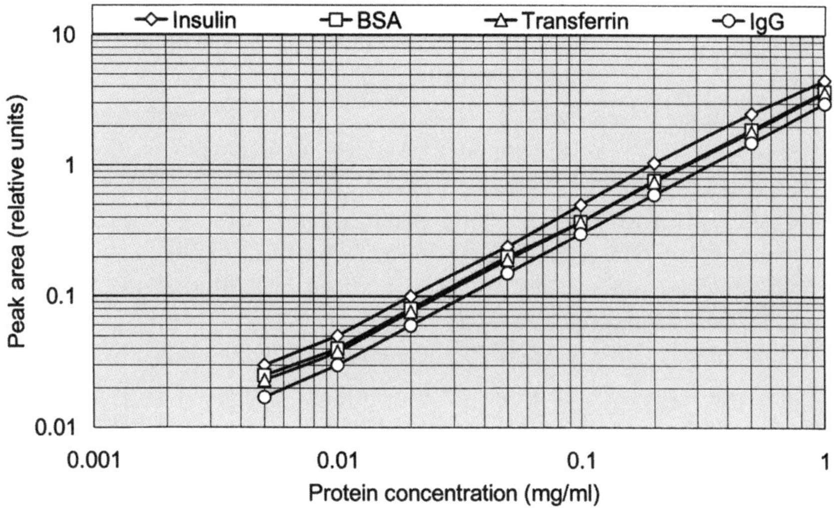

Fig. 16. Calibration curve of protein concentration. Injection 40 s by pressure. The peak area is presented in relative units. Other conditions as in **Fig. 12**.

Table 2
Initial and Final Concentration of Main Protein Components in the Cell Culture Broth

Protein	Initial protein concentration in the medium (mg/L)	Final protein concentration in the cell broth (mg/L)
Insulin	1	Lower
BSA	100	80–90
Transferrin	4.3	4
IgG	1–5	10–200
Other proteins and peptides	0	0–20–100

Previously, the IgG concentration in cell culture was determined by affinity HPLC and there were some problems evident with measurements carried out under different conditions. SDS-polymer CE usually gives higher values as it does not discriminate between native and denatured forms of the protein. During IgG production, both the media proteins and IgG are partially degraded. In addition, a small amount of the cell proteins and peptides are released in the cell culture broth. These new components could also be detected by SDS-polymer CE. Two techniques for sample preconcentration were compared, using identical samples: (1) ultrafiltration of microfiltered culture by centrifugation by Centricon-10 filters, and (2) precipitation in an ethanol-chloroform solution.

Concentration by Centricon filters was performed using centrifugation, according to the instructions of the supplier (Amicon Division, W. R. Grace & Co., Beverly, MA). Pefablock (Pentapharm, Basel, Switzerland) and EDTA (Merck, Darmstadt, Germany) were used for protease inhibition. For further protein concentration, we have used ethanol-chloroform precipitation: 1 vol of sample was added to 1 vol of methanol and one-quarter vol of chloroform (relatively to sample volume), mixed by vortexing and centrifuged for 10 min at 12,000*g*. After centrifugation, the protein should be in interphase in the form of a film between two liquids. The upper phase should be removed by a Hamilton syringe or pipet and one vol of methanol added for mixing and centrifugation under the same conditions. The supernatant was removed and the sediment was dried in air and dissolved in 0.1 *M* Tris-HCl, 1% SDS, pH 6.6. Before analysis, all buffers and samples were degassed and filtered through 0.22–0.45 μm filters.

There are two important aspects of sample concentration by ethanol-chloroform precipitation: (1) after the first centrifugation, care should be taken not to damage the protein film formed at the interphase upon removal of the top phase; (2) after the second centrifugation and liquid removal, before addition of SDS buffer, the protein sediment should not contain residual liquid or be over dried.

Usually it is very difficult and time consuming to dissolve dried protein, even in SDS solution. For better protein dissolution in SDS buffer, the sample may be heated up to 95°C for 5–30 min in a tightly closed tube. After cooling in ice, the tube should be centrifuged for 1 min at 2500*g* in order to collect all small droplets that have evaporated and condensed on the walls liquid. Without the heating step large proteins may not be completely dissolved. The final solution should be clear and easily passed through a microcellulose filter (0.22 μm).

Using unconcentrated cell culture we observed a BSA peak and only traces of IgG (*see* **Fig. 17A**). In front of BSA, a broad peak of a probably nonprotein nature is observed. After concentration of sample four times by ultrafiltration, resolution between peaks is decreased, the peaks become wider, noise is growing, and it becomes difficult to analyze small peaks. The concentrated solution was found to be unstable, so in a few hours of storage it was impossible to obtain meaningful results, even at higher concentration of SDS and in the presence of such protein stabilizers as EDTA and Pefablok. Precipitation technique with ethanol-chloroform was more effective. The resolution did not decrease, and we could analyze small peaks. Accurately concentrated solution was stable, with reproducible patterns after storage for 4 d (*see* **Fig. 17B**). The most effec-

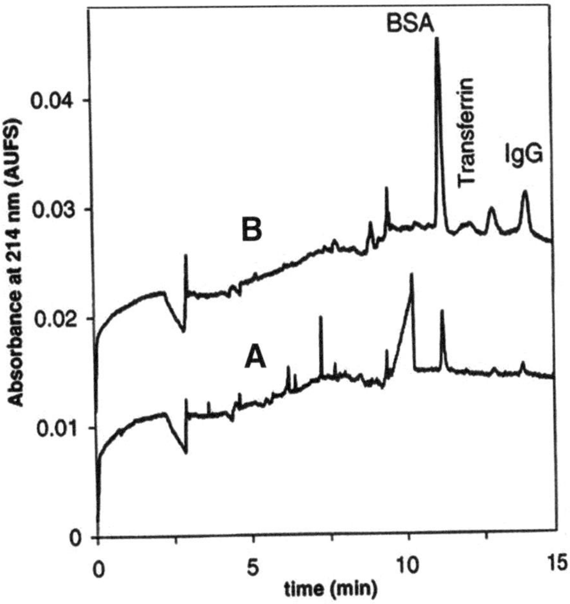

Fig. 17. Separation of proteins from hybridoma cell cultivation. (**A**) unconcentrated culture medium. (**B**) culture broth, concentrated five times with ethanol-chloroform precipitation, after 4 d of cell cultivation.

tive level of concentration by this technique is 5–10 times, depending on initial protein concentration. It should noted, that even under these conditions the initial concentration of insulin is not high enough for accurate quantitative analysis.

3.6. Analysis of Cell Cultivation

3.6.1. Batch Cultivation

After development of analytical methodology for SDS-polymer CE we have applied this tool for analysis of hybridoma cell cultivation (*see* **Fig. 18**). After 12 d of cultivation, the concentration of IgG was much higher and many low molecular weight proteins were detected (*see* **Fig. 18C**).

During a batch cell-cultivation process, the samples were analyzed each day by SDS-gel CE and by affinity chromatography (*see* **Fig. 19**) to quantify the mAb by an independent method. Cell-growth started after inoculation with 2.5×10^5 cells/mL and the 24-h lag phase (*see* **Fig. 19B**). In the next 3 d, cells grew exponentially with low cell-specific Ab production. After 96 h of cultivation, the cells entered the stationary phase, which lasted for 5 d. During this time, the cell-specific Ab production was high and the Ab concentration

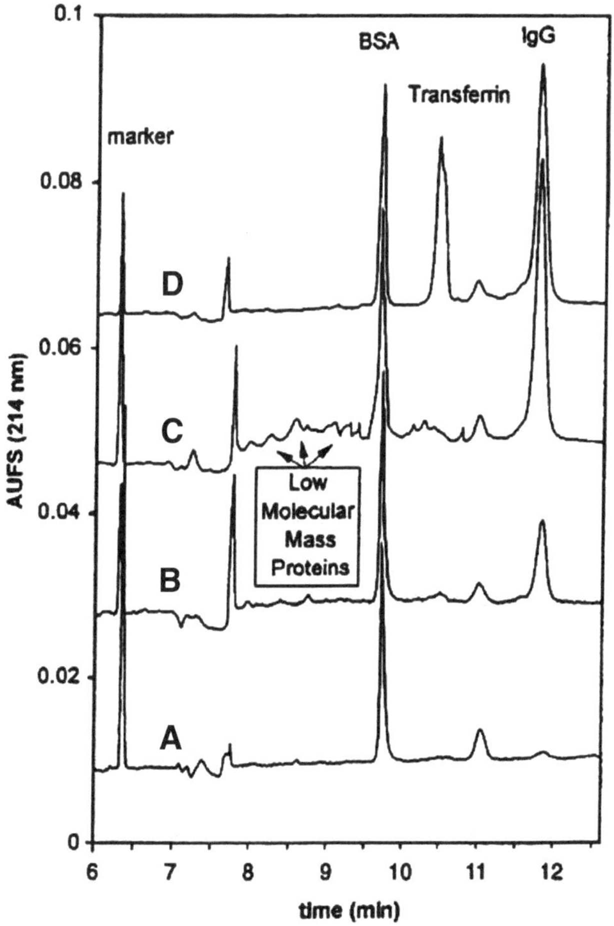

Fig. 18. SDS-gel CE analysis of protein compounds in the batch cell culture broth: (**A**) 1st, (**B**) 5th, (**C**) 12th d, (**D**) separation of the model proteins. Conditions of separation: coated capillary 27-cm length, 8.1 kV; electrokinetic injection 40 s, 5 kV. Reprinted with permission from **ref.** ***53***.

exceeded 80% of its maximum value. In the last 3 d, of culture, the starving phase, cell viability decreased dramatically. Within this period of time, Ab concentration increased, mainly by the release of preformed Ab from the cells.

All three phases (exponential growth, stationary phase, and starving phase) could be distinguished in the CE measurements. There was a slight increase of Ab concentration during the first 4 d of cultivation (lag and exponential phase) followed by 5 d of exponential increase of Ab concentration (stationary phase). During the last 3 d of cultivation, Ab concentration increased only slightly (starving phase). The BSA concentration decreased slightly during the station-

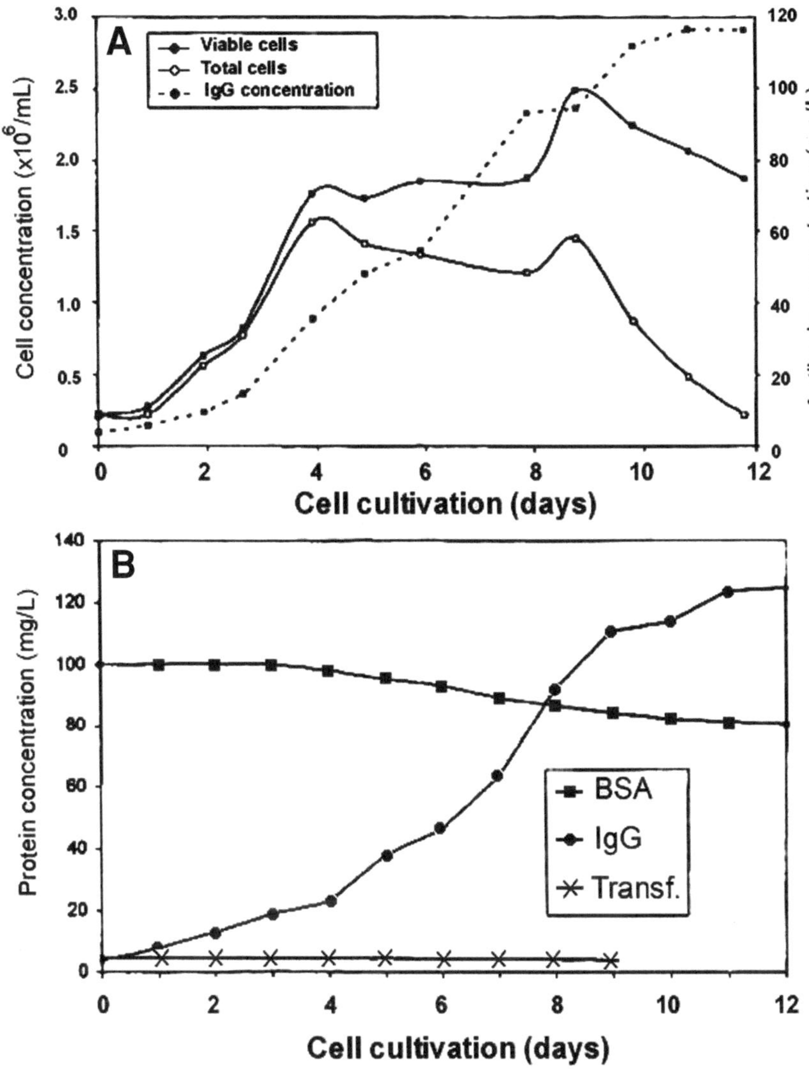

Fig. 19. Analysis of hybridoma cell cultivation in batch mode: (**A**) viable and total cell concentration and IgG concentration measured by affinity HPLC; (**B**) SDS-gel CE analysis of whole broth. Conditions as in **Fig. 17**. Reprinted with permission from **ref.** ***53***.

ary phase, whereas the transferring concentration stayed constant up to the ninth day. Because of the new peaks appearing near the transferrin position, it was difficult to quantify the transferrin during the later stages of growth. The protein concentration profile, measured by CE was similar to one measured by affinity HPLC (*see* **Fig. 19**).

3.6.2. Continuous Cultivation

The same analytical methods were employed for the analysis of the cell cultivation in continuous mode (*see* **Fig. 20**). The bioreactor was inoculated with 5×10^5 cells per mL reactor volume. After 9 d of growth, cell density reached a steady-state concentration of 1.7×10^8 cells per mL carrier. During this period of cultivation, antibody concentration increased. Owing to the exponential increase of the medium feed rate, there is a decrease in IgG concentration after d 3. After the addition of an inhibitor to the media (4 m*M* lactate), the cell density dropped to reach another steady-state value on d 17. Addition of even higher concentrations of the inhibitor (8 m*M*) led to a steep decrease in cell numbers at the end of the cultivation. These two steps of inhibition are clearly observed by SDS-gel CE analysis. The Ab concentration decreased after additions of inhibitor by a factor of two, whereas the concentration of BSA and transferrin did not change.

3.7. Materials

1. 1.0 *M* HCl was used for the rinsing of capillary.
2. 0.1 *M* Na-borate buffer, pH 9.0, was used for CZE separation.
3. Proteins for SDS-polymer capillary calibration (14–200 kDa) were supplied by Beckman, Fullerton, CA).
4. Orange G (Beckman) was used as a low-molecular-weight marker.
5. IgG (150 kD) was purified at the Institute of Enzyme Technology, by standard methods.
6. Other proteins used in the experiment: insulin (5.6 kD), BSA (67 kD), and transferrin (78–85 kD) were obtained from Gibco-BRL (Eggenstein, Germany).
7. Before analysis, all buffers and samples were degassed and filtered through a 0.22–0.45-μm filter.

3.8. Methods

1. All CE separations were performed with a P/ACE 2100 device for CE (Beckman).
2. Prior to each analytical run a bare silica capillary with modifications according to Hjerten ([*48*]; 37- or 47-cm length [30- or 40-cm effective length] × 100 μm id × 375 μm od) was rinsed with 1.0 *M* HCl and water for 2 min each, followed by filling with the SDS-polymer solution (eCAP SDS 14-200, Beckman) for 5 min.
3. The column temperature was maintained at 20°C by circulating a coolant to minimize band diffusion and ensure effective size separation.
4. Electrophoretic runs were conducted at 300 V/cm using the SDS-polymer solution (Beckman).
5. For SDS-Polymer separations we used the following conditions:
   a. Capillary length, 27 cm; 20-cm effective length; voltage 8.1 kV; electrokinetic injection for 20 s, 5 kV;
   b. Coated capillary 37-cm length; 12 kV; electrokinetic injection 40 s, 5kV.

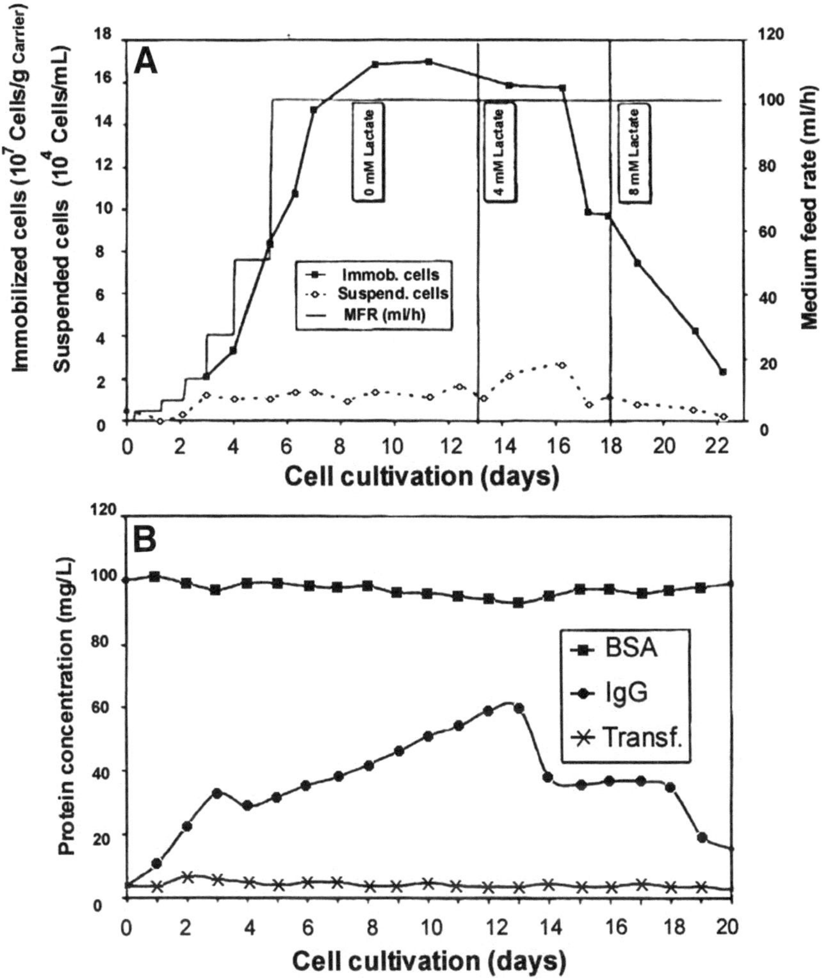

Fig. 20. Analysis of hybridoma cell cultivation in continuous mode: (**A**) immobilized suspended cells and flow-rate; (**B**) SDS-Gel CE analysis of whole broth. Conditions as in **Fig. 16**. Reprinted with permission from **ref.** ***53***.

6. For CZE analysis of IgG under basic conditions, an uncoated capillary of 57 cm (50-cm effective) length was used.
7. Reductive degradation of protein samples by 2ME was performed with heating at (100°C) for 1–3 min.

3.9. Cell Culture

The cell culture was performed as a batch with continuous fermentation in a 1-L spinner flask (Techne, Cambridge, GB) using a mouse–mouse hybridoma cell line secreting a mAb (IgG2A). At the start the vessel was inoculated with a cell density of 2×10^5 cells/mL. The culture medium consisted of a 3:1 mixture of DMEM and Ham's F-12 supplemented with various amino acids (all from Serva, Heidelberg, Germany), fatty acids, vitamins, trace-elements, 100 mg/L BSA, 1 mg/L insulin, 4.3 mg/L transferrin (all from Gibco-BRL, Eggenstein, Germany). Samples were taken daily. Cell counting and viability determination was done using a hemacytometer and Erytrosin-B staining. Glucose measurements were performed by Ebio-compact (Eppendorf, Heidelberg, Germany). The Ab concentration was also determined using a HPLC system (Pharmacia-LKB, Freiburg, Germany) with a ProAnaMabs column (Hyclone, Aalst, Belgium). Samples for hybridoma cell cultures were supplied by Mr. Holger Heine and Dr. Manfred Bizelli (IBT-II, KFA, Juelich, Germany). Conditions for cell cultivation have been described previously ***(49)***.

3.10. Notes

The major limitation for the analysis of cell culture by CE is the low protein concentration. There are several methods for sample concentration, each of which has its own limitations.

1. Proteins concentrated by adsorption on the column or filter may not be recovered properly. The elution buffer may affect the quality of the resulting CE separation.
2. Total concentration of cell culture extract by Centricon or Unicon concentration cells may cause the concentration of other high-molecular-weight compounds together with proteins, which affects the quality of signal background. The proper protein concentration should also separate other compounds, and eliminate high concentration of salt and organic solutions to simplify further CE separation.
3. The ethanol–chloroform precipitation is an effective means of protein concentration, although it affects the 3D structure of proteins.

4. Customized CE Techniques

4.1. Off-Line Coupling of the CZE and MALDI–TOF-MS in the Analysis of Cell Culture

Despite sharp peaks and high resolution of protein peaks, frequently the accuracy of CE is not enough. Even with SDS-polymer CE, it is hard to define exact molecular mass of the protein because of interaction of protein molecule with SDS. We have solved this problem by off-line coupling CE with MALDI–TOF-MS ***(50)***.

The harvest obtained at the end of the batch fermentation, discussed in the previous subsection, was analyzed by MALDI–TOF-MS (*see* **Fig. 21**). We can identify the peaks with molecular mass 66633 as BSA, 124993 as dimer of

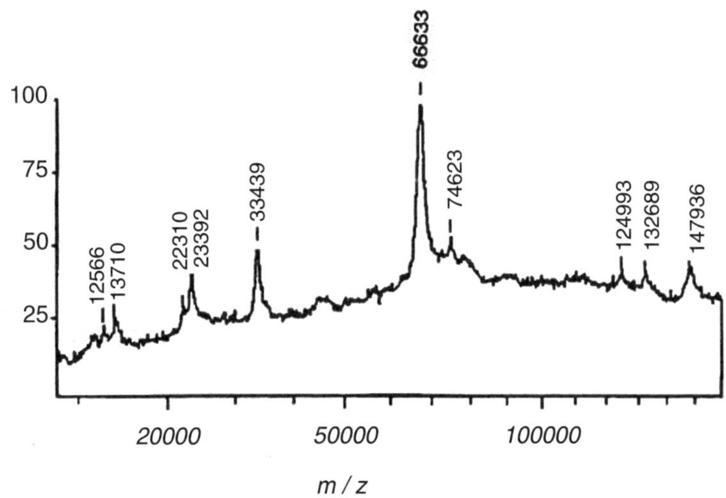

Fig. 21. MALDI–TOF-MS of the last harvest after batch cultivation. Reprinted with permission from **ref. *53***.

BSA, 33439 and 22310 as double- and triple-charged BSA; 147936 as IgG, 74623 as double-charged IgG, 124993 as IgG missing one L-chain, 23392 as L-chain, and 12566 and 13710 are unknown proteins. One of the difficulties using MALDI–TOF-MS analysis for mixtures is the different response of proteins during vaporization, which depends also on the molecular mass. By SDS-gel CE analysis, the concentration of IgG in the sample was higher than BSA (*see* **Fig. 18**, C), but in the MS plot the IgG appears as a small peak; whereas in the SDS-gel CE analysis many peaks with molecular mass 10–50 kD appear, which appear as baseline noise in the MALDI–TOF-MS scan. The problem of the difference in response could possibly be solved by changes in the sample preparation ***(51)***.

The main difficulty in the use of CE as SDS-gel CE or isoelectric focusing and MALDI–MS is the low compatibility of CE buffer additives, such as salts, gel, or ampholyte, with the operating conditions for MS ***(52)***. In order to avoid these effects, the separation of the harvest was performed by CZE in a coated capillary in 0.1 *M* NH_4OAc buffer (pH 3.5) (*see* **Fig. 22**). The proteins from the cell culture media were concentrated 10 times by ethanol–chloroform precipitation and dissolved in 50% CH_3CN, 0.1% TFA. The reasons to choose this buffer are: the protein was dissolved in acidic media with high concentration, the separation of the complex mixture in acidic buffer was better than under basic conditions, and the volatile acid could be evaporated after collection of the fraction during the preparation of the sample for MS.

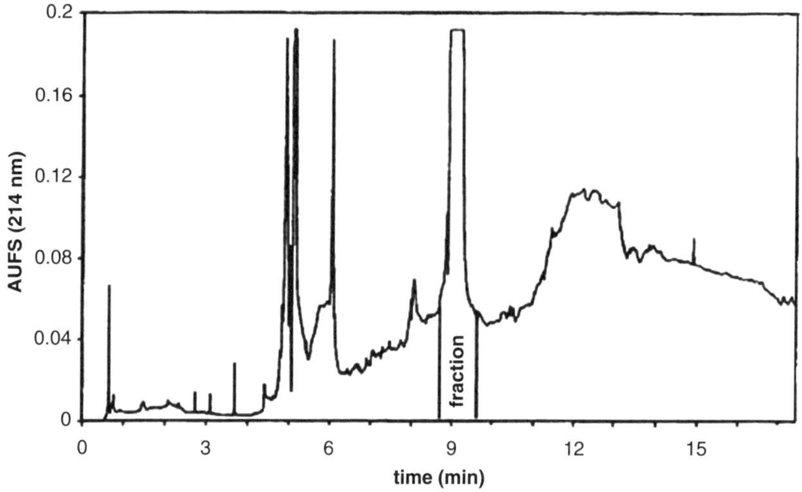

Fig. 22. Separation of proteins in the cell broth after batch cultivation by CZE. Conditions of separation: coated capillary 57-cm length, 0.1 *M* NH_4Oac, pH 3.5; injection by pressure 99 s. Reprinted with permission from **ref.** ***53***.

One of the main peaks, indicated in **Fig. 22**, was collected in a special vial in several consecutive separations from the same sample. The customized vial was constructed from the commercial one for protein injection (Beckman) by inserting a metal membrane in the center as shown in **Fig. 23**. The 0.1-mm stainless-steel membrane with platinum coating was permanently fixed in the vial with cyanoacrylate glue. This membrane separates the volume of electrolyte buffer from the electrode plus a minimum amount of buffer to allow fraction collection. There are two aims of the construction: to prevent dilution of the fraction and to allow adsorption of collected proteins on the electrode. Using the modified vial, the protein-containing fraction could be collected in 1–5 μL of the buffer. The short part of the capillary should be 3–5 mm longer than the electrode in order to reach the deepest point of the vial filled with 1–5 μL of the buffer. The addition of small droplet of mineral oil may also prevent the evaporation of the buffer with collected protein peak *(53)*.

Matrix (1 μL; 20 mg/mL in CH_3CN, 0.1% TFA) for the MALDI operation was added to the vial and after mixing the solution was used to carry out the MS analysis (*see* **Fig. 24**). The peak was identified as pure BSA (mol mass 66469). *M*/z 33367 and 132905 correspond to the double-charged species and the dimer of BSA, respectively. The results demonstrate that, in principle, a coupling of CZE and MALDI–TOF-MS is possible. Other peaks could be analyzed upon improved sensitivity of MALDI–MS and enhanced resolution of the CZE separation.

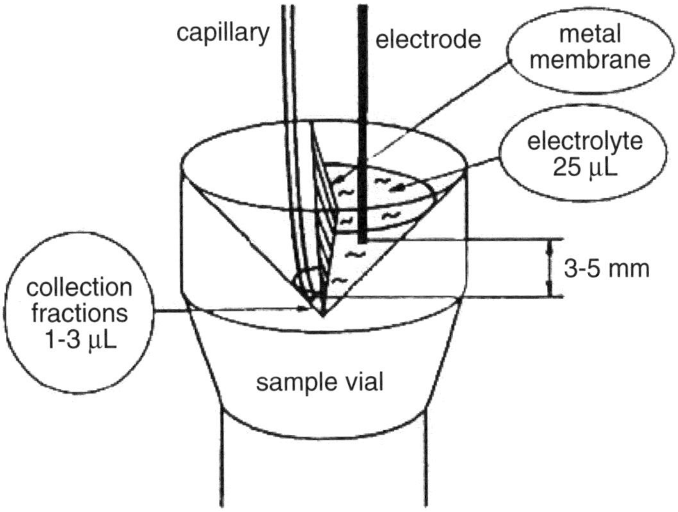

Fig. 23. Construction of the vial for the fraction collection. Reprinted with permission from **ref. *53***.

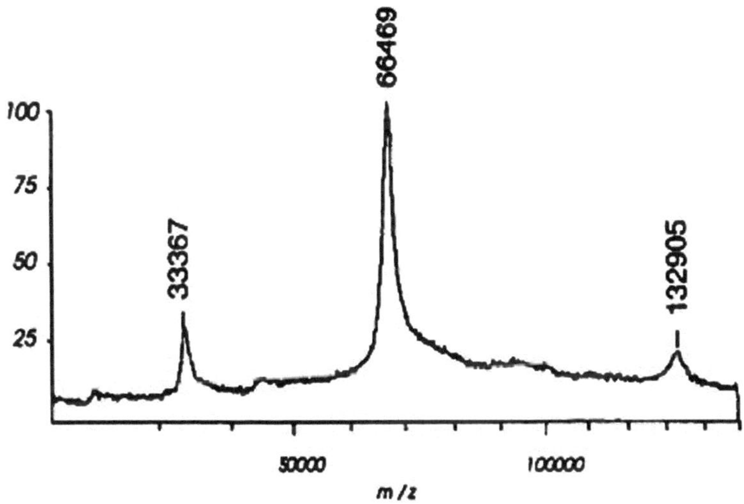

Fig. 24. MALDI–TOF-MS of one of the fractions separated by CZE (**Fig. 21**). Reprinted with permission from **ref. *53***.

The MALDI–TOF-MS measurements were performed with an instrument constructed in the Institute of Laser Medicine (Dusseldorf University) in the linear TOF mode with a laser beam wavelength $\lambda = 337.1$ nm ***(54)***. The samples were prepared in 2,5-dihydroxybenzoic acid with 10% of 2-hydroxy-5-methoxybenzoic acid as described by Karas ***(55)***. All chemicals used were at least analytical grade if not stated separately.

4.2. Rapid SDS Gel CE

In commercial instruments the capillary has long (20–50 cm) and short (2–10 cm) sections connected by a window for the detection, and the separation is usually performed in the long part of the capillary. In this case, the time of analysis is about 15–50 min. For accurate analysis of biotechnological processes and reactions, rapid or virtually on-line analyses are required. In the present work, the separation in the short section (7 cm) of the capillary was investigated as an alternative, and rapid separation was evaluated with respect to the resolution and number of theoretical plates ***(56)***.

Polymers or gels are useful media for electrophoretic separations mainly because they facilitate separations based on "molecular sieving effects." Furthermore, gels serve as anticonvective support and minimize analyte diffusion contributing to zone broadening. Gels prevent solute adsorption into the capillary walls and help to eliminate electro-osmosis ***(57)***. In the absence of electro-osmosis, the migration velocity (v) during electrophoresis is given by

$$v = \mu_{ep}E = \mu_{ep}V/L \tag{1}$$

where μ_{ep} is the electrophoretic mobility, E is the field strength (V/L), V is the voltage applied across the capillary, and L is the length of the capillary.

Assuming that the only contribution to band broadening arises from diffusion, the variance of the migrating zone width (σ^2) can be written as:

$$\sigma^2 = 2\,D\mathrm{t} = 2\,DL^2/\mu_{ep}V \tag{2}$$

where D is the diffusion coefficient of the solute.

The number of theoretical plates (N) is given by

$$N = L^2/\sigma^2 = \mu_{ep}\,V/2D. \tag{3}$$

Therefore, the separation efficiency is predominantly based on the voltage applied and not on the length of the capillary. Theoretically, maximum efficiency and short analysis times are obtained with high voltages and short columns, provided that there is efficient heat dissipation. In practice, the input of dispersive factors such as non-ideal sample injection, Joule heat, electromigration dispersion, diffusion, low electro-osmotic flow, and other factors lead to less-than-ideal results.

The separation in the normal mode was performed with two types of injection—pressure and electrokinetic. No significant differences in the efficiency of protein separation were detected between them, and only the data from the electrokinetic injection are shown in **Fig. 25**. As aforementioned, the main purpose of the current investigation is to develop a rapid analysis for biotechnological processes such as cell cultivation for the production of mAbs and other pharmaceutical proteins suitable for on-line monitoring. We analyzed hybridoma cell cultures containing BSA (66 kD), transferrin (82 kD), and IgG (150 kD), in addition to small amounts of insulin (5.6 kD).

The separation of the model proteins (BSA, transferring, and IgG) was achieved with good sensitivity, selectivity, and resolution (*see* **Fig. 25**). In the media, the concentration of BSA was 100 mg/L, whereas transferrin concentration was only 4 mg/L (*see* peak #3, **Fig. 25B**). Rapid separation was performed in the short part of the capillary reversing the polarity of the current and the direction of sample injection from the other (short) part of the capillary. No big differences were found in terms of the separation efficiency carried out in the long or short part of the capillary, but the separation time was decreased significantly (*see* **Fig. 26**).

The resolution (R) and number of theoretical plates (N) were calculated for these experiments according to

$$R = \Delta t / 4\sigma_t;\ N = (t/\sigma_t)^2$$

where Δt is the difference in the time of elution between two consecutive peaks and σ_t is the standard width of the single peak, and t is the total time for a given species to elute.

The resolution between peaks of BSA and transferrin in the normal mode is $R_{N(B/T)} = 5.8$; in the rapid mode it dropped slightly to $R_{R(B/T)} = 4.3$; analogously $N_{N(B/T)}$ was found as 64000, and $N_{R(B/T)}$ as 37000.

The ratio in length of the capillaries L_N/L_S is 3, similarly, the ratio in the time of analysis T_N/T_S is 3, but the relation between the resolution $R_{N(I/T)}/R_{R(I/T)}$ was found to be 1.34. The injection time (electrokinetic mode) for the protein separation in the long part of the capillary was 40 s, and in the short part, 20 s. At the shorter injection time, the sensitivity of the detection was not sufficient to quantify low protein–protein concentrations such as transferrin (*see* **Figs. 25B** and **26B**).

For injection by pressure in excess of 1 s, separation of the model proteins is not satisfactory, the protein peaks are bifurcates and very broad (*see* **Fig. 27**). Satisfactory results were achieved using the minimal possible time of injection for the device, which is 0.6 s. But under these conditions the reproducibility of the separation was low. The results demonstrate that the sample injection process is the major contributor to the dispersion factors in the equipment used.

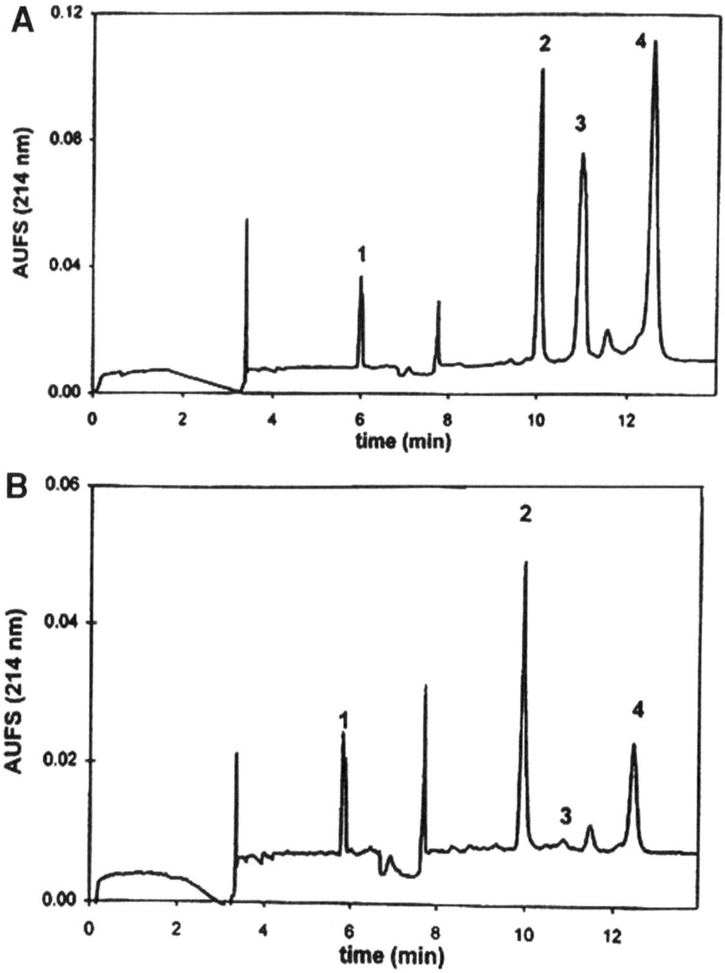

Fig. 25. (**A**) Separation of model proteins by SDS Gel CE; (**B**) Analysis of cell culture broth. Conditions: normal mode of separation, electrokinetic injection 40 s, 5 kV. 1, marker; 2, BSA; 3, transferring; 4, IgG.

4.3. Materials

1. 0.1 *M* NH_4OAc, pH 3.5, was used for CZE separation.
2. 1.0 *M* HCl was used for the rinsing of capillary.
3. Orange G (Beckman) was used as a low-molecular-weight marker.
4. Other proteins used in the experiment: insulin (5.6 kD), BSA (67 kD), and transferrin (78–85 kD) were obtained from Gibco-BRL (Eggenstein, Germany).
5. Before analysis, all buffers and samples were degassed and filtered through 0.22–0.45-μm filter.

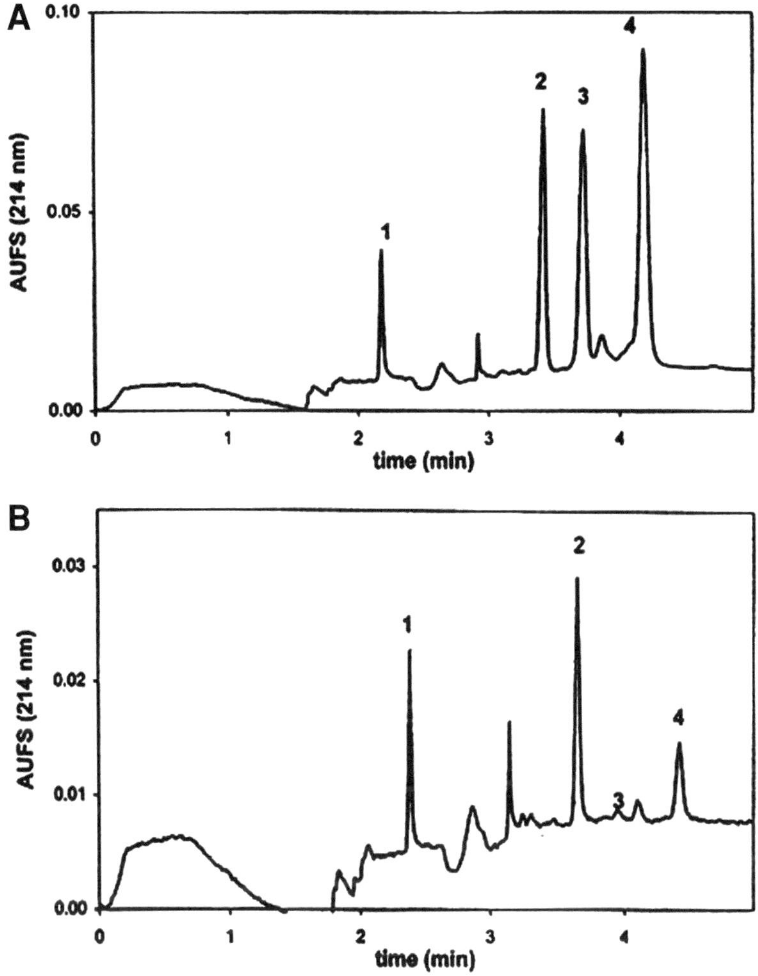

Fig. 26. (**A**) Rapid separation of model proteins by SDS Gel CE; (**B**) Analysis of cell culture broth. Conditions: reversed polarity, electrokinetic injection 20 s, 5 kV. Other conditions as in **Fig. 24**.

4.4. Methods

Conditions of separation by CZE:

1. Coated capillary 57 cm length, 0.1 *M* NH_4OAc (pH 3.5); injection by pressure 99 s.
2. The proteins from the cell culture media were concentrated 10 times by ethanol–chloroform precipitation and dissolved in 50% CH_3CN, 0.1% TFA.
3. The construction of the vial is shown in **Fig. 23** and described in **Subheading 4.1.**

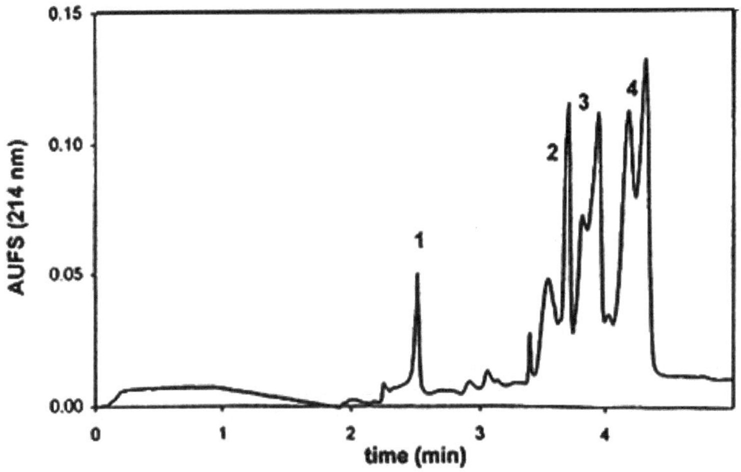

Fig. 27. Separation of model proteins by short-time mode of SDS-gel CE. Conditions: reversed polarity, injection by pressure 3 s. Other condition as in **Fig. 25**.

Conditions of separation by SDS-polymer CE:

1. Coated capillary 27 cm length, 8.1 kV.
   a. normal mode of separation, electrokinetic injection 40 s, 5 kV.
   b. reversed polarity, electrokinetic injection 20 s, 5 kV.
2. For separation under reversed-current polarity, we disconnected the electrodes and connected them in reverse order. The sample injection was performed from the short part of the capillary, followed by the regular procedure for the sample separation.
3. Because the proteins were separated in the short part of the capillary, the separation time was reduced from 12 to 4.5 min.

4.5. Notes

1. There are certain limitations for the fraction collection by CZE. The proper calculation of the separation time for the targeted fraction should be based on injection time and type, length of the separation zone (from injection end to detection window), and actual length of the capillary.
2. The concentration of the targeted protein can be increased by multiple separations; however, the accuracy of separation will be lower with each additional collection owing to small variations in separation time. The collected fraction may evaporate because of the small volume and long separation time of multiple injections.

3. Because this experiment was performed several years ago, there was no availability of fraction collection for CE. Currently, with rapid development of high-throughput-screening (HTS) technologies, fraction collection can be performed with higher accuracy and simplified significantly.

5. Rapid SDS-Polymer CE for the Analysis of Recombinant NADP$^+$-Dependent Formate Dehydrogenase During Expression in *E. coli* Cells and Purification

5.1. Overview

Another application of SDS-polymer CE was done for fast monitoring of the protein expression of NAD$^+$-dependent formate dehydrogenase (FDH) from methanol-utilizing microorganisms. This enzyme catalyses the oxidation of formate to carbon dioxide with the corresponding reduction NAD$^+$ to NADH. It provides the best NADH regeneration systems used in the processes of enzymatic synthesis of chiral and physiologically active compounds ***(58)***. Large-scale production of NAD$^+$-specific FDH from the yeast *Candida boidinii* was developed a few years ago ***(59)*** and this enzyme was used for the industrial production of *tert*-leucine and can be applied together with many other dehydrogenases ***(60)***. Normally, in nature, in coupled enzymatic systems NAD(H) is involved in biodegradation processes and NADP(H)—in biosynthesis. Unfortunately, a NADP$^+$-specific FDH not requiring complex metal clusters has not been found in living cells yet. A NADP$^+$-dependent enzyme has been generated, however, by the method of site-directed mutagenesis of the FDH gene from the bacterium *Pseudomonas* sp.101 ***(61)*** and first model experiments of coupled enzyme-catalyzed synthesis with NADPH regeneration have been reported ***(62)***. Overexpression and large-scale productions of this mutant FDH are necessary for industrial realization of such processes. Here, we describe rapid SDS-Gel CE as a method to analyze FDH expression in *E. coli* cells during cultivation and in further enzyme purification steps ***(63)***.

5.2. Results and Discussion

For short-duration SDS-polymer CE we have used the technique described in **Subheading 4.2.** The separation of the standard proteins for the molecular mass estimation was performed before measurements (*see* **Fig. 28A**). The protein peaks with different molecular masses were separated with high resolution. The separation of proteins occurs in the interval between 3 and 4.5 min. The peaks before 3 min arise from baseline noise and low-molecular-mass impurities. The calibration curve was generated as a function of lgMM vs the elution time. It should be noted, that the calibration curve is semilinear in the molecular mass range 14–150 kD.

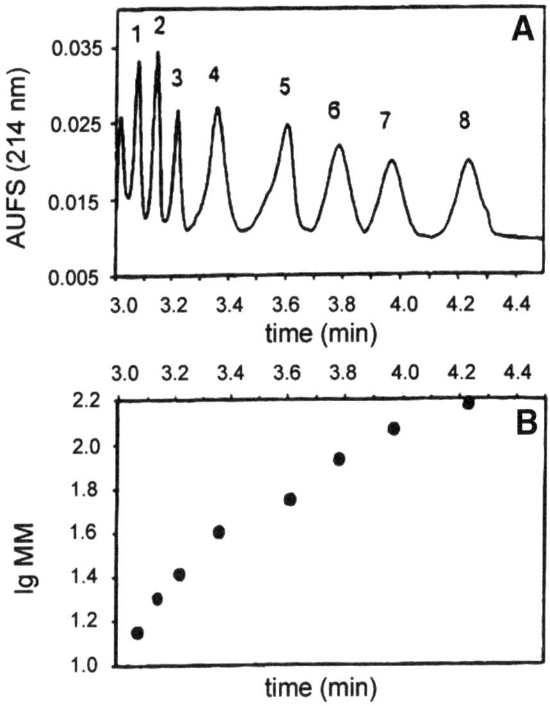

Fig. 28. (**A**) Separation of standard proteins by SDS-Gel CE, 27 cm coated capillary, 8.7 kV, 20 s electrokinetic injection at 5 kV. (**B**) Calibration curve of the separation of the standard proteins: 1, lysozyme (14 kD); 2, trypsin-inhibitor (20 kD); 3, triose phosphate isomerase (26.6 kD); 4, aldolase (39 kD); 5, gluramate dehydrogenase (55.5 kD); 6, fructose-6-phosphate kinase (85.2 kD); 7, β-galactosidase (116.3 kD); 8, IgG (150 kD). Reprinted with permission from **ref. *62***.

The fractions during the cultivation of *E. coli* containing the FDH gene were taken and analyzed in parallel by PAGE (*see* **Fig. 29**) and SDS-gel CE (*see* **Fig. 28**). The probes were analyzed in the CE at two wavelengths in the UV region, at 214 and 280 nm. At the time of analysis, the cells had entered the stationary phase of growth and the concentration of FDH was increasing rapidly or almost stepwise between 13 and 14.5 h of the cell cultivation. After 14.5 h the FDH content became constant and the cultivation was stopped. Data in Table 3 show that there are no big differences in protein content between the times of cultivation in the 14.5–17.5-h range. The recombinant mutant FDH is stable in the *E. coli* cells. As the protein content remained at plateau values, a proteolytic degradation of the proteins was not observed. The protein folding in the cells occured fast, as evidenced by the constant relation of the FDH specific activity at the different times (*see* **Table 3**).

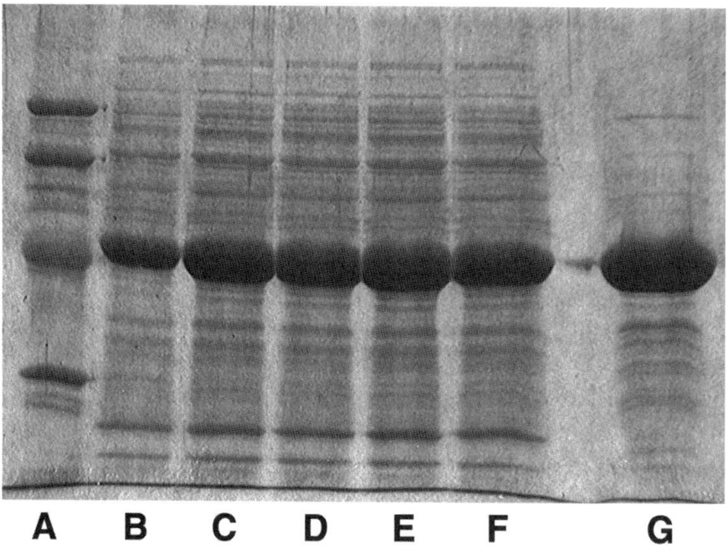

Fig. 29. SDS-PAGE separation of fractions during the cell cultivation and final purified FDH: (**A**) markers, (**B**) 13.5 h, (**C**) 14.5 h, (**D**) 15.5 h, (**E**) 16.5 h, (**F**) 17.5 h, (**G**) purified FDH. Reprinted with permission from **ref. *62***.

Table 3
Analysis of Protein Concentration, FDH Activity and Specific Activity at Different Times of the Cell Cultivation

Time (h)	FDH[a] concentration in probe (mg/mL)	FDH activity in probe (U/mL)	Specific activity (U/mg)	Cell absorbance A_{560} (optical units)
13.5	0.257	0.644	2.51	13.0
14.5	0.543	1.355	2.49	13.9
15.5	0.567	1.400	2.47	14.2
16.5	0.556	1.376	2.48	14.7
17.5	0.561	1.389	2.48	14.5

[a]Concentration of FDH in probe was determined from DC experiments as 214 nm (**Fig. 2**) using a calibration curve with BSA as a standard (data not shown). The specific activity of FDH was calculated as the ratio of enzyme activity to FDH concentration in probe (columns 3 and 2, respectively).

Using SDS-PAGE to separate proteins with large differences in concentration, the problem of quantification of the main protein vs impurities arises, which is illustrated in **Fig. 29**. Even if the content of the main protein is about 50–60%, the remaining 40–50% of the other proteins are distributed over different molecular masses (*see* **Fig. 29**, lanes B–F). If one wants to see a main protein and impurities with close molecular masses, as products of its degradation, or other impurities, different amount of samples should be applied. The registration and quantification is difficult even employing a laser gel scanner. In our case, we have overloaded the SDS-PAGE in order to show the impurities.

Compared to the SDS-PAGE separation, the concentration range of the SDS-gel CE and sensitivity are much higher. Otherwise, there is a full correspondence of the main peak and the impurities between the SDS-PAGE and the SDS-gel CE separations (*see* **Figs. 29–31**). The detection at 214 nm is unique for proteins and it is possible to compare all steps of the fermentation process registered by both methods (**Fig. 29**, lanes B–F, **Fig. 30[right]**, a–e). Even with detection at 280 nm and an overloading of the capillary resulting in an asymmetric peak of FDH, it is possible to quantify main impurities (3.1–3.3 min), and their strong reduction in the process of purification (*see* **Fig. 31A**) is evident. For the analysis of the FDH after different steps of purification with detection at 280 nm, a higher voltage was applied during the injection because of the lower sensitivity (compared to 214 nm). At 214 nm, all impurities were registrated and corresponded with the SDS-PAGE. We have overloaded the capillary with the purified FDH (*see* **Fig. 31**, b–e) at 214 nm in order to show the protein purity and the presence of impurities, if any. The little peak at 3.85 min corresponds with the thin high-molecular mass-band (about 90 kD) visible on SDS-PAGE (*see* **Fig. 29**, lane G)

A summary of the fermentation and purification process is shown in **Table 4**. The initial cell extract contained usually about 56% of FDH (*see* **Fig. 31**, a). After addition of 30% of ammonium sulfate, some protein impurities are precipitated and removed (*see* **Fig. 31**, b). At 75% saturation with ammonium sulfate, the precipitate contained 77% FDH (*see* **Fig. 31**, c). After redissolving, the protein mixture was applied to the hydrophobic interaction and size-exclusion chromatography. The final purity of the protein was better than 98%.

The purified protein was analyzed by MALDI–TOF-MS (*see* **Fig. 32**). The molecular mass of the FDH was found to be 44078 Da, which corresponds to the expected value, based on DNA sequence information. In the scan, some additional peaks correspond to the double and triple-charged molecules and to the single-charged dimer of FDH, respectively. The error in the single measurement was about 0.5%, and the average value of five measurements is quoted.

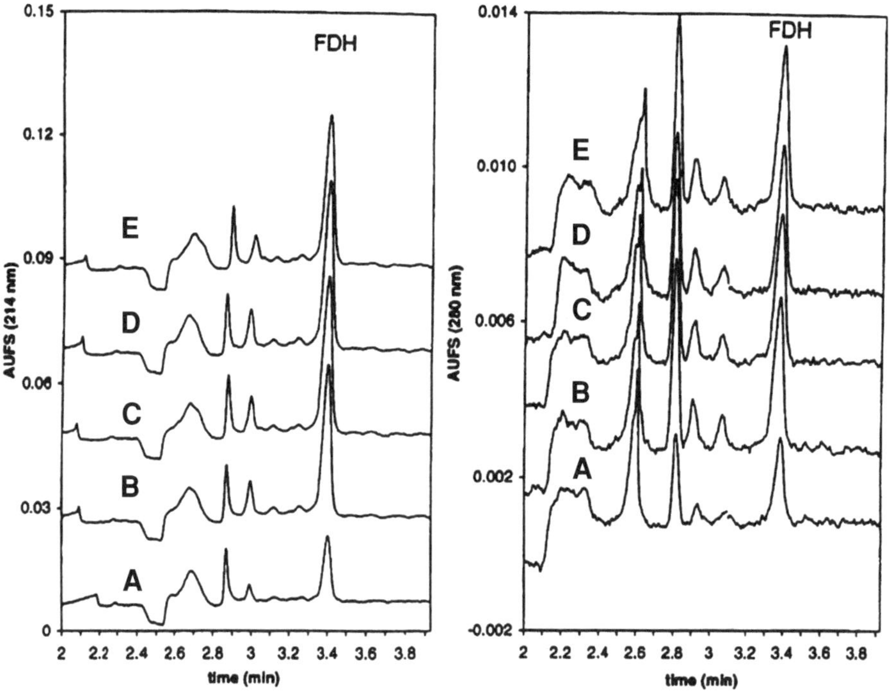

Fig. 30. Analysis of expression of FDH at different times of cell cultivation (left) 214 nm, (right) 280 nm: a, 13.5 h; b, 14.5 h; c, 15.5 h; d, 16.5 h; e, 17.5 h. Conditions: separation: 8.1 kV, 40–50 μA, Injection time 40 s, injection voltage 1 kV. Reprinted with permission from **ref. *62***.

5.3. Materials

1. 1.0 *M* HCl was used for the rinsing of capillary.
2. 0.1 *M* Tris-HCl, 1% SDS, pH 6.6, was used for protein sample preparation.
3. For calibration and analysis, the following proteins were used: lysozyme (14 kD), trypsin-inhibitor (20 kD), triose phosphate isomerase (26.6 kD), aldolase (39 kD), glutamate dehydrogenase (55.5 kD), fructose-6-phosphate kinase (85.2 kD), β-galactosidase (116.3 kD) (Boehringer Mannheim) and IgG (150 kD; Institute of Enzyme Technology, University of Dusseldorf, KFA Juelich, Germany).
4. Before analysis, all buffers and samples were degassed and filtered through 0.22–0.45-μm filters.

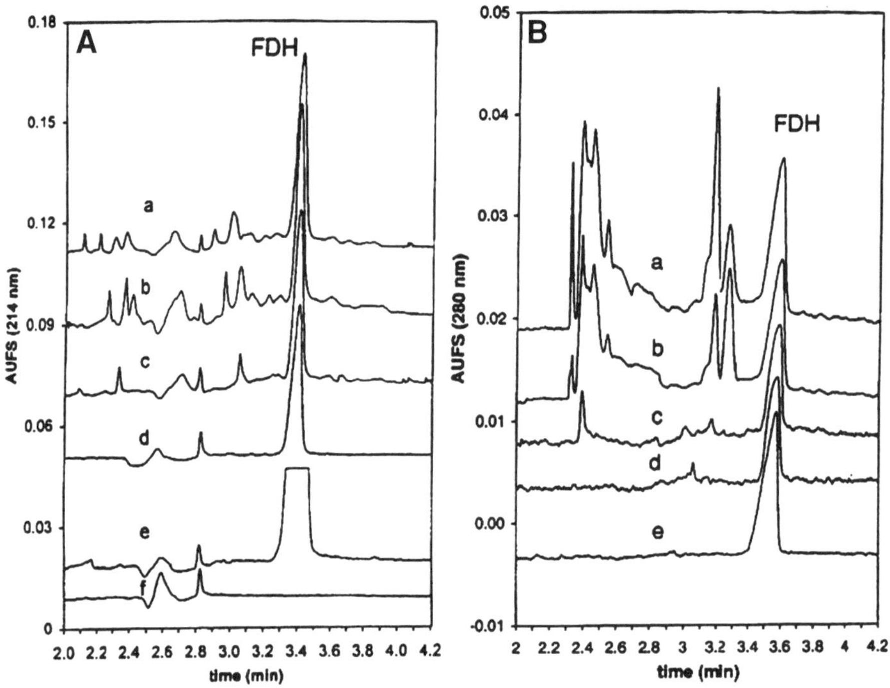

Fig. 31. Analysis of FDH samples at different stages of purification: a, cell-free extract; Conditions see **Fig. 3**; b, ammonium sulfate, 30% of saturation; c, ammonium sulfate, 75% of saturation; d, after purification on Phenyl-Sepharose; e, after gel-filtration through Superdex G 200; f, base-line. Reprinted with permission from **ref. *62***.

Table 4
Results of FDH Purification

No.	Purification step	Total protein (mg)	Total activity (U)	Specific activity (U/mg)	Purification fold	Purity by SDS-Gel CE %	Yield (%)
1	Cell-free extract	2380	3100	1.30	1	56	100
2	Ammonium sulfate, 30% of saturation	2140	3100	1.45	1.12	67	100
3	Ammonium sulfate, 75% of saturation	1570	2880	1.83	1.41	77	93
4	Phenyl Sepharose	1060	2510	2.37	1.82	94	81
5	Superdex G-200	856	2140	2.50	1.92	98	69

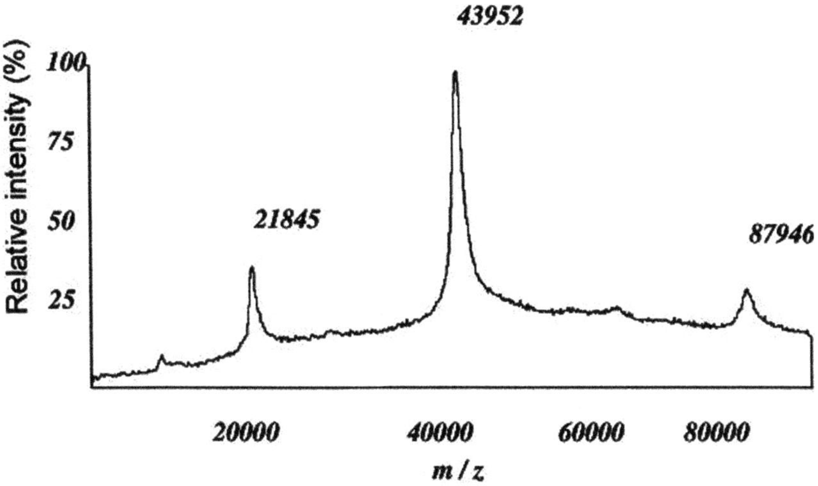

Fig. 32. MALDI–TOF-MS of purified FDH. Reprinted with permission from **ref.** ***62***.

5.4. Methods

1. All CE measurement were performed with a P/ACE 2100 device for capillary electrophoresis (Beckman).
2. Prior to each analytical run, a bare silica capillary (27-cm length, [7-cm effective length] × 100-µm id × 375-µm od) was rinsed with 1.0 *M* HCl water for 2 min followed by a filling with the SDS polymer solution (eCAP SDS 14-200; Beckman) for 2 min.
3. The column temperature was maintained at 20°C by circulating a coolant to minimize band diffusion and ensure effective size separation.
4. An electrophoretic run was conducted at 300 V/cm (8.1 kV for the capillary) using the SDS-polymer solution (Beckman). The molecular mass protein standard was injected for 20 s (5 kV) using the electrokinetic mode into the SDS-polymer-filled capillary column.
5. Model proteins were dissolved in the buffer for SDS gel CE: 0.1 *M* Tris-HCl, 1% SDS, pH 6.6.
6. Conditions of separation by SDS-polymer CE: coated capillary 27-cm length, 8.1 kV; reversed polarity, electrokinetic injection 20 s, 5 kV.
7. For the separation under reversed-current polarity, we have disconnected the electrodes and connected them in reverse order. The sample injection was performed from the short part of the capillary, followed by the regular procedure for the sample separation.
8. SDS-PAGE was performed in 10% polyacrylamide (0.8% *bis*-acrylamide) according to Lemmli ***(64)***.
9. For MALDI–TOF-MS conditions, see p. 104, first paragraph.

5.5. Notes

1. Another limitation for the in-process analysis by CE is the quantification of purity by detection in UV part of the spectrum. The absorption of unknown proteins or other charged high molecular weight compounds may differ, so the calculated purity of target protein may differ as well. However, this problem is common for other types of separations with UV detection.

6. Conclusion

The rapid growth and complexity of current therapeutic proteins produced by recombinant DNA technology requires fast and quantitative analysis, and monitoring of large-scale applications. We have shown successful application of CE for in-process analysis and final purity testing of the recombinant proteins from different sources. The rapid development of CE methodology and its versatility can be enhanced with complementary analytical separation techniques. In addition, CE–MS techniques provide higher accuracy in identification of main recombinant products and impurities.

Acknowledgments

The author gratefully acknowledges A. N. Wulfson, D. M. Koulich, S. A. and I. V. Nazimov from Shemyakin and Ovchinnikov Institute of Bioorganic Chemistry, V.Tishkov from Moscow State University, Moscow, Russia, J. Thömmes and M.-R. Kula from the Institute of Enzyme Technology, Dusseldorf University (Juelich, Germany), for helpful discussions and technical assistance; also the Alexander von Humboldt Foundation (Bonn, Germany) for a Postdoctoral Fellowship grant and additional financial support for scientific collaboration.

References

1. Tran, A., Parker, H., Levi, V., and Kunitani,M. (1998) Analysis of recombinant human platelet-derived growth factor by reversed-charge capillary zone electrophoresis. *Analyt. Chem.* **70,** 3809–3817.
2. Bergmann, J., Jaehde, U., and Schunack, W. (1998) Quantitative trace analysis of interleukin-3, interleukin-6, and basic model proteins using isotachophoresis-capillary zone electrophoresis with hydrodynamic counterflow. *Electrophoresis* **19,** 305–310.
3. Pantazaki, A., Taverna, M., and Vidal-Madjar, C. (1999) Recent advances in the capillary electrophoresis of recombinant glycoproteins. *Analyt. Chim. Acta* **383,** 137–156.
4. Michalski, W. P. and Shiell, B. J. (1999) Strategies for analysis of electrophoretically separated proteins and peptides. *Analyt. Chim. Acta* **383,** 27–46.

5. Hooker, A. D. and James, D.C. (2000) Analysis of glycoprotein heterogeneity by capillary electrophoresis and mass spectrometry. *Molec. Biotechnol.* **14,** 241–249.
6. Goldman, M. H., James, D.C., Rendall, M., Ison, A.P., Hoare, M., and Bull, A.T. (1998) Monitoring recombinant human interferon-gamma N-glycosylation during perfused fluidized-bed and stirred-tank batch culture of CHO cells. *Biotechnol. Bioeng.* **60,** 596–607.
7. Lagu, A. L. (1999) Applications of capillary electrophoresis in biotechnology. *Electrophoresis* **20,** 3145–3155.
8. Hooker, A. D. and James, D. C. (1999) Monitoring recombinant glycoprotein heterogeneity. *Methods Biotechnol.* **8,** 277–288.
9. Liu, T., Shao, X.-X., Zeng, R., and Xia, Q.-C. (1999) Analysis of recombinant and modified proteins by capillary zone electrophoresis coupled with electrospray ionization tandem mass spectrometry. *J. Chromatogr. A* **855,** 695–707.
10. Strege, M. A. and Lagu, A. L. (1997) Capillary electrophoresis of biotechnology-derived proteins. *Electrophoresis* **18,** 2343–2352.
11. Kaiser, R. E., Strege, M. A., and Lagu, A. L. (1997) Assessment of protein drugs obtained by recombinant DNA technology. *Process Control Quality* **10,** 205–228.
12. Yeung, B., Porter, T. J., and Vath, J. E. (1997) Direct isoform analysis of high-mannose-containing glycoproteins by online capillary electrophoresis electrospray mass spectrometry. *Analyt. Chem.* **69,** 2510–2516.
13. Gysler, J., Helk, B., Dambacher, S., Tjaden, U. R., and Van der Greef, J. (1999) Characterization of recombinant cytokine fragments using isotachophoresis-capillary zone electrophoresis, reversed-phase high performance liquid chromatography, and mass spectrometry. *Pharmaceut. Res.* **16,** 695–701.
14. van de Goor, T. (1997) Peptide mapping using HPCE: use of hyphenated methods. Book of Abstracts, 213th ACS National Meeting, San Francisco, CA, Apr. 13–17.
15. Dittmann, M. M. and Rozing, G. P. (1999) High-sensitivity separations of sodium dodecyl sulfate-protein complexes with capillary gel electrophoresis. *LC-GC* **17,** 132–137.
16. Bergmann, J., Jaehde, U., Mazereeuw, M., Tjaden, U. R., and Schunack, W. (1996) Potential of online isotachophoresis-capillary zone electrophoresis with hydrodynamic counterflow in the analysis of various basic proteins and recombinant human interleukin-3. *J. Chromatogr. A* **734,** 381–389.
17. Vo, H. C., Britz-Mckibbin, P., Chen, D. D. Y., and MacGillivray, R. T. A. (1999) Undercarboxylation of recombinant prothrombin revealed by analysis of γ-carboxyglutamic acid using capillary electrophoresis and laser-induced fluorescence. *FEBS Lett.* **445,** 256–260.
18. Schlenck, A., Schiele, F., Barbier, A., Shuvaev, V. V., Visvikis, S., and Siest, G. (1999) Capillary electrophoretic analysis of recombinant human apolipoprotein E. Calibration mode of a protein reference material. *J. Chromatogr. A* **853,** 237–241.
19. Taverna, M., Tran, N. T., Merry, T., Horvath, E., and Ferrier, D. (1998) Electrophoretic methods for process monitoring and the quality assessment of recombinant glycoproteins. *Electrophoresis* **19,** 2572–2594.

20. Kundu, S., Fenters, C., Lopez, M., et al. (1997) Capillary electrophoresis for purity estimation and in-process testing of recombinant GB virus-C proteins. *J. Capillary Electrophoresis* **4,** 7–13.
21. Facchetti, I. (1998) Determination of aggregates in a formulated recombinant fusion protein by SDS-capillary electrophoresis. *Biomed. Chromatogr.* **12,** 138–139.
22. Denton, K. A. and Tate, S. A. (1997) Capillary electrophoresis of recombinant proteins. *J. Chromatogr. B Biomed. Sci. Applicat.* **697,** 111–121.
23. Patrick, J. S. and Lagu, A. L. (2001) Review applications of capillary electrophoresis to the analysis of biotechnology-derived therapeutic proteins. *Electrophoresis* **22,** 4179–4196.
24. Ovchinnikov, Yu. A., Efimov, V. A., and Chakhmakhcheva, O. G. (1983) Chemical-enzymic synthesis and cloning of a human proinsulin gene. *Doklady Akademii Nauk SSSR* **270,** 743–747.
25. Klyushnichenko, V. E., Koulich, D. M., Yakimov, S. A., et al. (1994) Recombinant human insulin. III. High-performance liquid chromatography and high-performance capillary electrophoresis control in the analysis of step-by-step production of recombinant human insulin. *J. Chromatogr. A* **661,** 83–92.
26. Patrick, J. S. and Lagu, A. L. (1992) Determination of recombinant human proinsulin fusion protein produced in Escherichia coli using oxidative sulfitolysis and two-dimensional HPLC. *Analyt. Chem.* **64,** 507–511.
27. United States Pharmacopoeia, 1984 Revision XX, United States Pharmacopoeia Convention, Rockville, MD, pp. 2177–2179,
28. British Pharmacopoeia 1988, H.M. Stationary Office, London, pp. 312–313.
29. Sergeev, N. V., Gloukhova, N. S., Nazimov, I. V., et al. (2001) Monitoring of recombinant human insulin production by narrow-bore reversed-phase high-performance liquid chromatography, high-performance capillary electrophoresis and matrix-assisted laser desorption ionisation time-of-flight mass spectrometry. *J. Chromatogr. A* **907,** 131–144.
30. Felten, C., Quan, C. P., Chen, A. B., et al. (1999) Use of acidic and basic pH and calcium ion addition in the capillary zone electrophoretic characterization of recombinant human deoxyribonuclease, a complex phosphoglycoprotein. *J. Chromatogr. A* **853,** 295–308.
31. Radko, S.P., Stastna, M., Buzas, Z., Kingsley, D., and Chrambach, A. (1999) Charge heterogeneity of commercial, red-shifted recombinant green fluorescent protein, revealed by capillary zone electrophoresis under nondenaturing conditions. *Analyt. Biochem.* **274,** 146–148.
32. St'astna, M., Radko, S. P., and Chrambach, A. (2000) Separation efficiency in protein zone electrophoresis performed in capillaries of different diameters. *Electrophoresis* **21,** 985–992.
33. Stastna, M., Radko, S. P., and Chrambach, A. (2001) Discrimination between peak spreading in capillary zone electrophoresis of proteins due to interaction with the capillary wall and due to protein microheterogeneity. Electrophoresis 22, 66–70.

34. Kulish, D. M., Klyushnichenko, V. E., and Vul'fson, A. N. (1994) Separation of proteins by liquid chromatography and capillary electrophoresis. *Zhurnal Fizicheskoi Khimii* **68,** 1841–1847.
35. Grady, J. K., Zang, J., Laue, T. M., Arosio, P., and Chasteen, N. D. (2002) Characterization of the H- and L-subunit ratios of ferritins by sodium dodecyl sulfate-capillary gel electrophoresis. *Analyt. Biochem.* **302,** 263–268.
36. Kinghorn, N. M., Norris, C. S., Paterson, G. R., and Otter, D. E. (1995) Comparison of capillary electrophoresis with traditional methods to analyze bovine whey proteins. *J. Chromatogr. A* **700,** 111–123.
37. Tsuji, K. (1994) Sodium dodecyl sulfate polyacrylamide gel- and replacable polymer filled electrophoresis for molecular mass determination of proteins of pharmaceutivcal interest. *J. Chromatogr.* **662,** 291–299.
38. Tsuji, K. (1993) Evaluation of sodium dodecyl sulfate non-acrylamide, polymer gel-filled capillary electrophoresis for molecular size separation of recombinant bovine somatotropin. *J. Chromatogr.* **652,** 139–147.
39. Werner, W. E., Demorest, D. M., Stevens, J., and Wiktorovicz, J. E. (1993) Size-dependent separation of proteins denatured in SDS by capillary electrophoresis using a replaceable sieving matrix. *Anal. Biochem.* **212,** 253–258.
40. Nakatani, M., Shibukawa, A., and Nakagawa, T. (1993) Sodium dodecyl sulfate polyacrylamide solution filled capillary electrophoresis of proteins using linear polyacrylamide coated capillary. *Biol. Pharm. Bul.* **16,** 1185–1188.
41. Hjerten, S. (1985) High-performance electrophoresis. Elimination of electroendosmosis and solute adsorption. *J. Chromatogr.* **347,** 191–198.
42. Cobb, K. A., Dolnik, V., and Novotny, M. (1990) Electrophoretic separations of proteins in capillaries with hydrolytically-stable surface structures. *Anal. Chem.* **62,** 2478–2483.
43. Nakatani, M., Shibukawa, A., and Nakagawa, T. (1994) High-performance capillary electrophoresis of SDS-proteins using pullulan solution as separation matrix. *J. Chromatogr.* **672,** 213–218.
44. Ganzler, K., Greve, K. S., Cohen, A. S., Karger, B. L., Guttman, A., and Cooke, N. C., (1992) High-performance capillary electrophoresis of SDS-protein complexes using UV-transparent polymer networks. *Anal. Chem.* **64,** 2665–2671.
45. Tsuji, K. (1994) Factors affecting the performance of sodium dodecyl sulfate gel-filled capillary electrophoresis. *J. Chromatogr.* **661,** 257–264.
46. Klyushnichenko, V. and Kula, M.-R. (1997) Sodium dodecyl sulfate-polymer capillary electrophoresis for the analysis of cell culture proteins. *Electrophoresis* **18,** 2019–2023.
47. Werner, W. E., Demorest, D. M., and Wiktorovicz, J. E. (1993) Automated Ferguson analysis of glycoproteins by capillary electrophoresis using a replaceable sieving matrix. *Electrophoresis* **14,** 29–38.
48. Hjerten, S. (1985) *J. Chromatogr.* **347,** 191–198.
49. Thoemmes, J., Gaethgens, J., Bizelli, M., Rundstadler, P.W.and Wandrey, C. (1993) The influence of dissolved oxygen tension on the metabolic activity of an immobilized hybridoma population. *Cytotechnol.* **13,** 29–38.

50. Klyushnichenko, V., Rodenbrock, A., Thömmes, J., Kula, M.-R., Heine, H., and Biselli, M. (1998) Analysis of hybridoma cell culture processes by SDS/gel capillary electrophoresis and matrix-assisted laser desorption ionization-time-of-flight MS. *Biotechnol. Appl. Biochem.* **27,** 181–188.
51. Kallweit, U., Börnsen, K. O., Kresbach, G. M., and Widmer, M. (1996) Matrix compatible buffers for analysis of proteins with matrix-assisted laser desorption/ionization mass spectrometry. *Rapid Comm. Mass Spectrom.* **10,** 845–849.
52. Winkler, M. A., Kundu, S., Robey, T. E., and Robey, W. G. (1996) Comparative peptide mapping of a hepatitis C viral recombinant protein by capillary electrophoresis and matrix-assisted laser desorption time-of-flight mass spectrometry. *J. Chromatogr. A* **744,** 177–185.
53. Shimura, K., Uchiyama, N., and Kasai, K.-I. (2001) Prevention of evaporation of small-volume sample solutions for capillary electrophoresis using a mineral-oil overlay. *Electrophoresis* **22,** 3471–3477.
54. Spengler, B., Kirsch, D., Kaufmann, R., Karas, M., Hillenkamp, F., and Giessmann U. (1990) The detection of large molecules in matrix-assisted UV laser desorption. *Rapid Commun. Mass Spectrometry* **4,** 301–305.
55. Karas, M., Ehring, H., Nordhoff, E., et al. (1993) Matrix-assisted laser desorption/ionization mass spectrometry with additives to 2,5-dihydroxybenzoic acid. *Org. Mass Spectrom.* **28,** 1476–1481.
56. Klyushnichenko V. and Kula, M.-R. (1997) Rapid SDS gel capillary electrophoretic analysis of proteins. *J. Cap. Elec.* **4,** 61–64.
57. Li, S. F. Y. (1993) *Capillary Electrophoresis. Principles, Practice and Applications.* Elsevier Science, Amsterdam, pp. 9–13.
58. Hummel, W. and Kula M.-R. (1989) Dehydrogenases for the synthesis of chiral compounds. *Eur. J. Biochem.* **184,** 1–13.
59. Wuester-Botz, D., Pascold, H., Striegel, B., Gieren, H., Kula, M.-R., and Wandrey C. (1994) Continuous computer controlled production of formate dehydrogenase and isolation on a pilot scale. *Chem. Eng.Technol.* **17,** 131–137.
60. Kragl, U., Kruse, W., Hummel, W., and Wandrey C. (1996) Enzyme engineering aspects of biocatalysis: cofactor regeneration as example. *Biotechnol. Bioengin.* **52,** 309–319.
61. Tishkov, V.I., Matorin, A.D., Rojkova, A.M., et al. (1996) Site-directed mutagenesis of the formate dehydrogenase active centre: role of the His332-Gln313 pair in the enzyme catalysis. *FEBS Lett.* **390,** 104–108.
62. Seelbach, K., Riebel, B., Hummel, W., et al. (1996) A novel, efficient regeneration method of NADPH by using new formate dehydrogenase. *Tetrahedron Lett.* **37,** 1377–1380.
63. Klyushnichenko, V., Tishkov, V. and Kula, M.-R. (1997) Rapid SDS-Gel capillary electrophoresis for the analysis of recombinant NADP+-dependent formate dehydrogenase during expression in *Escherichia coli* cells and its purification. *J. Biotechnol.* **58,** 187–195.
64. Laemmli, U.K. (1970) Cleavage of structural proteins during the assembly of the head of bacteriophage T4. *Nature* **227,** 680–685.

6

Capillary Electrophoresis of Proteins in a Quality Control Environment

David L. Good, Stacey Cummins-Bitz, Raeann M. Fields, and Brian K. Nunnally

Summary

A method for determining the purity of recombinant carboxypeptidase B utilizing CE-SDS has been developed and validated for use in a manufacturing quality control laboratory. The method was optimized, prior to validation, to allow for the lowest limit of quantitation. The method was validated with the typical ICHQ2A parameters, including accuracy, linearity, LOQ, precision, robustness/ruggedness, and specificity. All validation parameters met the acceptance criteria defined in the validation protocol.

Key Words

Capillary; electrophoresis; enzyme purity; protein; quality control; sodium dodecyl sulfate (SDS); validation.

1. Introduction

After a decade of development and refinement, the capillary electrophoresis (CE) of proteins has moved beyond the R&D laboratory into the manufacturing quality control laboratory (QCL). CE methods have historically been slow in gaining acceptance in a QCL environment because the ruggedness and cost-effectiveness requirements were so high. Now that CE has gained a foothold in QCL, its inherent advantages over traditional technologies, such as sodium dodecylsulfate-polyacrylamide gel electrophoresis (SDS-PAGE), have been demonstrated. Several companies have reported implementing CE for development applications or batch release for their biopharmaceutical products. CE has had a fundamental positive impact on protein products such as Inerfergen®

From: *Methods in Molecular Biology, vol. 276: Capillary Electrophoresis of Proteins and Peptides*
Edited by: M. A. Strege and A. L. Lagu © Humana Press Inc., Totowa, NJ

(Amgen) ***(1)***, Enbrel® (Immunex) ***(2)***, and Synagis® (MedImmune) ***(3)***. In addition, Genentech has reported the use of CE for several of their therapeutic antibodies ***(4,5)***. Today, CE can be considered a "routine analytical tool in pharmaceutical analysis" ***(6)***.

Significant advances in the design and engineering of CE instruments have been made in recent years. Improvements in cooling systems and voltage control and the introduction of instruments designed to be more easily maintained have been major advances and have been especially important to the analysis of proteins because of the sensitivity of these analytes to their environment. Further, the addition of more powerful software and flexibility in modes of separation has increased the utility of the instruments. Corresponding advances in capillaries and gel buffers have mirrored the advances in the instruments. The batch-to-batch variability of these components has greatly decreased and the results have been apparent. For example, Schenerman and Bowen have reported migration time intermediate precision (%CV) as low as 1.2% and repeatability equal to 0.15% ***(3)***. Hunt and Nashabeh reported overall precision repeatability and intermediate precision (RSD) of 0.9% for the assay of their therapeutic proteins ***(5)***.

SDS-based gel analysis (CE-SDS) is the most common CE application for protein pharmaceuticals. Method development is simple and the method itself can be quite robust. The advantages of this technique over traditional SDS-PAGE applications in a QCL environment are profound. Staining is not necessary (i.e., on-column detection), which allows for significant time savings. Significant improvements in linearity are observed, with correlation coefficients greater than 0.998 being possible for CE methods ***(7–9)***. The technique provides superior accuracy as excellent recoveries have been obtained for CE-SDS methods reported in the literature ***(7,9)***. Direct comparisons of CE-SDS and SDS-PAGE demonstrate a significant decrease in variability in the former method ***(10)***. The instrumentation has become more automated, facilitating easy integration into the modern pharmaceutical laboratory. Finally, CE has software that is easily validatable per the 21 Code of Federal Regulations Part 11. Instrument companies such as Beckman and Agilent have produced software systems that can comply with FDA expectations. The validation data reported for CE-SDS methods are able to stand on their own merit but appear especially profound when compared to SDS-PAGE methods ***(1,4,7,10)***. Additional strategies for the method development and validation of CE methods have been reviewed previously ***(11)***.

The development and validation of a purity method for recombinant carboxypeptidase B, described in **Subheading 3**, demonstrates the process used to define a CE method suitable for use in a QCL.

2. Materials

1. Capillary electrophoresis system (Beckman, P/ACE MDQ System).
2. Vortex mixer.
3. Pipets able to dispense accurately from 10 to 5000 µL.
4. Dry bath incubator (e.g., Fisher, cat. no. 11-718).
5. Polypropylene microcentrifuge tubes.
6. Spectrophotometer.
7. Purified water.
8. 5.0-mL disposable syringes.
9. 0.20-µm low protein binding syringe filters.
10. 100-µm id capillary, SDS-coated, 65 cm, cat. no. 241521).
11. SDS sample buffer (Beckman, cat. no. 241525 or Bio-Rad, cat. no. 148-5033).
12. SDS gel buffer (Beckman, cat. no. 477416 or Bio-Rad, cat. no. 148-5032).
13. Blank cartridge kit (Beckman P/ACE MDQ cartridge, cat. no. 144738).
14. Internal reference marker (e.g., Orange G, Beckman, cat. no. 241524).
15. Carboxypeptidase B reference material.
    a. Albumin standard (bovine serum albumin [BSA]) (e.g., Pierce, cat. no. 23209).
    b. Purified water (e.g., Milli-Q® purified water).
    c. 1 *N* HCl (reagent grade).
    d. Tris-HCl buffer: 0.025 *M* Tris-HCl, 0.1 *M* NaCl, pH 7.65. Weigh approx 3.04 g of Tris-HCl and 5.8 g of NaCl and dissolve in approx 950 mL Milli-Q purified water. Add 5 *N* HCl (to pH of 7.65 ± 0.1). Bring final volume to 1 L. May be used for 1 mo.
    e. 5 *N* HCl (reagent grade).

3. Methods

The CE-SDS method was developed to replace an existing SDS-PAGE method. The two methods were developed to determine the overall purity of carboxypeptidase B. The SDS-PAGE method will be discussed in **Subheading 3.1.** The CE-SDS method will be the focus of the remainder of the chapter.

3.1. SDS-PAGE Method

The current purity analysis technique for carboxypeptidase B is a validated SDS-PAGE method. The accuracy (*see* **Fig. 1**) of the method was determined by spiking a range of concentrations of 3-phosphoglyceric phosphokinase, which was used as an internal standard owing to the inherent nonlinearity of stain uptake for SDS-PAGE methods. The accuracy, by spike recovery, was determined to be 87% for the 5% spike and 97% for the 10% spike. Carryover between lanes can be seen in the gel in **Fig. 1.** Blank lanes must be included to prevent sample crosscontamination. This limits the throughput of the method.

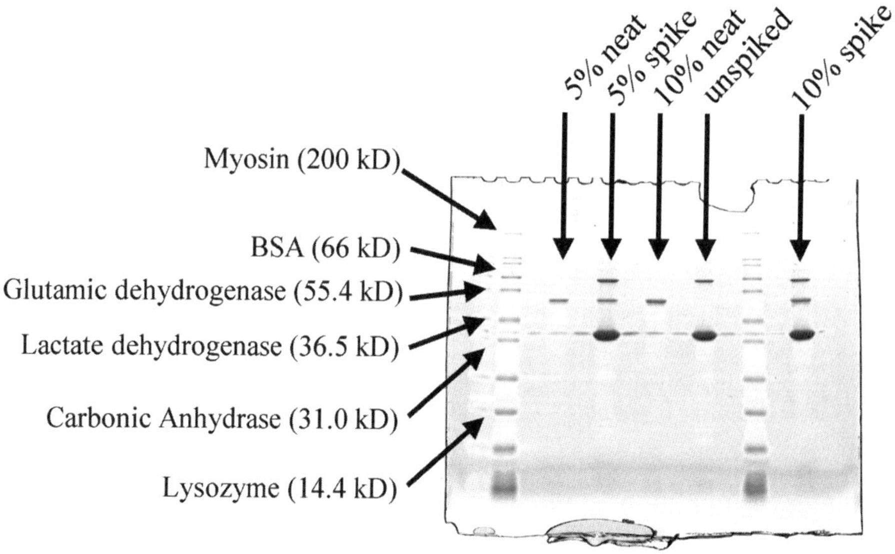

Fig. 1. Accuracy of the SDS-PAGE method. The accuracy of the method was determined by spiking 3-phosphoglyceric phosphokinase into the sample matrix and measuring the recovery.

The linearity of the SDS-PAGE method was demonstrated by fitting the absorbance signal versus concentration to a line. A sample gel is shown in **Fig. 2**. The linear fit is shown in **Fig. 3**. The correlation coefficient (r) for the line was 0.9850, typical for an SDS-PAGE method. The use of densitometry to determine the absorbance limits the ability of the method to show a high degree of linearity.

The precision of the SDS-PAGE method was better than expected. Monomer (band located at ~32 kD) RSD of 3.7% was demonstrated during validation of the method. An example of a SDS-PAGE gel showing the precision experiment is shown in **Fig. 4**. The dimer (band located at ~64 kD) RSD was determined to be 25.9%, which is not high considering the low levels of dimer observed in most samples.

3.2. CE-SDS Method

The methods here describe the steps to follow to develop the optimal, robust method for purity determinations by CE-SDS. The elements of the process are sample preparation, electrophoretic separation, validation, and system suitability.

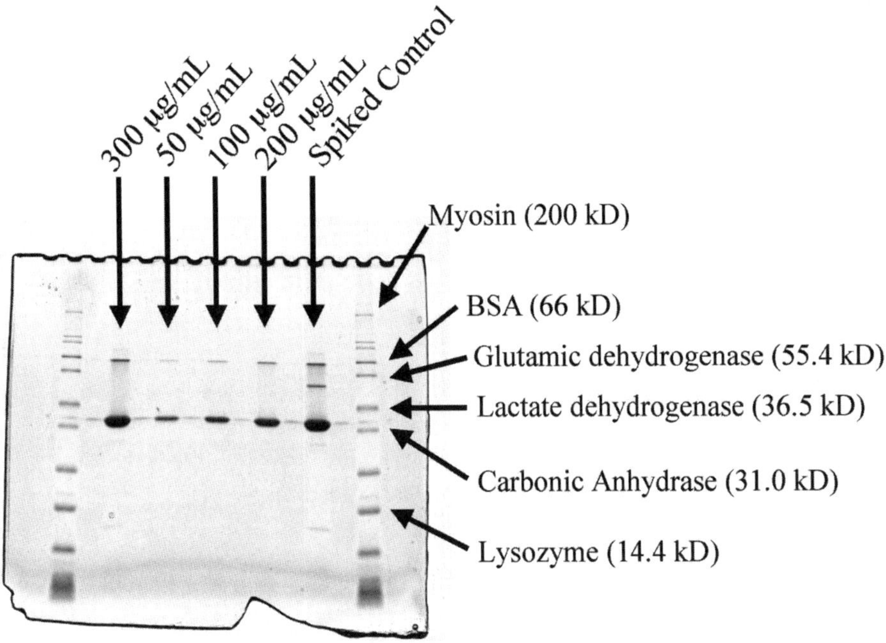

Fig. 2. Linearity of the SDS-PAGE method.

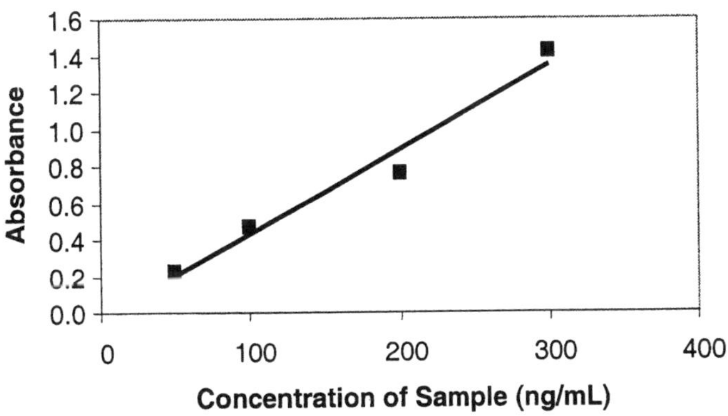

Fig. 3. Linear fit for the SDS-PAGE method. The correlation coefficient (r) is 0.9850.

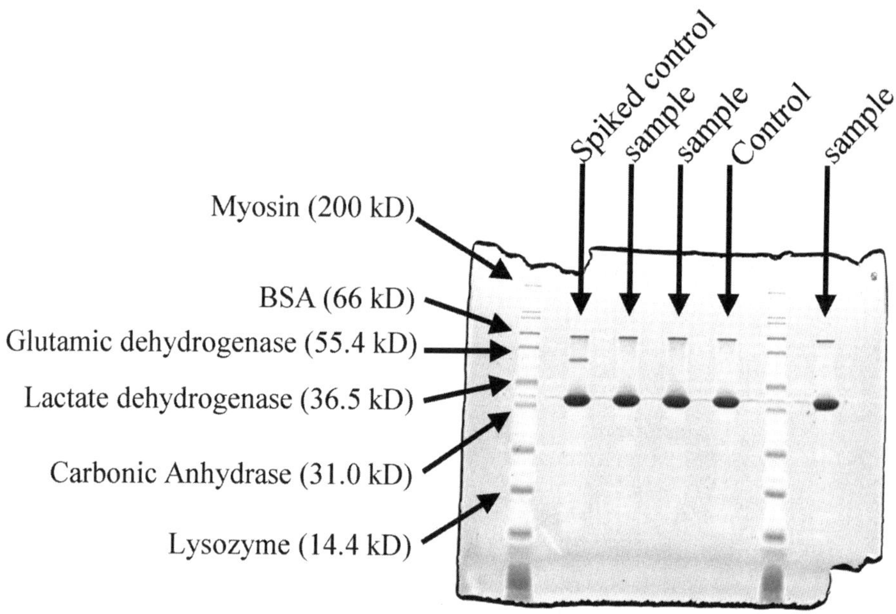

Fig. 4. Example gel demonstrating the precision of the SDS-PAGE method.

3.2.1. Sample Preparation

Sample preparation is a key step in achieving the best resolution in CE-SDS. The goal is to find conditions that robustly denature the protein without inducing the formation of degradation products. The experiments described in the next section represent the optimization of the sample preparation in this method.

3.2.2. Optimization

To achieve the lowest possible limit of quantitation, it is also necessary to use the highest concentration of the sample permitted by the method and analyte. Empirically, the concentration of protein in the sample is limited by the solubility of the protein, the solubility of SDS, and the capacity of the capillary. The upper protein concentration is often practically limited at 2 mg/mL. The detailed sample prep is described.

1. Dilute the sample solution, at least 1:1, to 1.0 mg/mL with syringe-filtered (0.20 μm) SDS sample buffer in a 2.0-mL microcentrifuge tube. Into the diluted sample, add a 5% spike, by volume, of Orange G reference marker. Mix thoroughly.
2. Place tubes in the dry bath incubator set at approx 100°C (± 5°C) for approx 3 min.
3. Allow samples to cool, transfer to 0.5-mL CE sample vials, and cap.

Table 1
Sample Prep Optimization

Time (minutes)	Temperature (°C)
2	60
5	60
10	60
2	100
5	100
10	100
2	boiling
5	boiling
10	boiling

The preparation of the sample is further optimized by testing at different temperatures and times as shown in **Table 1**.

Figure 5 shows sample that has been overheated. Note the formation of a degradation peak just in front of the main peak and the other new peaks, when compared to **Fig. 6**. This is not ideal and represents an analytical modification of the sample, rather than the inherent nature of the sample being analyzed. **Figure 6** represents material that has been prepared by the optimized method. The electropherogram demonstrates no unexpected peaks and no analytical artifacts. Parameters such as peak shape and the appearance of new peaks were considered to determine the optimal conditions. Peak areas suggested complete protein recovery.

3.2.3. Protocol

1. Dilute the sample solution, at least 1:1, to 1.0 mg/mL with syringe-filtered (0.2 μm) SDS sample buffer in a 2.0-mL microcentrifuge tube.
2. Add a 5% spike, by volume, of Orange G reference marker into the diluted sample. For example, spike 15 μL to a total sample volume of 300 μL. Mix thoroughly.
3. Place tubes in the dry bath incubator set at 100°C for approx 3 min (*see* **Notes 1** and **2**).
4. Allow samples to cool to ambient temperature, by refrigeration if necessary, and then transfer to 0.5-mL CE sample vials, and cap.

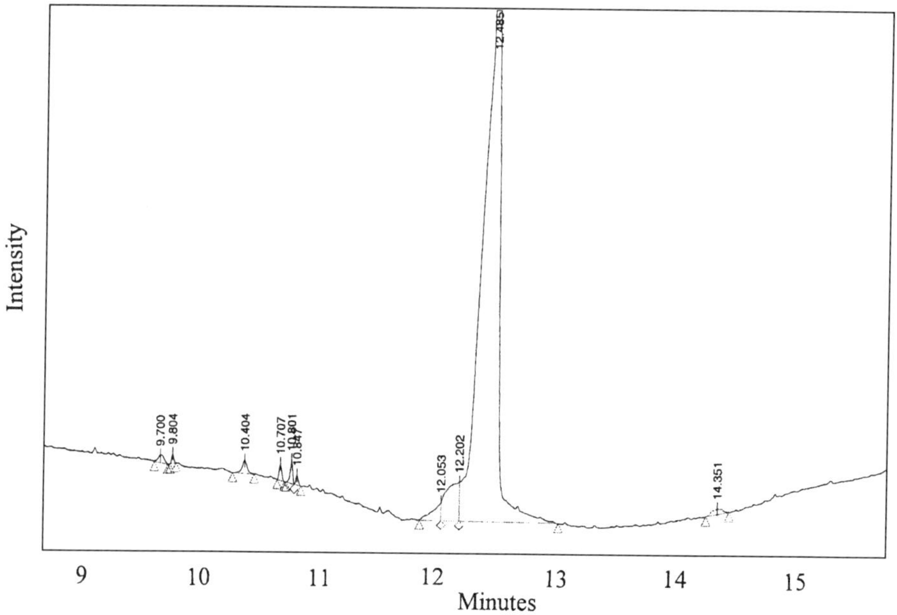

Fig. 5. Electropherogram of sample that has been overheated.

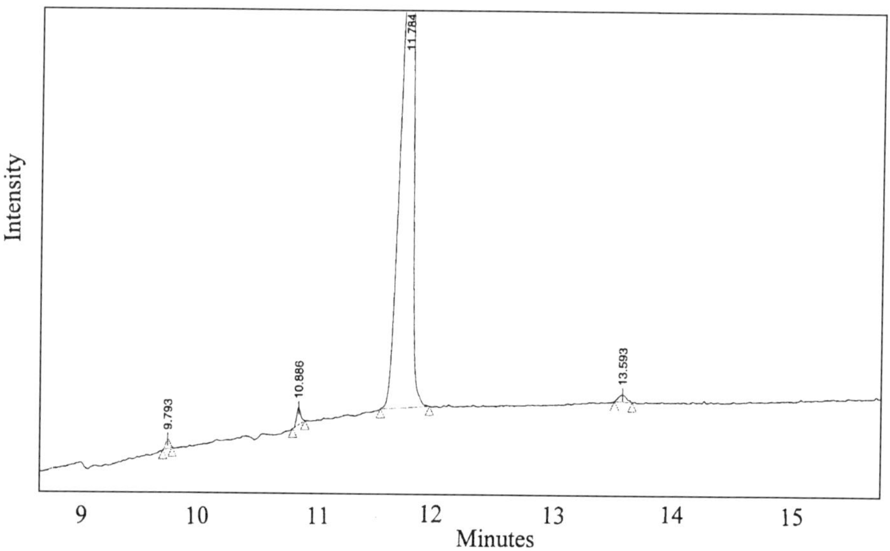

Fig. 6. Electropherogram of an optimally prepared sample.

3.3. Electrophoretic Separation

The electrophoretic conditions are optimized for the protein being tested. Voltage, capillary length, and buffer type are adjusted as necessary to optimize the separation. For the Beckman SDS gel buffer, the optimum conditions were 300 V/cm (e.g., 40-cm capillary using 12 kV of voltage). For the Bio-Rad gel buffer, the optimum conditions were 600 V/cm (e.g., 40-cm capillary using 24 kV of voltage). *See* **Notes 3** and **4**.

Set the UV detector at 214 nm, the capillary temperature to 20°C, and the autosampler temperature between 2–8°C. The method parameters are used for the PACE/MDQ system are detailed in **Table 2**.

The voltage needed depends upon the field strength desired with respect to the capillary length and SDS gel buffer used. **Figure 7** shows the final electropherogram with the optimized conditions detailed in **Subheading 3.2.3.**

3.4. Validation

To demonstrate that the CE-SDS method is providing reliable data, a validation exercise is undertaken. The CE-SDS method was validated independent of the SDS-PAGE method. The parameters studied during the exercise are shown in **Table 3**.

3.4.1. Accuracy, Linearity, and Limit of Quantitation

As with any purity method where standard impurities are not available, accuracy is inferred from specificity and precision (ICHQ2A). To demonstrate accuracy throughout the range of the method, the protein is diluted to a series of concentrations and each dilution is introduced into the capillary. The area of the main peak is then plotted as a function of the concentration, as shown in **Fig. 8**. The validation criteria for accuracy, linearity, and limit of quantitation (LOQ) was as follows. The r-value for the lines fit through the monomer and dimer peaks must be greater than 0.99. For accuracy, the mean bias must not be more than 20%. If all points on the standard curve did not meet these criteria, the lowest point that met the criteria was considered to be the LOQ of the method.

The area at each concentration is plotted and a linear fit was performed on the data (*see* **Fig. 8**). The data showed a linear relationship with a correlation coefficient of 0.9998.

Although the correlation coefficient is expected to be greater than 0.999 (meeting the linearity acceptance criterion), the accuracy is confirmed by plotting a line through the undiluted material and zero and measuring the bias between the predicted and measured values for each dilution. For the enzyme, these data are shown in **Table 4**. The low level of bias in the data indicates that the CE-

Table 2
Method Settings for the P/ACE MDQ System

Step	Time	Event	Value	Duration	Inlet	Outlet	Summary
1		Rinse, pressure	50.0 psi	1.00 min	1 *N* HCl	Empty	Fwd
2		Rinse, pressure	20.0 psi	1.00 min	1 *N* HCl	Empty	Fwd
3		Rinse, pressure	20.0 psi	1.00 min	1 *N* HCl	Empty	Fwd
4		Rinse, pressure	20.0 psi	3.00 min	Gel Buffer	Empty	Fwd
5		Wait		0.00 min	Water	Water	
6		Inject, pressure	0.5 psi	60.0 s	Sample	Gel buffer	Override o.k., fwd
7		Wait		0.00 min	Water	Water	
8	0.00	Separate, voltage	12 or 24 kV*	30.00 min	Gel buffer	Gel buffer	0.17 min Ramp, reverse polarity, 20.0 psi, both
9	0.00	Relay on			N/A	N/A	01:30.0
10	1.00	Autozero			N/A	N/A	
11	30.00	Stop data			N/A	N/A	
12		End			N/A	N/A	

SDS method is accurate. The acceptance criterion for accuracy is met. The data generated in this validation study show that the CE-SDS method provides greater linearity than the SDS-PAGE method (*see* **Subheading 3.1.**), as expected.

To test the limit of quantitation, standard material is spiked with known levels of another protein to simulate an impurity. For example, the recovery of BSA spiked at 1.0, 0.5, 0.25, 0.1, and 0.05% shows acceptable recovery at 0.1% and above.

3.4.2. Precision

Repeatability is tested across both preparations and injections with a typical study shown in **Table 5**. Both relative migration times (RMT) and peak area are expected to be repeatable. The precision acceptance criteria were related to repeatability, intermediate precision, and total purity variability. The repeatability must be less than or equal to 2%. The acceptance criteria for intermediate precision and total purity variability were that they each must be less than or equal to 10%.

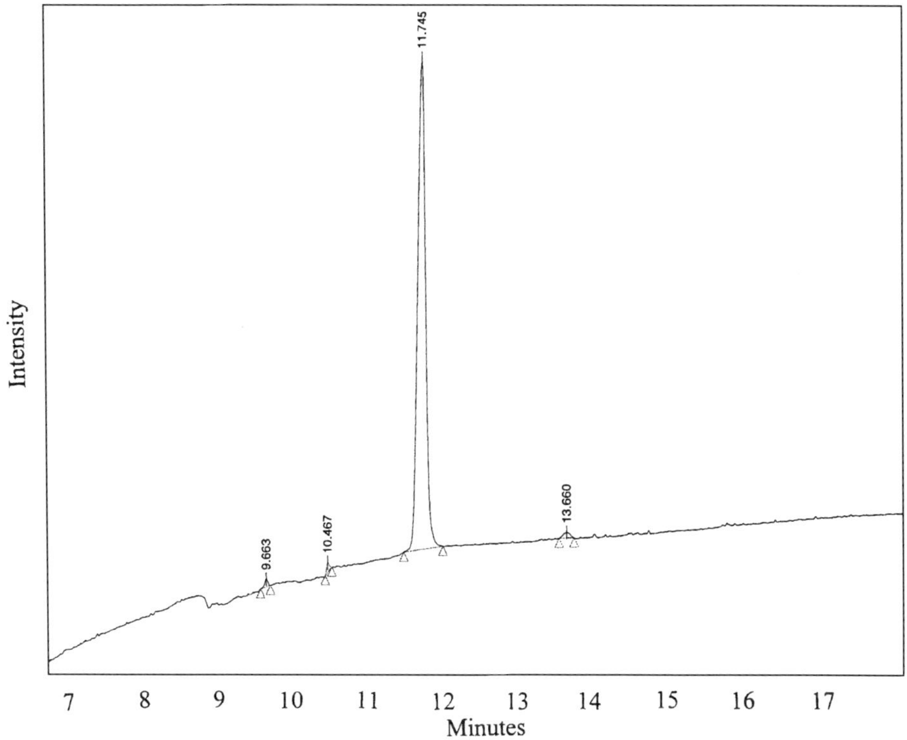

Fig. 7. Electropherogram of optimized method. Optimization included sample preparation and electrophoretic conditions.

Table 3
Validation Parameters Selected for the CE-SDS Method Validation

Parameter	Description
Accuracy/Linearity/LOQ	Defines the range in which data may be generated
Precision	Estimates for introduction, preparation, and setup variability.
Robustness/Ruggedness	Describes the extent to which method parameters can be adjusted without affecting data outcomes.
Specificity	Assures the method is measuring what and only what it is intended to measure.

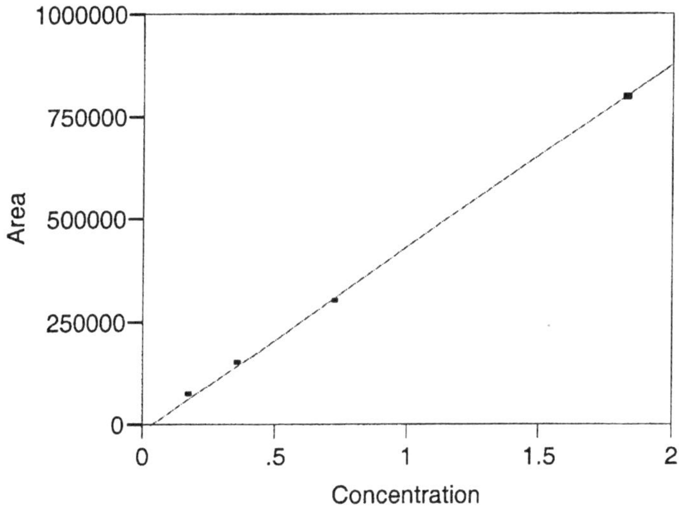

Fig. 8. Accuracy of the CE-SDS method. The correlation coefficient (r) is 0.9998.

Table 4
Linearity of the CE-SDS Method

Concentration mg/mL	Area	Percent accuracy
1.828	804783	0.00
0.7312	305425	-5.12
0.3656	150130	-6.73
0.1828	76031	-5.53

The relative standard deviation for 1/RMT monomer was 0.085% for injections and 0.092% for preps. The relative standard deviation for monomer area was 1.50% for injections and 3.57% for preps. These data compare favorably to the SDS-PAGE method and underscores the importance of the preparation to the variability of the technique. Overall, the method demonstrated very low variability and met all applicable acceptance criteria. Repeatability for the entire study was 0.88%, whereas intermediate precision and total purity variability were 2.18 and 2.33%, respectively.

Table 5
Repeatability of the CE-SDS Method

Prep	Injection	1/RMT Monomer	1/RMT Dimer	1/RMT unknown	Area Monomer	% Purity Monomer
1	1	0.679	0.590	0.811	752382	99.11
1	2	0.679	0.593	0.811	743520	99.31
1	3	0.680	0.592	0.811	764539	99.29
2	1	0.679	0.592	0.810	718019	99.29
2	2	0.678	0.591	0.810	725156	99.19
2	3	0.678	0.591	0.809	702729	99.22

Table 6
Ruggedness Design for the CE-SDS Method Validation

Setup	Analyst	Capillary	SDS-sample buffer vendor	SDS gel buffer vendor
1	1	2	Biorad	Biorad
2	1	1	Beckman	Beckman
3	2	2	Beckman	Beckman
4	2	1	Biorad	Beckman
5	1	1	Beckman	Biorad
6	2	2	Biorad	Beckman
7	2	2	Beckman	Biorad
8	1	1	Biorad	Biorad

3.4.3. Robustness/Ruggedness

To test the ruggedness and robustness of the method, the parameters that may be expected to vary over time are purposely adjusted. Specifically, an experimental design, shown in **Table 6**, is used to insure that different analysts, capillaries, sample buffers, and gel buffer do not give significantly different results (i.e., using analysis of variation, the *p*-value is equal or less than 0.05). No parameters tested gave significantly different results. This model accounted for more than 80% of the variability encountered during the study indicating that the parameters chosen were the largest contributors to the overall variability of the method.

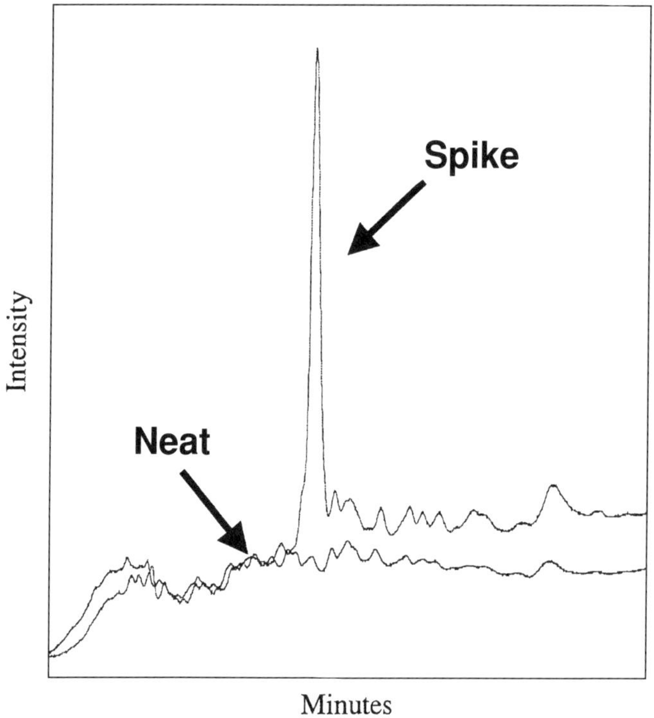

Fig. 9. Electropherogram showing the specificity of the CE-SDS method. The spike sample is easily differentiated from the host cell protein milieu. The approximate migration time of the main peak was 12 min.

3.4.4. Specificity

For recombinant proteins, specificity is demonstrated by showing the resolution of the protein of interest from the host cell protein milieu. The acceptance criterion for this parameter was that the spike recovery must be between 95 and 105%. Spike recoveries of 96.2 and 97.6% were observed in the validation study, meeting the acceptance criterion. These spike recoveries were better than those obtained from the SDS-PAGE experiments. As shown in **Fig. 9**, the CE-SDS method was capable of resolving the protein of interest from the host cell proteins.

3.5. System Suitability

System suitability is the set of analyses that are performed to ensure that the instrument is functioning properly prior to running the samples to be tested. To ensure that the method performance continues to be reliable, a system suitability strategy is developed for the method. For example, if the current is stable and the migration time does not vary significantly, the instrument can be considered suitable for sample analysis. System suitability is an evolving factor that should be reevaluated as additional capabilities of the method are determined.

3.5.1. CE System

A stable current (± 5 µA) is achieved throughout the run after the migration time of the reference marker.

3.5.2. Migration Time

Bracket samples, up to 10 injections in a sequence, with carboxypeptidase B standard injections. The main peak in each injection must be within a migration time of ± 0.1 minutes.

3.6. Conclusion

The CE-SDS method is a rugged, reliable method with excellent accuracy, linearity, precision, and specificity for the analysis of carboxypeptidase B. The CE-SDS purity method met all of the acceptance criteria in the validation protocol and demonstrated that it represents an improvement over the current SDS-PAGE method. We consider this the first step in our efforts to replace SDS-PAGE methods with CE-SDS for obvious technical and business reasons.

4. Notes

1. The total volume while heating samples should be at least 300 µL.
2. To prevent the tubes from opening while boiling, create holes into the caps of the tubes using a scalpel or other sharp, fine-point instrument.
3. If the capillary is new, conditioning of the capillary may be necessary. Using 45.0 psi pressure, rinse with 1 *N* HCl for 5 min, followed by purified water for 3 min, and then SDS gel buffer for 2 min. Repeat as necessary.
4. Sharp spikes and/or significant jumps in the baseline are indicators that an air bubble may have entered the system.

Acknowledgment

The authors would like to acknowledge Mark Strege and Ray Kaiser for their advice and counsel on the preparation of this chapter.

References

1. Park, S. S., Cate A., and Chang, B. S. (2001) Use of capillary electrophoresis to determine the dilute protein concentration in formulations containing interfering excipients. *Chromatographia* **53,** S34–S38.
2. Jochheim, C., Novick, S., Balland, A., et al. (2001) Separation of Enbrel® (rhuTNFR:Fc) isoforms by capillary isoelectric focusing. *Chromatographia* **53,** S59–S65.
3. Schenerman, M. A. and Bowen, S. H. (2001) Optimization, validation, and use of capillary gel electrophoresis for quality control testing of Synagis®, a monoclonal antibody. *Chromatographia* **53,** S66–S72.
4. Ma, S. and Nashabeh, W. (2001) Analysis of protein therapeutics by capillary electrophoresis. *Chromatographia* **53,** S75–S89.
5. Hunt, G. and Nashabeh, W. (1999) Capillary electrophoresis sodium dodecyl sulfate nongel sieving analysis of a therapeutic recombinant monoclonal antibody: a biotechnology perspective. *Anal. Chem.* **71,** 2390–2397.
6. Altria, K. D., Chen, A. B., and Clohs, L. (2001) Capillary electrophoresis as a routine analytical tool in pharmaceutical analysis. *LCGC* **19,** 972–985.
7. Izco, J. M., Ordóñez, A. I., Torre, P., and Barcina, Y. (1999) Validation of capillary electrophoresis in the analysis of ewe's milk casein. *J. Chromatogr. A* **832,** 239–246.
8. Schenerman, M. A. and Bowen, S. H. (2001) Optimizing sample preparation for capillary gel electrophoresis. *LCGC 19,* 190–198.
9. Hwang, K.-H., Lee, K.-W., Kim, C.-S., Han, K., Chung, Y.-B., and Moon, D.-C. (2001) Determination of recombinant human Epidermal Growth Factor (rhEGF) in a pharmaceutical preparation by capillary electrophoresis. *Arch Pharm Res* **24,** 601–606.
10. Bowen, S. H. and Schenerman, M. A. (1998) Replacing slab gel electrophoresis methods with capillary electrophoresis for quality control and stability testing. *BioPharm* **11,** 42–50.
11. Wätzig, H., Degenhardt, M., and Kunkel, A. (1998) Strategies for capillary electrophoresis: method development and validation for pharmaceutical and biological applications. *Electrophoresis* **19,** 2695–2752.

7

Analysis of Neutral N-Linked Oligosaccharides From Antibodies Using Free-Solution Capillary Electrophoresis in Bare Fused-Silica Capillaries

Jeffrey S. Patrick, Brenda P. Rener, Gregory S. Clanton, and Avinash L. Lagu

Abstract

Conditions for the enzymatic release, chemical derivatization, and analysis of oligosaccharides from the consensus glycosylation sites on antibodies are described. Release of the oligosaccharides is from the native protein. The APTS derivatives of the released oligosaccharides are then analyzed by capillary electrophoresis (CE) using a free solution separation in a bare fused silica capillary. An example of the application of the method to the analysis of the oligosaccharide population from antibodies obtained from different cell lines is provided. The separation conditions provide for resolution of various galactose positional isomers, including those derived from different linkage configurations.

Key Words

Antibodies; capillary zone electrophoresis; derivatization, fluorescence, fused-silica; LIF; N-linked; oligosaccharides; qualitative; quantitation; antibodies.

1. Introduction

The development and approval of protein therapeutics has seen exponential growth over the past decade. This rapid increase in the investigation and development of proteins of recombinant origin has created significant challenges for the protein analytical scientist. This has required the continuing development of new tools with which to address issues in a reliable and rigorous fashion. Among the tools which have seen rapid and prolific development have been mass spectrometry (MS) ***(1–3)*** and capillary electrophoresis (CE) ***(4–6)***. Within the various classes of therapeutic proteins, one of the most challenging molecu-

From: *Methods in Molecular Biology, vol. 276: Capillary Electrophoresis of Proteins and Peptides*
Edited by: M. A. Strege and A. L. Lagu © Humana Press Inc., Totowa, NJ

lar properties to quantify is the broad and heterogeneous population of glycoforms that are often present. The complexity and impact on bioactivity have been documented thoroughly for numerous complex glycoproteins. The glycosylation of these compounds has been demonstrated to affect bioactivity, pharmacokinetics, and other biophysical properties. However, it has also been demonstrated that conditions during cell culture, among others, can influence the glycosylation observed *(7)* making it important to have reliable methods with which to monitor this property.

A subclass of therapeutic proteins which has dominated development efforts is monoclonal antibodies (MAbs). Current estimates indicate that these molecules, which occur in several classes, occupy more than half of the development efforts and new drug product submissions. As with complex glycoproteins, the oligosaccharides present on antibodies (Abs) can significantly affect the activity observed. An Ab is represented in (**Fig. 1**) and is composed of light (L) and heavy (H) chains, two each, which then give rise to Fab and Fc regions on the molecule. All Abs of animal origin have a common site of N-linked glycosylation located on each of the two Fc portions. Occupancy of this site has significant impacts on effector functions that are the mechanisms by which the cell initiates killing functions (so-called cellular- and Ab-mediated cytotoxicity) *(8,9)*. These sites are consistently occupied by biantennary structures such as those in **Fig. 2**. These structures are neutral in nature with 0, 1, or 2 terminal galactose residues. These structures are referred to as G0, G1, and G2, respectively. Dependent upon the source of the Ab (i.e., cell-line, genetic construction, and culture conditions) other glycoforms may also be present, including fucosylation, sialylation, secondary galactose structures, and even remote complex glycosylation sites. The nature of each of these entities is determined in part by the host system in which the Ab is expressed.

Common methods of oligosaccharide analysis are high-pH anion exchange chromatography (HPAEC) of the free oligosaccharide *(10,11)* or weak anion exchange HPLC of the derivatized oligosaccharide *(12)*. These charge-based separations are not productive for neutral oligosaccharides where the difference in the oligosaccharide-based charge is negligible. Other efforts have utilized MS analysis of the intact Ab or its partially reduced form, which provides mass information but does not readily provide information on positional isomers or other isobaric or nominally isobaric species. Capillary electrophoresis (CE) of derivatized oligosaccharides using gels or coated capillaries and laser-induced fluorescence (LIF) have been applied to effect the separation and quantitation of neutral oligosaccharides such as those found on Abs *(13)*. This type of technology is presently commercially available from Beckman Coulter *(14)*. These approaches provide useful separation and quantitative population

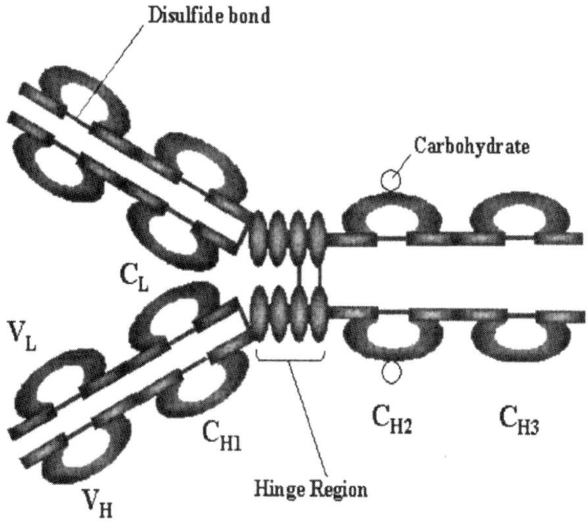

Fig. 1. Depiction of a typical IgG. The location of the carbohydrate is indicated in the figure.

estimates but restrict the user to capillaries provided by a single vendor or to the use of complex additives, which may introduce long-term variability in the separation from one batch to another batch. In the **Subheading 3.** conditions for the separation of the APTS-derivatives (9-aminopyrene-1,4,6-trisulfonic acid; *see* structure in **Fig. 3**) of neutral oligosaccharides released from Abs are discussed. The release of the oligosaccharide is from the native protein using PNGase F, and the separation is effected using a bare-fused silica capillary. The most commonly used conditions are provided along with experiments to define the identity of typical and atypical structures observed. Both the analysis and release of the oligosaccharides are generic and have been applied to a range of Abs. The release of oligosaccharides from the native Ab is intended to provide some selection for the accessible Fc structure in the case where secondary sites exist. The conditions established have been optimized for the analysis of neutral oligosaccharides but are amenable to the analysis of acidic variants (e.g., sialylated or sulfated) as well. The conditions provide separation of positional isomers as well as linkage (α vs. β) isomers.

2. Materials

1. 9-Amino-1,4,6-trisulfonic acid (APTS; part no. 501309, Beckman Coulter, Fullerton,CA).
2. Peptidyl-*N*-Glycanase F (PNGase F; QA-Bio, part no. E-PNG01, San Mateo, CA).
3. Solutions of Abs at 0.5–12 mg/mL in bicarbonate, acetic acid, or other volatile buffer system.

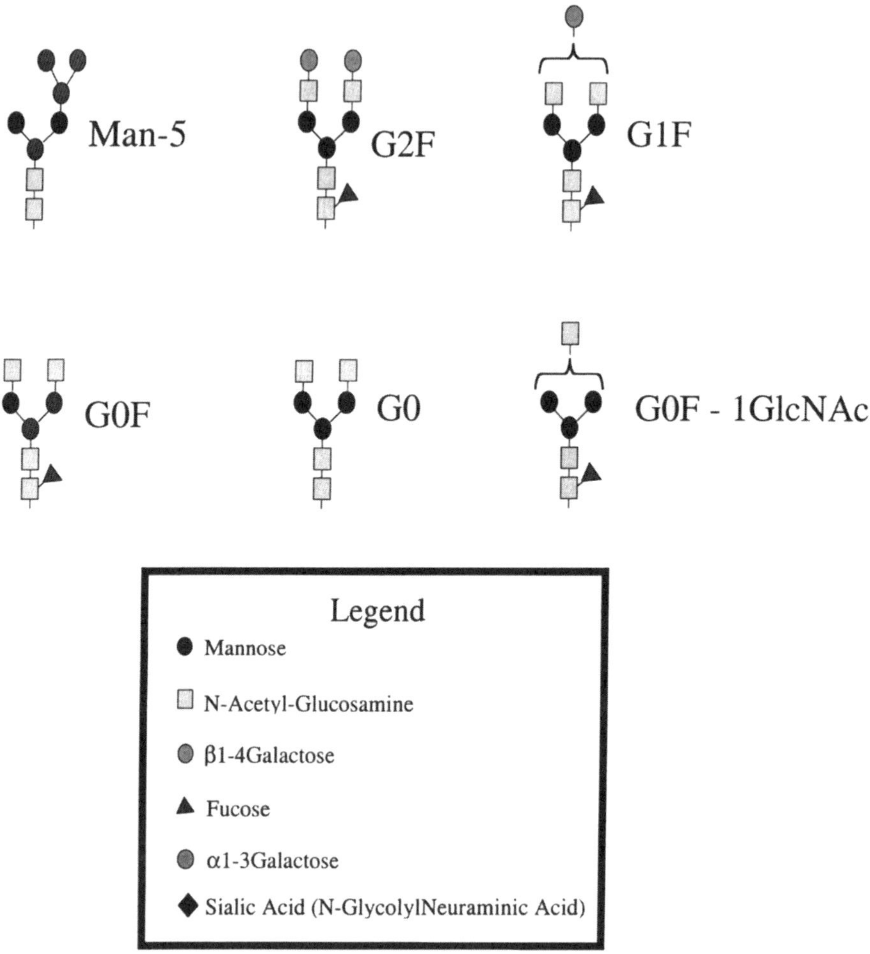

Fig. 2. Structures of typical Asparagine-linked (N-linked) biantennary oligosaccharides present on the antibodies.

4. α-Galactosidase selective for 1–3 bond (from green coffee bean).
5. β-Galactosidase.
6. Neuraminidase (a.k.a. sialidase).
7. Oligosaccharide reference materials including G0, G0F, Man-5, G2, and G2F.
8. Centrifugal evaporator (e.g., SpeedVac, Thermo-Savant).
9. Water bath capable of control between 35 and 60°C (target temperatures of 37 and 55°C).
10. Milli-Q water or equivalent 18 MΩ water.
11. Fused silica capillary (50 µm id at least 60 cm in length; available from Beckman Coulter [Fullerton, CA] with detection window in place).

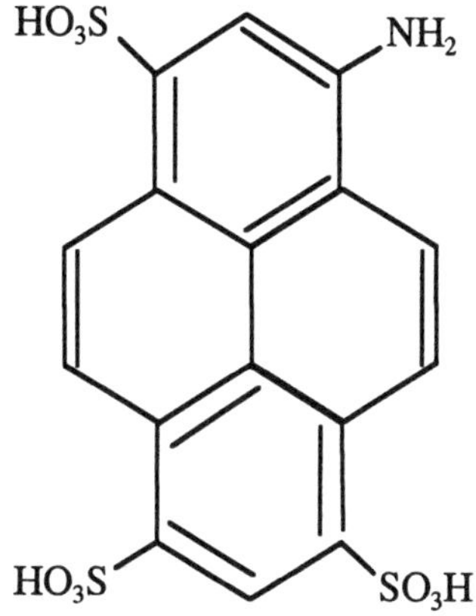

Fig. 3. Structure of 9-amino-1,4,6-trisulfonic acid (APTS).

12. 250 m*M* Sodium phosphate buffer, pH 7.5.
13 0.22 µm Nylon filters adaptable to leur-lock syringes.
14. 0.22 µm Nylon filters for filtration of CE running buffer.
15. 10,000 mol wt CO (molecular weight cut-off) Microcon ultracentrifugation device (Millipore, Billerica, MA).
16. 1 *M* Sodium cyanoborohydride in THF (available from Aldrich, part no. 29,681-3).
17. 500 L Conical polypropylene centrifuge tubes compatible with 14,000*g* centrifugation.
18. Centrifuge capable of generating 14,000*g* centrifugal force.
19. HPLC-grade triethanolamine.
20. Glycerol (reagent grade or better).
21. 1/10 Running buffer: 10 m*M* triethanolamine (HCl), pH 7.5, with 1% glycerol.
22. CE running buffer: 100 m*M* triethanolamine (HCl), pH 7.5, with 10% glycerol.
23. 0.1 *N* Sodium hydroxide.
24. 0.1 *M* Hydrochloric acid.
25. 0.5 *N* Sodium hydroxide.
26. Capillary electrophoresis system equipped with a laser-induced fluorescence detector having an argon laser or one capable of excitation at 488 nm and detection of emitted light at nominally 520 nm. An example of one such system is a Beckman Coulter P/ACE MDQ CE system (Beckman Coulter, Fullerton, CA).
27. LIF: 520-nm bandpass filter with the excitation provided by a 488-nm argon laser.

3. Methods

3.1. Sample Preparation

A schematic of the sequence for the release and derivatization of oligosaccharides is shown in **Fig. 4**, and detailed preparation instructions follow.

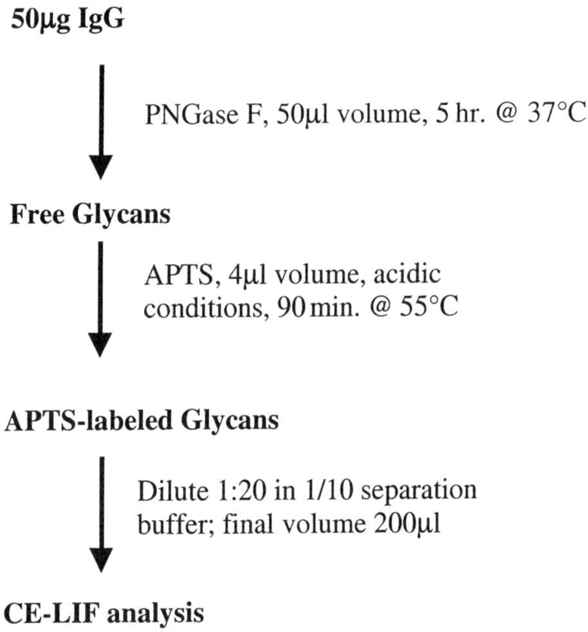

Fig. 4. The procedure for the release and derivatization of oligosaccharides.

The sample of intact Ab can be obtained in any buffer but will require exchange into a moderately dilute volatile buffer such as water, 10 mM ammonium bicarbonate, or 1% acetic acid. The tendency of Abs to aggregate and be retained at surfaces makes it critical that some measurement be used to track protein recovery prior to release of the oligosaccharides until the specific Ab of interest has been handled extensively (e.g., absorbance at 278 nm). Although written with intact Ab in mind, the procedure is amenable to Ab fragments such as "half antibodies" (one light chain joined to one heavy chain), free heavy chain, and the Fc portion produced by enzymatic treatment with papain.

The volumes of reagent and sample listed here are based on IgG protein concentrations between 0.5 and 12 mg/mL. For differing sample concentrations the aliquoted volumes of sample should be adjusted to provide approx 50 μg of Ab. This amounts to approx 0.3 nmoles of Ab, which will provide approx 0.6 nmoles of released oligosaccharide. Rigorous experimentation has demonstrated that complete release of the oligosaccharides present in the Fc region is complete under the conditions described here without reduction or alkylation. Particularly in the instance of recombinantly engineered Abs, other glycosylation sites may be present on the Ab. The release of oligosaccharide from these positions may or may not occur without reduction and alkylation and may require additional experimentation. Regardless of the details, it is recommended that the molecule of interest be evaluated for release kinetics during initial studies or that complementary techniques including matrix-assisted laser desorption/ionization (MALDI)–MS be used to evaluate the oligosaccharide release.

The steps to release the oligosaccharide and generate the APTS derivative of the released oligosaccharide are as follows (approximate time is 2 h hands-on activity plus enzyme incubation; reasonable throughput of 10–15 samples per sample set):

1. Transfer a volume corresponding to approx 50 ± 15 μm of IgG to a conical polypropylene tube such as a 500-μL (or smaller) Eppendorf tube. Evaporate the sample to dryness in a vacuum centrifuge, such as a Speed-Vac (*see* **Note 1**).
2. An amount of protein lower than this may be used provided that the reaction volumes described below are adjusted in proportion. Set up digestion mixture by adding 38 μL purified water to the dried pellet. Add 10 μL 250 m*M* sodium phosphate, pH 7.5 (referred to as 5X reaction buffer by enzyme sources and typically supplied with the enzyme). The final buffer in the reconstituted sample is now nominally 50 m*M* sodium phosphate, pH 7.5. Mix the sample by gentle vortexing and assure that the dried residue is dissolved.
3. Add 2 μL PNGase F (10 mU). The PNGase F used in the investigational studies for these conditions is supplied by QA-Bio. The specific activity of the enzyme is >20 U/mg, and the activity concentration is >5 U/mL (*see* **Note 2**).
4. Mix gently and leave in a water bath set at 37°C for 5–18 h (*see* **Note 3**).
5. Centrifuge the sample briefly to get any condensate that may have formed during the incubation out of the lid of the tube. Transfer the sample, which contains the deglycosylated Ab and the released oligosaccharides to a 10,000 mol wt CO. Microcon ultracentrifugation device. (The Microcon should be prerinsed with water per manufacturer's instructions.)
6. Centrifuge the Microcon containing the underivatized oligosaccharides at a relative centrifugal force (rcf) of 14,000*g* for about 9 min or as directed in the manufacturer's instructions. (*see* **Notes 4**, **5**, and **6**).
7. Transfer the oligosaccharide solution to a 500-μL tube. Evaporate to dryness. (Again, do not use heat to dry samples.)
8. Add 2 μL APTS labeling reagent (*see* **Note 7**). Mix well by vortexing. In fume hood, add 2 μL 1 *M* sodium cyanoborohydride in THF (*see* **Note 8**). Mix the sample well by vortexing gently and centrifuge briefly to make sure all liquid is at bottom of tube. Incubate in a water bath set at 55°C for 90 min (*see* **Note 9**).
9. Remove the reaction tube from the water bath and centrifuge briefly. Add 46 μL purified water to stop additional reaction and mix well. The samples should be diluted 1:20 in 10 m*M* triethanolamine (HCl), pH 7.5, with 1% glycerol, which is 1/10 strength separation buffer (*see* **Note 12**).

3.2. Instrumental Conditions

The experiments described were performed using the conditions listed below on a Beckman Coulter P/ACE MDQ CE system (Beckman Coulter, Fullerton, CA) with a LIF detector. Although not used to obtain the data presented in this manuscript, an Agilent CE system has been used to successfully acquire comparable data. Either system was operated under the same nominal conditions within the constraints of the system design.

3.3. Separation Conditions

CE was achieved as follows.

1. A bare fused-silica capillary (Beckman Coulter) with a length of 50 cm and an id of 50 μm was installed in the Beckman Coulter cartridge. (Other suppliers of fused-silica capillaries have also been used successfully.) The detection window was effectively 46 cm from the sampling end of the capillary. The capillary was conditioned using 0.1 *N* sodium hydroxide to assure full and uniform activation of the capillary walls and removal of trace metals that might deleteriously affect the separation. (This is critically important to proper capillary performance and reproducible application from capillary to capillary.) A typical conditioning setup will be as follows.

Time (min)	Reagent	Duration (min)	Pressure (psi)
0.0	Water	1.00	50.0
1.0	1 *N* HCl	5.00	20.0
6.0	Water	1.00	50.0
7.0	0.1 *N* NaOH	25.00	20.0
32.0	Water	1.00	50.0
33.0	Buffer	10.00	20.0

2. The running electrolyte of 100 m*M* triethanolamine, pH 7.5, with 10% glycerol was prepared by dispensing the appropriate weight of HPLC-grade triethanolamine and the appropriate weight of glycerol to provide 100 m*M* and 10% (w/v) into purified water then adjusting the pH down with the addition of 1 *N* hydrochloric acid. The solutions are filtered through 0.22-μm nylon membranes (*see* **Note 11**).
3. The sample is injected using pressure at 0.5 psig for 10 s. *See* **Note 12**.
4. The separation is achieved under constant voltage conditions using 22 kV while the capillary is maintained at 18°C (*see* **Note 13**). Data are acquired for 50 min, then the system is recycled using a 0.5-N sodium hydroxide wash to remove compounds adsorbed to the walls of the capillary.

A detailed time program for sample analysis is shown in **Table 1**.

Table 1
Program for Sample Analysis

Time	Reagent	Action	Duration	Pressure/Voltage
0.0 min	0.5 *N* NaOH	Rinse	5.00 min	40 psi
5.0 min	Water	Rinse	1.00 min	40 psi
6.0 min	Buffer (TEA)	Rinse	5.00 min	40 psi
11.0 min	Sample	Injection	10.0 s	0.5 psi
11.1 min	Buffer (TEA)	Separate	50.0 min	22 kV

3.4. Results and Discussion

The oligosaccharide profiles obtained from some typical recombinant MAbs are shown in **Fig. 5** and **6**. The Abs were isolated from the cell culture supernatant from two different cell lines identified here as cell lines 1 and 2. The protein portion of the Ab is identical in the two samples. The samples were analyzed using the conditions described previously. The electropherogram from sample 1 (*see* **Fig. 5**) shows four prominent and not less than 10 minor components. The electropherogram from sample 2 (*see* **Fig. 6**) shows the same four prominent components but significantly fewer (~5) components of lesser intensity. The oligosaccharide derivatives observed were identified by standard addition experiments when reference materials were available. These include G0, G0F, Man-5, G2, and G2F. (Note that G1 and G1F isomer mixtures may be generated from the G2 and G2F materials by treatment with α-galactosidase. This reaction is rapid and may generate a mixture that includes levels of G0 and G0F in addition to the desired G1 and G1F molecules). Alternatively, the oligosaccharide is identified through the observation of peak behavior and intensities upon treatment with exoglycosidases of known specificity. An example of this experiment is shown in **Fig. 7**, in which unknown peaks were suspected to be associated with α (1–3) galactose. The G1F positional isomers are well resolved in both of the electropherograms in **Figs. 5** and **6**. Two additional pairs of positional isomers [Galα(1–3) G3F and Galα(1–3) G2F] are baseline-separated in the cell line 1 sample (*see* **Fig. 4**; sample 1). However, one of the Galα(1–3) G2F appears as a shoulder preceding the G2F peak. To confirm the identity of the Galα(1–3) isomeric oligosaccharides the samples were treated using α(1–3) galactosidase from green coffee beans. A sample treated in this way and prior to treatment are shown in **Fig. 7A** (before enzyme) and **B** (after enzyme). The only notable differences in the electropherograms are the absence of peaks at nominally 21- and 28-min migration time and the small relative increase in the abundance of the G2F and G1F

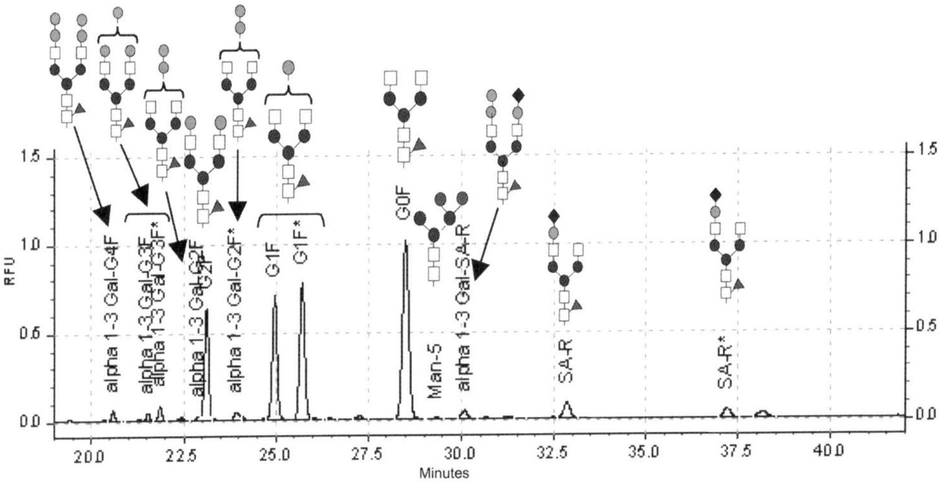

Fig. 5. Electropherogram produced from the APTS-derivatized oligosaccharides released from an Ab using cell line 1 (sample 1). The symbols are as described in the legend to **Fig. 2**. Common nomenclature is also included in the figure.

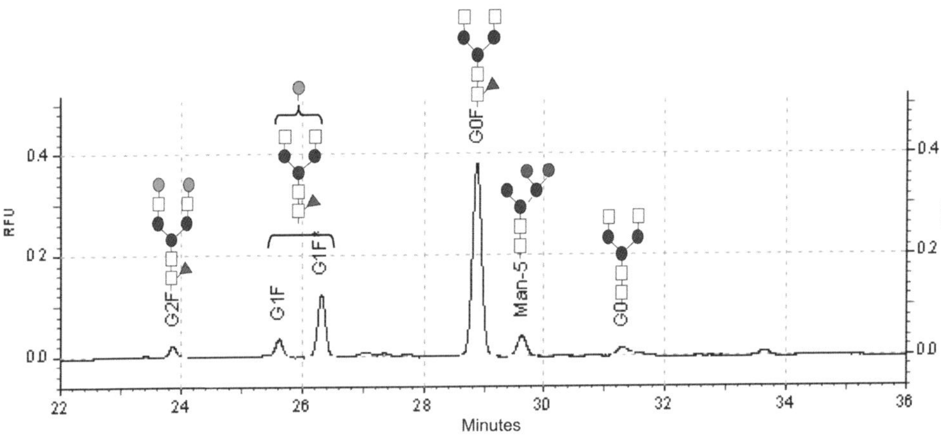

Fig. 6. Electropherogram produced from the APTS-derivatized oligosaccharides released from an Ab using cell line 2 (sample 2). The symbols are as described in the legend to **Fig. 2**. Common nomenclature is also included in the figure.

peaks. In a similar fashion, the presence of sialic acid at the terminus of the structures is confirmed using neuraminidase. This is shown by comparison of the electropherograms in **Fig. 7A** (before neuraminidase treatment) and **C** (after treatment with neuraminidase). As with the galactosidase, the disappearance of small peaks and slight increases in the relative intensities of neutral oligosaccharides is observed. To further support the other identifications, the released oligosaccharides were also treated with α-galactosidase. This shows (*see* **Fig. 7A** and **D**) the collapse of the majority of the peak area to a single peak corresponding to G0F.

In comparing these two electropherograms in **Figs. 5** and **6**, one notes small differences in relative and absolute migration. Typically, within-run migration time variations are less than 1 min. This is most pronounced when a new capillary is involved. As the capillary is conditioned through sample analysis, the migration time variability becomes minimized. It is strongly recommended that either a control sample or standard cocktail be analyzed with some frequency during a series of samples.

3.5. Conclusion

The conditions for the release, derivatization, and separation of neutral oligosaccharides from native Abs have been presented. Conditions have been developed which demonstrate the capability of a free solution CE–LIF method to:

Resolve all major neutral oligosaccharides from typical IgG from different hosts, including positional isomers not distinguishable by LC–MS or MS alone.

Use a simple buffer and a bare fused-silica capillary.

Confirm the identification of the oligosaccharide using comparison with reference materials or treatment with appropriate exoglycosidases.

Provide selectivity and specificity beyond LC–MS or weak anion exchange HPLC.

Provide quantitative information using only 3–6 fmoles of oligosaccharide from IgG on the capillary.

4. Notes

1. Do not dry the samples under heat as this may degrade some of the carbohydrates.
2. If the PNGase F is obtained from a source other than QA-Bio be aware that different vendors may determine enzyme activity differently. The conditions described herein may not work with enzymes from other vendors (i.e., the enzymes may not be equally active).
3. A water bath is strongly recommended for all incubations to maintain a uniform temperature, minimize "hotspots" that may occur in heating blocks, and minimize degradation and excessive evaporation that may occur during high temperature excursions.

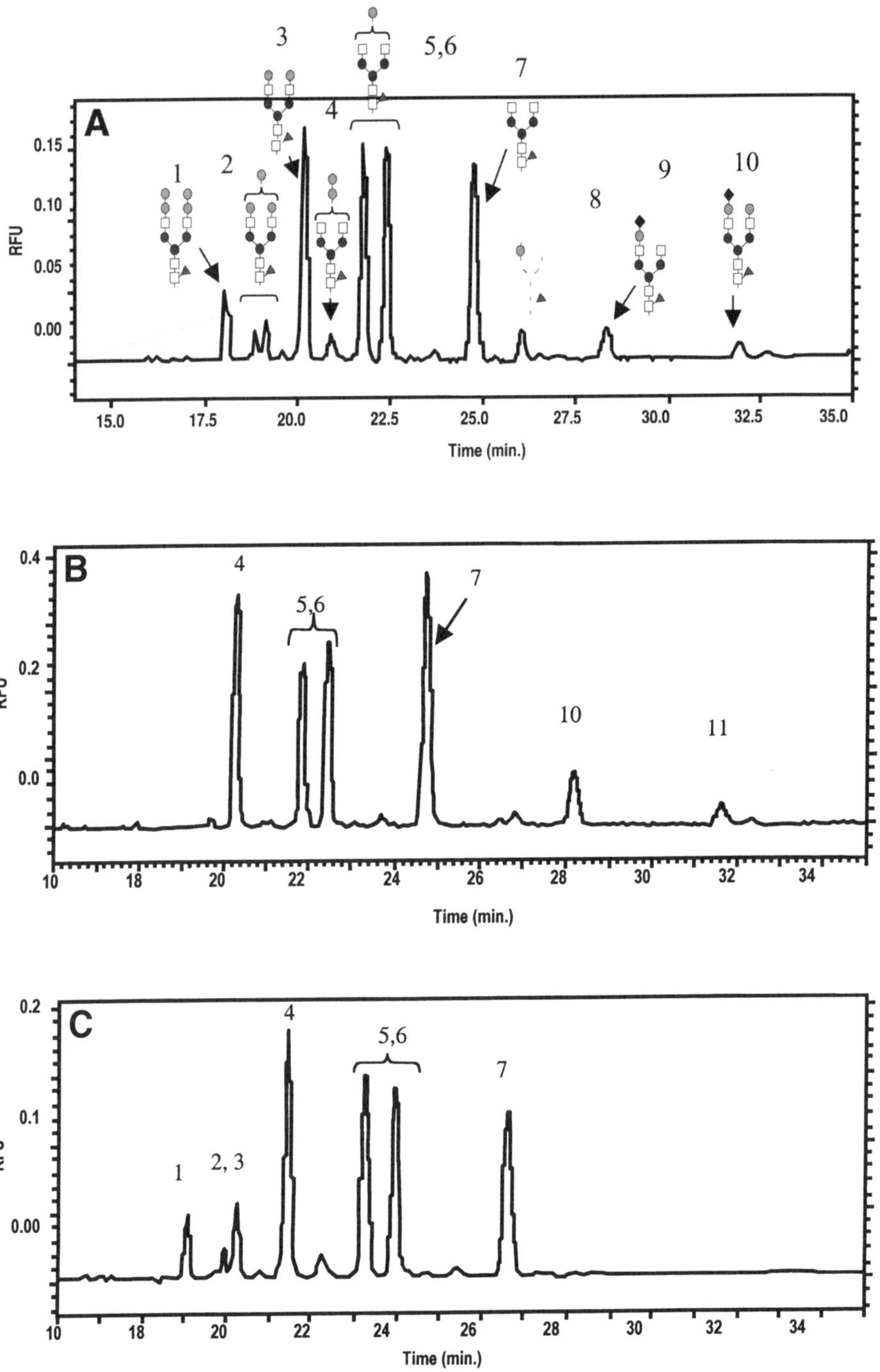
A
1
2
3
4
5,6
7
8
9
10
RFU
0.15
0.10
0.05
0.00
15.0
17.5
20.0
22.5
25.0
27.5
30.0
32.5
35.0
Time (min.)
B
4
5,6
7
10
11
RFU
0.4
0.2
0.0
10
18
20
22
24
26
28
30
32
34
Time (min.)
C
1
2, 3
4
5,6
7
RFU
0.2
0.1
0.00
10
18
20
22
24
26
28
30
32
34
Time (min.)

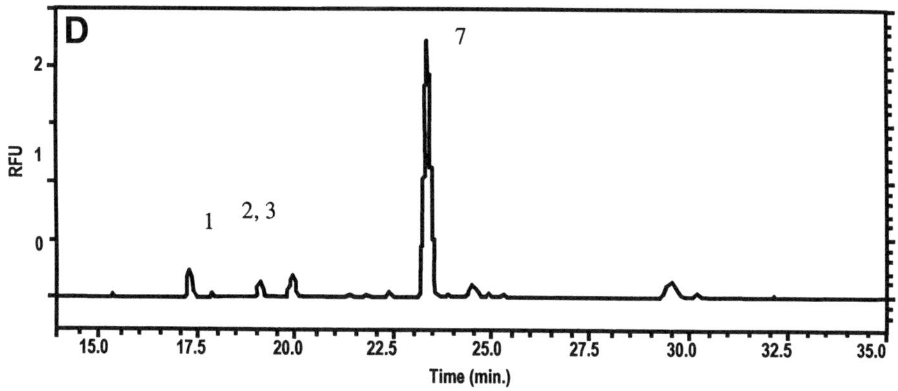

Fig. 7. Identification of oligosaccharides using treatment with α(1–3)galactosidase and neuraminidase. **(A)** Untreated sample of derivatized oligosaccharides released from an antibody; **(B)** Derivatized oligosaccharides from (A) after treatment with α(1–3)galactosidase; **(C)** Derivatized oligosaccharides from (A) after treatment with Neuraminidase; **(D)** Derivatized oligosaccharides from (A) after treatment with α-galactosidase. The peaks are identified as follows: 1, G4F (α); 2, G3F (α); 3, G2F; 4, G2F* (α); 5, G1F; 6, G1F*; 7, G0F; 8, SAR (α); 9, SAR; 10, SAR* in which α designates the presence of 1 or more alpha-1,3 galactose residues. The structures are shown in 7(A), but only numbers provided in B–D. **(A)**, Oligosaccharides released from sample without treatment. **(B)**, Oligosaccharides released from sample and treated with α-galactosidase. **(C)**, Oligosaccharides released from sample and treated with neuraminidase. **(D)**, Oligosaccharides released from sample and treated with α-galactosidase.

4. When performing the separation of the protein from the released oligosaccharides it is important to notice whether the entire volume of sample liquid passes through. It is important to use a centrifugation time where this DOES NOT occur. The retention of some liquid on the upper side of the membrane provides assurance that the membrane remained intact and the proteins, which may interfere with the electrophoresis, were removed from the oligosaccarides. The released oligosaccharides will pass through the membrane into the receiving tube, whereas the protein backbone and any residual PNGase F will be retained by the membrane.
5. If alternate devices are used to separate the glycans from the residual protein it is necessary to run controls (complete system blanks) to monitor other aldehydes or carbohydrate-polymers which may become derivatized and interfere with the analysis. The use of Dextran and other carbohydrate-based sizing columns is discouraged because of the potential for oligosaccharide leaching from these materials, which may react with APTS.
6. During the separation of the oligosaccharides from the deglycosylated protein, the protein-containing portion should be retained until the oligosaccharide analy-

sis is completed. If there are no oligosaccharides detected in the analysis, the release of the carbohydrate may be attempted again or the protein may be interrogated using alternative methods including mass spectrometry to determine whether the oligosaccharides are present or not on the residual protein.

7. The APTS reagent (available in solid form from Beckman Coulter, part no. 501309) is prepared by dissolving the solid in a single vial in 48 µL of 15% acetic acid in water. Mix well and store at -20°C up to 2 wk in the dark.
8. The cyanoborohydride reagent is flammable, and is an HCN source. Use extreme caution.
9. Appropriate precautions should be used if an amount of protein lower than the amount described in this procedure is used. A simple adjustment of reaction and dilution volumes may not result in the analytical performance. The derivatization reaction is most efficient at lower amounts of sample but reduction of sample volumes may compromise effective handling and sample injection on the CE system. Sample handling, control of evaporation, and appropriate mixing become significantly more difficult as the volumes below are reduced to accommodate reduced amounts of antibody.
10. Dilution of the derivatized oligosaccharides into the 1/10 buffer provides for a more robust analysis with improved peak shape and more precise migration times. It is recommended that the samples be filtered through 0.22-µm nylon membranes to ensure that all particulates are removed. Samples should be stored at –20°C when not in use and should be discarded after 3 mo.
11. The pH of the CE running buffer should be between 7.5 and 7.6 to assure proper selectivity. Weighing of the triethanolamine and glycerol provides a more robust buffer preparation owing to the viscous nature of these chemicals. High-quality glycerol should be used. Impurities in the glycerol may react to produce fluorescent derivatives which behave similarly to the oligosaccharides.
12. As in the case of sialylated, sulfated, or phosphorylated oligosaccharides, there may be some discrimination toward potential-driven sample injections. These should, therefore, be avoided. For this reason, pneumatic injections, either pressure or gravity driven, should be used.
13. It is very important that the capillary be maintained at subambient temperature as the migration and, more importantly, the selectivity and resolution are affected by temperature shifts.

References

1. Patrick, J. S. (2002) Advances in analytical chemistry for biotechnology—mass spectrometry of peptides, proteins and glycoproteins, in *Handbook of Industrial Cell Culture: Mammalian, Microbial and Plant Cells*, (Vinci, V. A. and Parekh, S. R., eds.), Humana, Totowa, NJ.
2. Chapman, J. R., ed. (2000) *Mass Spectrometry of Proteins and Peptides—Methods in Molecular Biology*, vol. 146. Humana, Totowa, NJ.

3. Snyder, A. P., ed. (1995) Biochemical and biotechnological applications of electrospray ionization mass spectrometry, *ACS Symposium Series*, vol. 619, American Chemical Society, Washington, DC.
4. Patrick, J. S. and Lagu, A. L. (2001) Review: applications of capillary electrophoresis to the analysis of biotechnology-derived therapeutic proteins. *Electrophoresis* **22,** 4179–4196.
5. Taverna, M., Tran, N. T., Merry, T., Horvath, E., and Ferrier, D. (1998) Electrophoretic methods for process monitoring and the quality assessment of recombinant glycoproteins. *Electrophoresis* **19,** 2572–2594.
6. Kakehi, K., Kinoshita, M., and Nakano, M. (2002) Analysis of glycoproteins and the oligosaccharides thereof by high-performance capillary electrophoresis-significance in regulatory studies on biopharmaceutical products. *Biomed. Chrom.* **16,** 103–115.
7. Patel, T., Parekh R., Moellering, B., and Prior, C. (1992) Different culture methods lead to differences in glycosylation of a murine IgG monoclonal antibody. *Biochem. J.* **285,** 839–845.
8. Tao, M.-H. and Morrison, S.L. (1989), Studies of aglycosylated chimeric mouse-human IgG. Role of carbohydrate in the structure and effector functions mediated by the human IgG constant region. *J. Immunol.* **143,** 2595–2601.
9. Dwek, R. (1995) Glycobiology: Towards understanding the function of sugars. *Biochem. Soc. Trans.* **23,** 1–25.
10. Townsend, R. R., Hardy, M. R. and Lee, Y. C. (1989) Separation of oligosaccharides using high-performance anion-exchange chromatography with pulsed amperometric detection. *Methods in Enzymology* **179,** 66–76.
11. Basa, L. J. and Spellman, M. W. (1990) Analysis of glycoprotein-derived oligosaccharides by high-pH anion exchange chromatography, *J. Chromatogr.* **499,** 205–220.
12. Maness, N. O., Miranda, E. T., and Mort, A. J. (1991) Recovery of sugar derivatives from 2-aminopyridine labeling mixtures for high-performance liquid chromatography using UV or fluorescence detection. *J. Chromatogr.* **587,** 177–183.
13. Ma, S. and Nashabeh W. (1999) Carbohydrate analysis of a chimeric recombinant monoclonal antibody by capillary electrophoresis. *Anal. Chem.* **71,** 5185–5192.
14. http://www.beckman.com/products/instrument/analytical/ce/pacemdq_carbohydratesystem_inst_dcr.asp

8

Affinity Capillary Electrophoresis to Examine Receptor–Ligand Interactions

Maryam Azad, John Kaddis, Valerie Villareal, Lili Hernandez, Catherine Silverio, and Frank A. Gomez

Summary

Afffinity capillary electrophoresis (ACE) is a new analytical technique that has been shown to be an efficient and accurate tool in studying biomolecular noncovalent interactions and determining binding and dissociation constants of formed complexes. ACE uses as its basis the change in migration time of a receptor upon binding to a ligand found in the electrophoresis buffer. Subsequent Scatchard analysis using noninteracting markers realizes a binding constant. Herein, ACE and three modifications in the technique, partial-filling ACE (PFACE), flowthrough PFACE (FTPFACE), and multiple-step ligand injection ACE (MSLIACE) are used to probe the binding of ristocetin A (Rist A) and vancomycin (Van) from *Streptomyces orientalis* to D-Ala-D-Ala terminus peptides and carbonic anhydrase B (CAB, E.C.4.2.1.1) to arylsulfonamides.

Key Words

Affinity capillary electrophoresis, binding constants, carbonic anhydrase B, receptor–ligand interactions, ristocetin, scatchard plot, vancomcyin.

1. Introduction

During the past decade advances in molecular biology have helped in determining a myriad of biological interactions. Recognition of one molecule by another is the key event of biological life and the specificity of these interactions is its most important aspect. By elucidating how and to what extent molecules interact, the development of treatments for human diseases including Parkinson's, Alzheimer's, and cancer might be expedited.

From: *Methods in Molecular Biology, vol. 276: Capillary Electrophoresis of Proteins and Peptides*
Edited by: M. A. Strege and A. L. Lagu © Humana Press Inc., Totowa, NJ

Over the past few years, affinity capillary electrophoresis (ACE) has been shown to be a versatile technique to study a variety of receptor–ligand interactions including protein–protein, protein–DNA, protein–drug, protein–carbohydrate, peptide–peptide, peptide–carbohydrate, peptide–dye, carbohydrate–drug, and antibody–antigen ***(1–28)***. In a typical form of ACE, a sample of receptor and noninteracting standard(s) is exposed to an increasing concentration of ligand in the running buffer, thereby, causing a shift in the migration time of the receptor relative to the standard(s). The change in migration time is then used for Scatchard analysis to obtain a value for K_b.

In this chapter, we describe the use of several different modes of ACE in examining receptor–ligand interactions using as model systems ristocetin A, carbonic anhydrase B (CAB, E.C.4.2.1.1), and vancomycin (Van) from *Streptomyces orientalis*. Emphasis is placed on describing the unique capabilities and advantages and disadvantages of each technique.

2. Materials

2.1. Standard ACE

1. 0.025 *M* Tris, 0.192 *M* glycine, pH 8.3 (*see* **Note 1**).
2. 1.0 mg Ristocetin A (Rist A) (Biodata Corporation, Horsham, PA) in 1.0 mL 0.025 *M* Tris, 0.192 *M* glycine buffer.
3. 50 μL Mesityl oxide (MO) (Calbiochem, San Diego, CA) in 1.0 mL 0.025 *M* Tris, 0.192 *M* glycine buffer (*see* **Note 2**).
4. 1.0 mg Carbonic anhydrase B (CAB, EC 4.2.1.1, containing CAA and CAB isozymes, from bovine erythrocytes) (Sigma-Aldrich, St. Louis, MO) in 1.0 mL 0.025 *M* Tris, 0.192 *M* glycine buffer (*see* **Note 3**).
5. Electrophoresis sample: 30 μL Rist A solution, 100 μL CAB solution, 4 μL MO solution, in 200 μL sample vial.
6. 5.0 mg Fluorenylmethoxy carbonyl (Fmoc)-Gly-D-Ala-D-Ala, 1 in 5.0 mL 0.025 *M* Tris, 0.192 *M* glycine buffer. Make up eight solutions (10.0 mL total with addition of electrophoresis buffer) of the following concentrations of 1: 20, 40, 50, 75, 100, 150, 200, 300 μM and divide into two sample vials (4.2 mL each; *see* **Note 4**).

2.2. Partial-Filling Affinity Capillary Electrophoresis (PFACE)

1. 0.025 *M* Tris, 0.192 *M* glycine, pH 8.3 (*see* **Note 1**).
2. 1.0 mg Van from *S. orientalis* (Sigma-Aldrich) in 1.0 mL 0.025 *M* Tris, 0.192 *M* glycine buffer.
3. 50 μL MO (Calbiochem) in 1.0 mL 0.025 *M* Tris, 0.192 *M* glycine buffer (*see* **Note 2**).
4. Electrophoresis sample: 20 μL Van solution, 4 μL MO solution, 140 μL electrophoresis buffer in 200 μL sample vial.

5. 5.0 mg *N*-Succinyl-D-Ala-D-Ala, 2, in 5.0 mL 0.025 *M* Tris, 0.192 *M* glycine buffer. Make up eight solutions (10.0 mL total) of the following concentrations of 2:50, 100, 175, 300, 425, 600, 825, 1150 μ*M* and divide into two sample vials (4.2 mL each; *see* **Note 5**).

2.3. Flowthrough Partial-Filling Affinity Capillary Electrophoresis (FTPFACE)

1. 0.025 *M* Tris, 0.192 *M* glycine, pH 8.3 (*see* **Note 1**).
2. 1.0 mg Carbonic anhydrase B (CAB, EC 4.2.1.1, containing CAA and CAB isozymes, from bovine erythrocytes),) (Sigma-Aldrich) in 1.0 mL 0.025 *M* Tris, 0.192 *M* glycine buffer.
3. 1.0 mg Horse heart myoglobin (HHM) (Sigma-Aldrich) in 1.0 mL 0.025 *M* Tris, 0.192 *M* glycine buffer (*see* **Note 6**).
4. 50 μL MO (Calbiochem) in 1.0 mL 0.025 *M* Tris, 0.192 *M* glycine buffer (*see* **Note 2**).
5. Electrophoresis sample: 30 μL CAB solution, 30 μL HHM solution, 4 μL MO solution, 120 μL electrophoresis buffer in 200 μL sample vial.
6. 5.0 mg [4-(Aminosulfonyl)phenyl-methyl-amino]-6-oxohexanoic acid, 3, in 1.0 mL 0.025 *M* Tris, 0.192 *M* glycine buffer. Make up 10 solutions (10.0 mL total) of the following concentrations of 3: 2.0, 3.0, 4.0, 6.0, 8.0, 14, 22, 30, 60, 80 μ*M* and divide into two sample vials (4.2 mL each; *see* **Note 7**).

2.4. Multiple-Step Ligand Injection ACE (MSLIACE)

1. 0.025 *M* Tris, 0.192 *M* glycine, pH 8.3 (*see* **Note 1**).
2. 1.0 mg Van from *S. orientalis* (Sigma-Aldrich) in 1.0 mL 0.025 *M* Tris, 0.192 *M* glycine buffer.
3. Carbonic anhydrase B (CAB, EC 4.2.1.1, containing CAA and CAB isozymes, from bovine erythrocytes), (Sigma-Aldrich) in 1.0 mL 0.025 *M* Tris, 0.192 *M* glycine buffer (*see* **Note 3**).
4. 50 μL MO (Calbiochem) in 1.0 mL 0.025 *M* Tris, 0.192 *M* glycine buffer (*see* **Note 2**).
5. Electrophoresis sample: 20 μL Van solution, 30 μL CAB solution, 4 μL MO solution, 130 μL electrophoresis buffer in 200 μL sample vial.
6. 5.0 mg *N*-Acetyl-D-Ala-D-Ala, 4 (Sigma-Aldrich) in 5.0 mL 0.025 *M* Tris, 0.192 *M* glycine buffer. Make up eight solutions (10 mL total) of the following concentrations of 4: 50, 100, 150, 300, 400, 600, 800, 1200 μ*M* and divide into two sample vials (4.2 mL; each *see* **Note 8**).

2.5. Equipment

1. High-performance capillary electrophoresis (CE) system (Beckman Model P/ACE 5510; Fullerton, CA).
2. Uncoated fused silica capillaries (Polymicro Technologies, Inc., Phoenix, AZ) with an internal and external diameter of 50 and 360 μm, respectively, a length

from inlet to detector of 37.0 cm for standard ACE (57.0 cm for PFACE; 67.0 cm for FTPFACE; 87.0 for MSLIACE) and a length from detector to outlet of 6.5 cm (*see* **Note 9**).
3. Detection: 200 nm on-column.
4. Polyethylene solution vials.
5. pH meter.

3. Methods

The methods described herein outline the use of four ACE techniques to probe the binding of model biological systems to ligands: (1) standard ACE to examine the binding of Rist A to D-Ala-D-Ala terminus peptides; (2) PFACE to examine the binding of Van from *S. orientalis* to D-Ala-D-Ala terminus peptides; (3) FTPFACE to examine the binding of carbonic anhydrase B (CAB, EC 4.2.1.1) to arylsulfonamides, and; (4) MSLIACE to examine the binding of Van from *S. orientalis* to D-Ala-D-Ala terminus peptides.

Standard ACE has several advantages as a method for measuring affinity constants over assay techniques. First, it requires small quantities of both protein and ligand. Second, purification of the sample prior to injection is not necessary as long as the component to be analyzed can be separated from other species. Third, it does not require radiolabelled or chromophoric ligands. Fourth, the commercial availability of automated instrumentation, and the high reproducibility of data, make it experimentally convenient. PFACE and FTPFACE are more advantageous than standard ACE in that less quantities of material are needed for a given assay. MSLIACE requires even less material than the other ACE techniques and the binding assay can be conducted in less time.

3.1. Standard ACE (Fig. 1)

1. Electrophoresis sample to sample holder 21.
2. Peptide solution in vials 12 and 2, 13 and 3, respectively, in order of increasing concenration of 1 (*see* **Note 10**).
3. Electrophoresis buffer (2 × 4.2 mL) in sample holder positions 11 and 1.
4. 3.0 min Rinse at high pressure (20 psi) with electrophoresis buffer.
5. Instrument programmed to inject electrophoresis sample for 3.0 s at low pressure (0.5 psi) and to run increasing concentrations of 1 in electrophoresis buffer for 2.0 min (*see* **Note 11**).
6. The conditions used in CE were as follows: voltage, 25 kV; current, 7.7 μA depending on the capillary length; detection, 200 nm; temperature, 23 ± 0.5°C.
7. Run three to five repetitions of each concentration of 1.
8. For each new concentration of peptide fill the capillary column for 1.0 min (20 psi) with 1.

9. Upon completion of the electrophoresis runs, record the migration times of Rist A, MO, and CAB and compute the binding constant by Scatchard analysis.
10. In this form of Scatchard analysis K_b is estimated using a dual-marker form of analysis which we term the relative migration time ratio (RMTR) (*see* **Eq. 1**).

$$\mathrm{RMTR} = (t_r - t_s') / (t_s' - t_s) \quad (1)$$

of a receptor referenced to two noninteracting standards. Here, t_r, t_s, and t_s' are the measured migration times of the receptor peak, and the two noninteracting standard peaks, respectively. A Scatchard plot can be obtained via **Eq. 2**.

$$\Delta\mathrm{RMTR}_{R,L}/[\mathrm{L}] = \mathrm{K_b}\Delta\mathrm{RMTR}_{R,L}\mathrm{max} - \mathrm{K_b}\Delta\mathrm{RMTR}_{R,L} \quad (2)$$

Here, $\Delta\mathrm{RMTR}_{R,L}$ is the magnitude of the change in RMTR as a function of the concentration of ligand. **Eq. 2** allows for the estimation of $\mathrm{K_b}$ on a relative time scale using two noninteracting standards and compensates for fluctuations in voltage in the capillary column.
11. Upon electrophoresis, a dynamic equilibrium is achieved between the plug of Rist A and 1 resulting in a shift in migration time of the Rist-1 complex. The complexation between 1 and Rist A resulted in an increasing negative charge and the complex is detected later than the uncomplexed form.
12. **Figure 2A** shows a representative series of electropherograms of Rist A in capillaries partially filled with increasing concentrations (0–300 μ*M*) of 1.
13. **Figure 2B** is a Scatchard plot of the data for Rist A. A binding constant of $4.1 \times 10^4 M^{-1}$ was obtained for the binding of 1 to Rist A.

3.2. PFACE

1. Electrophoresis sample to sample holder 21.
2. Peptide solution in vials 12 and 2, 13 and 3, respectively, in order of increasing concenration of 2 (*see* **Note 10**).
3. Electrophoresis buffer (2 × 4.2 mL) in sample holder positions 11 and 1.
4. 3.0 min Rinse at high pressure (20 psi) with electrophoresis buffer.
5. Instrument programmed to inject solutions of 2 at low pressure (0.5 psi) for 15 s, electrophoresis sample for 3.0 s at low pressure (0.5 psi), and then in electrophoresis buffer for 4.0 min.
6. The conditions used in CE were as follows: voltage, 25 kV; current, 5.2 μA; detection, 200 nm; temperature, 23 + 0.5°C.
7. Run three to five repetitions of each concentration of 2.
8. Upon completion of the electrophoresis runs record the migration times of Van and MO and compute the binding constant by Scatchard analysis.
9. In this form of Scatchard analysis a non-interacting standard was used in estimating the binding constant using **Eqs. 3** and **4**. Here, t_{eo} and t_R

$$M = (t_{\mathrm{eo}} / t_R) + 1 \quad (3)$$

$$\Delta M_{R,L}/[\mathrm{L}] = \mathrm{K_b}\Delta M_{R,L}\mathrm{max} - \mathrm{K_b}\Delta \mathrm{M}_{R,L} \quad (4)$$

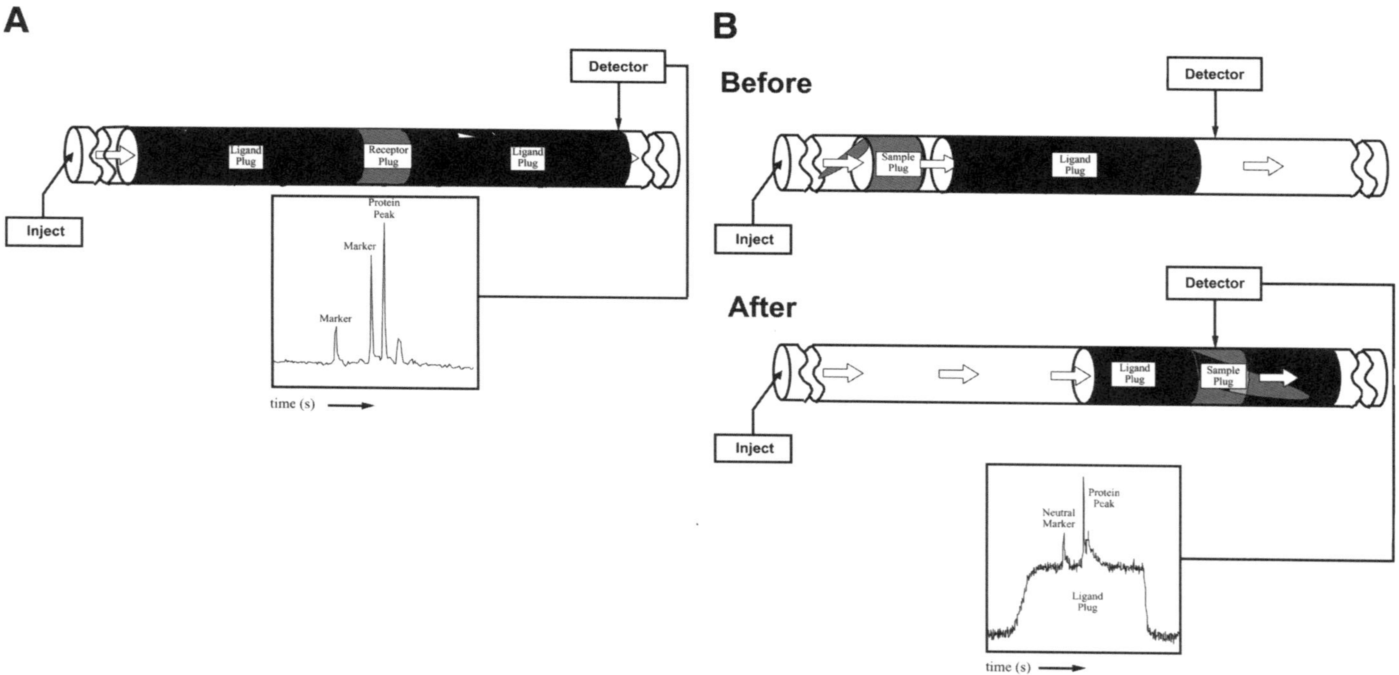
A
Inject
Ligand Plug
Receptor Plug
Ligand Plug
Detector
Marker
Marker
Protein Peak
time (s)
B
Before
Inject
Sample Plug
Ligand Plug
Detector
After
Inject
Ligand Plug
Sample Plug
Detector
Neutral Marker
Protein Peak
Ligand Plug
time (s)

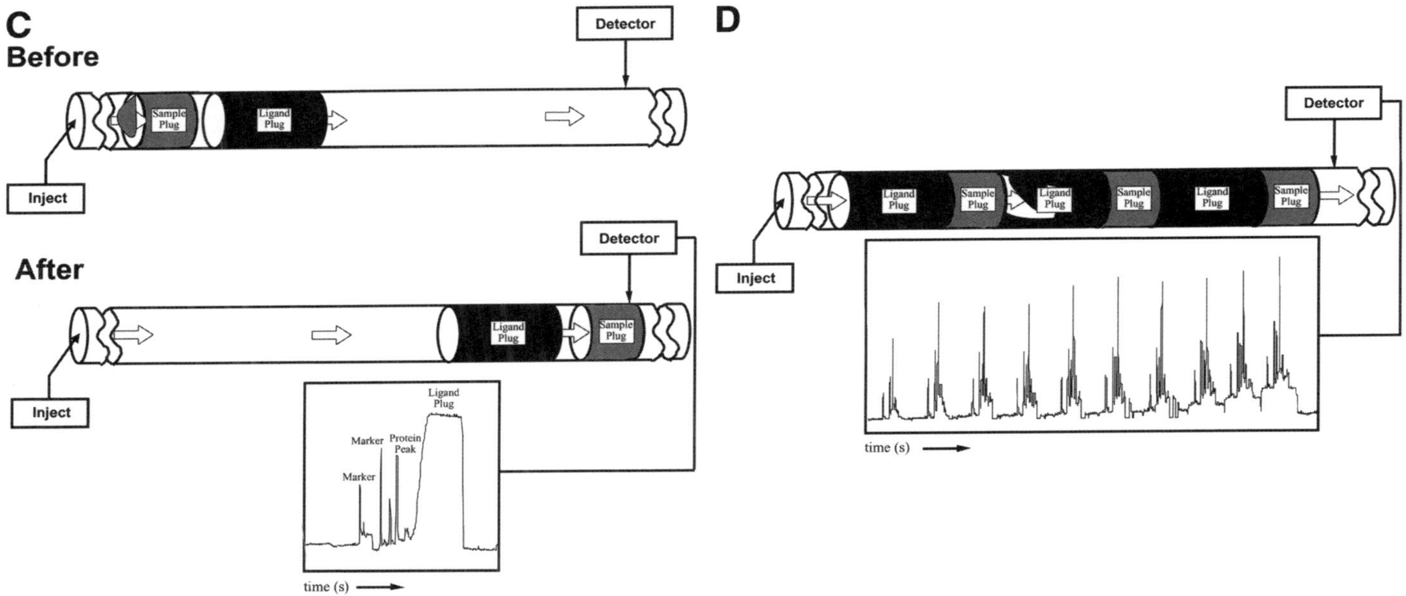

Fig. 1. (**A**) Schematic of a standard ACE experiment. (**B**) Schematic of a partial-filling ACE experiment. (**C**) Schematic of a flowthrough partial-filling affinity capillary electrophoresis experiment. Used with permission from **ref. *3***. (**D**) Schematic of a multiple-step ligand-injection ACE experiment. Used with permission from **ref. *4***.

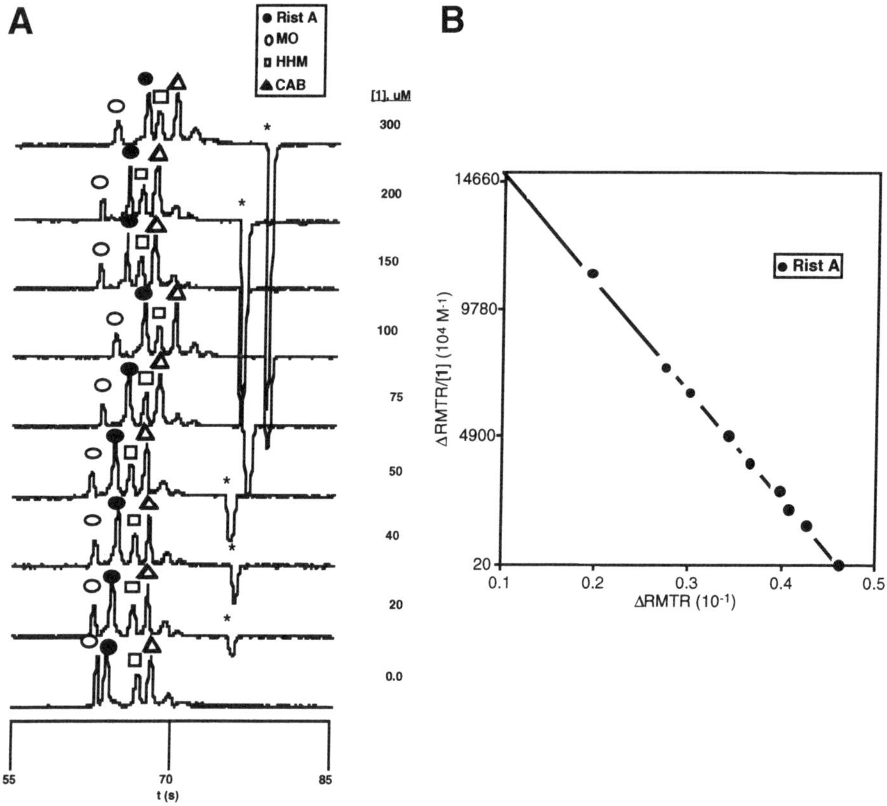

Fig. 2. (**A**) A representative series of electropherograms of Rist A in 0.192 *M* glycine-0.025 *M* Tris-HCl buffer, pH 8.3, containing various concentrations of 1 using the standard ACE technique. MO and CAB were used as internal standards. The total analysis time was 2.0 min at 25 kV (current: 7.7 μA) using a 30.5-cm (inlet to detector), 50-μm id open, uncoated quartz capillary. (**B**) Scatchard plot of the data for Rist A according to **Eq 2**. Used with permission from **ref.** ***1***.

are the measured migration times of the reference peak MO and Van, respectively. A Scatchard plot can be obtained using **Eq. 4**. $\Delta M_{R,L}$ is the magnitude of the change in the mobility ratio (*M*) as a function of the concentration of 2. **Eq. 4** allows for the estimation of K_b' on a relative time scale and compensates for fluctuations in voltage and/or capillary length.

10. Upon increasing the concentration of 2 in the capillary column a shift in the migration time of Van is observed. The Van-2 complex is more negative than Van and upon binding shifts to the right (longer migration time). The neutral

marker MO is unaffected by the change in the concentration of 2 and its migration time does not vary significantly during the course of the experiment, hence, MO can be used as a marker in the analysis of K_b. As shown in **Fig. 3A** the change in concentration of 2 in the column is visualized as an increased height in the ligand plateaus. The box-like structure of the ligand peak at all concentrations of ligand denotes both a uniform injection of ligand into the column and a stable concentration of peptide in the capillary column. For the sample plug to elute on top of the ligand boxes care was taken to ensure that a long enough time of ligand was injected into the capillary, otherwise, incomplete overlap will occur making analysis of the interaction problematic. Further proof that binding is occurring is shown as an increase in the peak area of the Van peak on increasing the concentration of 2 in the capillary column.

11. **Figure 3A** shows a representative series of electropherograms of Van in a capillary partially filled with 2.
12. **Figure 3B** is a Scatchard plot of Van using varying concentrations of 2 in the running buffer using *M* as the basis for the analysis.

3.3. Flowthrough Partial-Filling Affinity Capillary Electrophoresis (FTPFACE)

1. Electrophoresis sample to sample holder 21.
2. Peptide solution in vials 12 and 2, 13 and 3, respectively, in order of increasing concenration of 3 (*see* **Note 10**).
3. Electrophoresis buffer (2 × 4.2 mL) in sample holder positions 11 and 1.
4. 3.0 min Rinse at high pressure (20 psi) with electrophoresis buffer.
5. Instrument programmed to inject solutions of 3 at low pressure (0.5 psi) for 0.1 min, followed by electrophoresis sample for 3.0 s at low pressure (0.5 psi), and then electrophoresis buffer for 7.0 min.
6. The conditions used in CE were as follows: voltage, 28 kV; current, 5.8 μA; detection, 200 nm; temperature, 23 ± 0.5°C.
7. Run three to five repetitions of each concentration of 2.
8. Upon completion of the electrophoresis runs record the migration times of CAB, HHM, and MO and compute the binding constant by Scatchard analysis (same form of analysis as for Rist A).
9. Upon electrophoresis the sample plug flows into the domain of the ligand plug which is migrating at a slower velocity through the capillary column. A dynamic equilibrium is quickly reached between CAB and 3. Continued electrophoresis causes the sample plug to flow through the ligand plug and is detected first. The ligand plug is detected second as a rectangular shaped box. Complexation between CAB and 3 results in an increasing negative charge on CAB and, hence, it migrates later than the uncomplexed form.
10. **Figure 4A** shows a series of electropherograms of CAB in capillaries partially filled with increasing concentrations of 3.
11. **Figure 4B** is a Scatchard plot of the data for CAB obtained using the RMTR (**Eq. 2**) form of analysis.

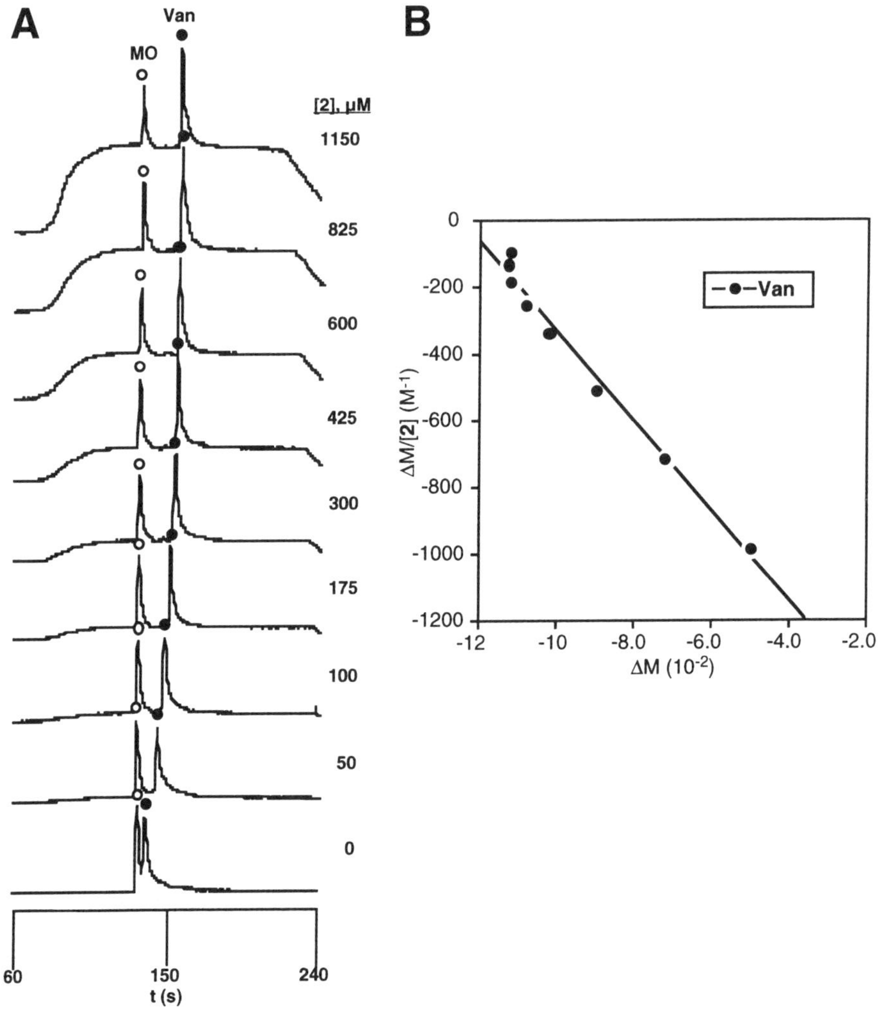

Fig. 3. (**A**) A representative set of electropherograms of Van in 0.192 *M* glycine-0.025 *M* Tris-HCl buffer, pH 8.3, containing various concentrations of 2 using the partial-filling ACE technique. The total analysis time in each experiment was 4.0 min at 25 kV (current: 5.2 μA) using a 60.5-cm (inlet to detector), 50-μm id open, uncoated quartz capillary. MO was used as an internal standard. (**B**) Scatchard plot of the data for vancomycin according to **Eq. 4**. Used with permission from **ref. *2***.

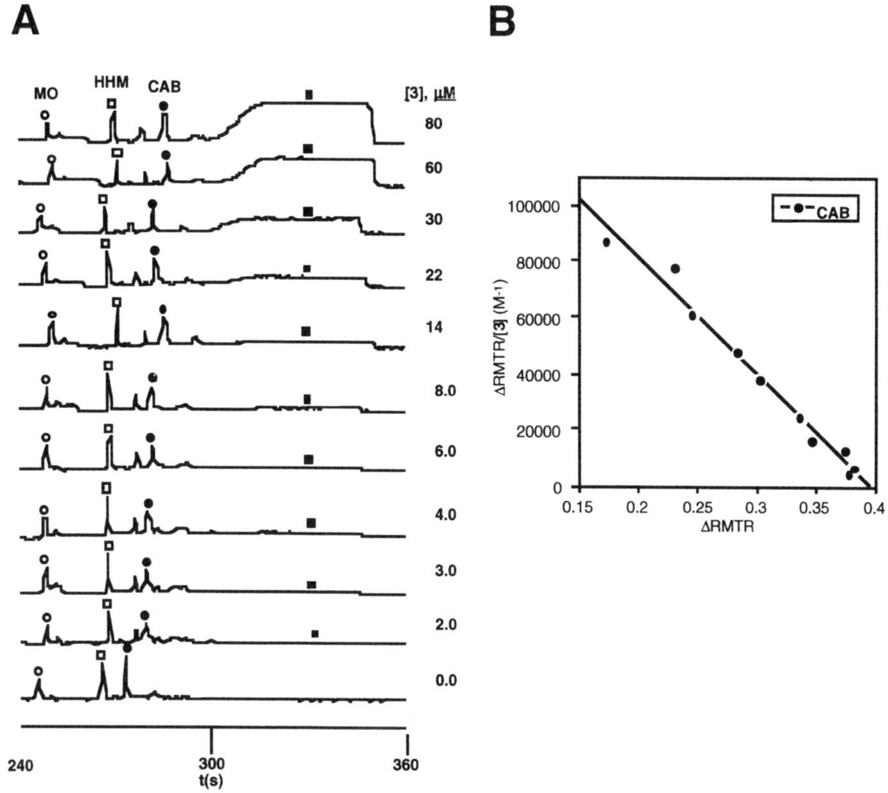

Fig. 4. (**A**) A representative set of electropherograms of CAB in 0.192 *M* glycine-0.025 *M* Tris-HCl buffer, pH 8.6 containing various concentrations of 3 using the flowthrough partial-filling affinity CE technique. The total analysis time in each experiment was 6.5 min at 28 kV (current: 5.8 µA) using a 60.5-cm (inlet to detector), 50-µm id open, uncoated quartz capillary. MO and HHM were used as internal standards. (**B**) Scatchard plot of the data for carbonic anhydrase B according to **Eq. 2**. Used with permission from **ref.** ***3***.

3.4. Multiple-Step Ligand Injection ACE (MSLIACE)

1. Electrophoresis sample to sample holder 21.
2. Peptide solution in vials 12 and 2, 13 and 3, respectively, in order of increasing concenration of 4 (*see* **Note 10**).
3. Electrophoresis buffer (2 × 4.2 mL) in sample holder positions 11 and 1.
4. 3.0 min Rinse at high pressure (20 psi) with electrophoresis buffer.
5. Instrument programmed to inject a solution of electrophoresis buffer (0.5 psi) not containing 4 (solution A) followed by a sample (7.2 nL; a 1-s time of injection at low pressure equates to 1.2 nL of volume of solution) of solution (solution B)

containing electrophoresis sample. The sample was subjected to electrophoresis in a solution (solution C) containing the first concentration (50 μ*M*) of 4 for 2.0 min at 24 kV. A second solution of electrophoresis sample (0.5 psi) (14.4 nL) (solution B) was injected for 12 s and subjected to electrophoresis in the next higher concentration of 4 (100–1200 μ*M*) for 2.0 min at 24 kV. The process of sample injection and ligand electrophoresis was repeated until all concentrations of ligand were run.

6. Run electrophoresis for 4.0 min.
7. The conditions used in CE were as follows: voltage, 24 kV; current, 4.0 μA; detection, 200 nm; temperature, 23 ± 0.5°C.
8. Run three to five repetitions of each concentration of 2.
9. Upon completion of the electrophoresis runs record the migration times of Van and MO and compute the binding constant by Scatchard analysis (same form of analysis as for Rist A).
10. In this technique, a plug of Van and noninteracting standards is injected and electrophoresed in buffer containing a given concentration of peptide. The sequence is repeated at increasing concentrations of peptide until all concentrations of ligand were run. Analysis of the change in the RMTR affords K_b. **Figure 5A** shows a representative set of electropherograms of Van in increasing concentrations of 4. Periodic injections of 4 at higher concentration result in the Van peak shifting to the right for any concentration of 4 in the buffer. Ligand 4 is a small negatively charged molecule and has a more negative electrophoretic mobility than both complexed and uncomplexed Van, hence, it elutes at a greater migration time than both Van and complexed Van. The complexation between 4 and Van resulted in an increasing negative charge and the peak for Van complexed to the ligand shifts to a longer migration time relative to the neutral marker MO increasing 4 in the running buffer. At any one time during the experiment, only three different ligand concentrations and three sample plugs are contained in the capillary column. The instrument was programmed in order to ensure all plugs of sample were contained in one single electropherogram. The total time for the experiment was approx 27 min.
11. **Figure 5A** is a single electropherogram is generated using the multiple-plug ligand injection ACE technique.
12. **Figure 5B** is a Scatchard plot of the data for Van using **Eq. 4**.

4. Notes

1. The use of Tris-gly buffer is neither critical for ACE analysis, nor is a pH of 8.3. Our labs use this buffer concentration and pH because at pH less than 8.0 the peak for CAB tends to broaden because of protein adsorption to the walls of the capillary column.
2. MO is a volatile organic compound and should be stored in sealed containers to maintain its concentration. Solution should be replaced daily as decomposition in buffers will occur. Other neutral markers can be used in the analysis.

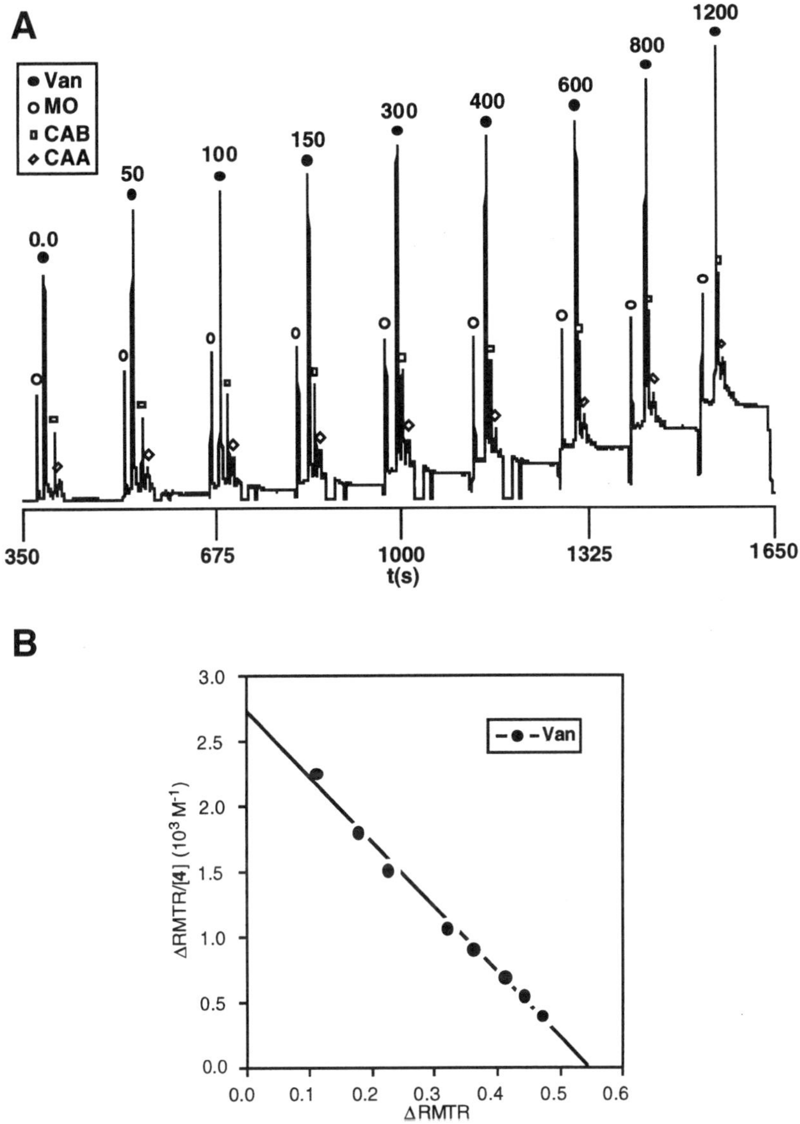

Fig. 5. (**A**) A representative electropherogram of Van in 0.192 *M* glycine-0.025 *M* Tris-HCl buffer, pH 8.3, containing various concentrations of 4 using the multiple-step ligand-injection ACE technique. The total analysis time in each experiment was 27 min at 24 kV (current: 4.0 μA) using a 80.5-cm (inlet to detector), 50-μm id open, uncoated quartz capillary. MO and CAB (containing CAA and CAB isozymes) were used as internal standards. The number above each set of sample peaks refer to the concentration of 4 in μ*M*. (**B**) Scatchard plot of the data for Van according to **Eq. 2**. Used with permission from **ref. *4***.

3. Carbonic anhydrase B is a noninteracting standard used in the data analysis. It has no affinity to either glycopeptide antibiotics or the peptide ligands. Other markers can be used as long as they do not migrate at or near the migration times of Rist A.
4. Compound 1 was synthesized in our laboratories based on standard peptide chemistries *(10)*. Other D-Ala-D-Ala terminus peptides can be used because they all have similar binding affinities to Rist A as 1.
5. Compound 2 was synthesized in our laboratories *(2)*. Other D-Ala-D-Ala terminus peptides can be used since they all have similar binding affinities to Van as 2.
6. Horse-heart myoglobin is a noninteracting standard used in the data analysis. It has no affinity to either CAB or 3. Other markers can be used as long as they do not migrate at or near the migration times of CAB.
7. Compound 3 was synthesized in our laboratories *(2)*. Other negatively charged arylsulfonamides can be used since they have similar binding affinities to CAB as 3.
8. Other D-Ala-D-Ala terminus peptides can be used since they all have similar binding affinities to Van as 4.
9. Length of column is not critical in standard ACE, PFACE, and FTPFACE, as long as a dynamic equilibrium is obtained prior to detection. For MSLIACE, length of column is critical as multiple injections are required in the experiment. The length of column depends on the number of ligand concentrations to be run during the experiment.
10. The instrument can be programmed to run the electrophoresis buffers at increasing concentrations.
11. High and low pressure settings can be programmed using the instrument software.

Acknowledgment

The authors gratefully acknowledge financial support for this research by grants from the National Science Foundation (CHE-0136724), Research Corporation (CC5293), and the Department of Defense (F49620-02-1-0445).

References

1. Azad, M., Hernandez, L., Plazas, A., Rudolph, M., and Gomez, F. A. (2003) Determination of binding constants between the antibiotic Ristocetin A and D-Ala-D-Ala terminus peptides by affinity capillary electrophoresis. *Chromatographia* **57,** 339–344.
2. Heintz, J., Hernandez, M., and Gomez, F. A. (1999) Use of a partial-filling technique in affinity capillary electrophoresis for determining binding constants of ligands to receptors. *J. Chromatogr. A* **840,** 261–268.
3. Mito, E., and Gomez, F. A. (1999) Flow-through partial-filling affinity capillary electrophoresis can estimate binding constants of ligands to receptors. *Chromatographia* **50,** 689–694.

4. Zhang, Y. and Gomez, F. A. (2000) Multiple-step ligand injection affinity capillary electrophoresis for determining binding constants of ligands to receptors. *J. Chromatogr. A* **897,** 339–347.
5. Mito, E., Zhang, Y., Esquivel, S., and Gomez, F. A. (2000) Estimation of receptor-ligand interactions by the use of a two-marker system in affinity capillary electrophoresis. *Analyt. Biochem.* **280,** 209–215.
6. Kawaoka, J. and Gomez, F. A. (1998) Use of mobility ratios to estimate binding constants of ligands to proteins in affinity capillary electrophoresis. *J. Chromatogr. B* **715,** 203–210.
7. Chu, Y.-H., Avila, L. Z., Gao, J., and Whitesides, G. M. (1995) Affinity capillary electrophoresis. *Acc. Chem. Res.* **28,** 461–468.
8. Qian, X. -H., and Tomer, K. B. (1998) Affinity capillary electrophoresis investigation of an epitope on human immunodeficiency virus recognized by a monoclonal antibody. *Electrophoresis* **19,** 415–419.
9. Kiessig, S., Bang, H., and Thunecke, F. (1999) Interaction of cyclophilin and cyclosporins monitored by affinity capillary electrophoresis. *J. Chromatogr. A* **853,** 469–477.
10. Zhang, Y., Kodama, C., Zurita, C., and Gomez, F. A. (2001) On-column ligand synthesis coupled to partial-filling affinity capillary electrophoresis to estimate binding constants of ligands to a receptor. *J. Chromatogr. A* **928,** 233–241.
11. Erim, F. B. and Kraak, J. C. (1998) Vacancy affinity capillary electrophoresis to study competitive protein-drug binding. *J. Chromatogr. B* **710,** 205–210.
12. Shimura, K. and Kasai, K. (1997) Affinity capillary electrophoresis: a sensitive tool for the study of molecular interactions and its use in microscale analyses. *Anal. Biochem.* **251,** 1–16.
13. Gomez, F. A., Avila, L. Z., Chu, Y.-H., and Whitesides, G. M. (1994) Determination of binding constants of ligands to proteins by affinity capillary electrophoresis: compensation for electroosmotic flow. *Anal. Chem.* **66,** 1785–1791.
14. Zhang, X., Davidson, E. W., Nguyen, T. H., Evans, R. W., Im, S. J., and Barker, G. E. (1996) Investigation of chiral resolution using displacement interactions with polymer networks in capillary affinity zone electrophoresis. *J. Chromatogr. A* **745,** 1–8.
15. Colton, J. J., Carbeck, J. D., Rao, J., and Whitesides, G. M. (1998) Affinity capillary electrophoresis: a physical-organic tool for studying interactions in biomolecular recognition. *Electrophoresis* **19,** 367–382.
16. Heegaard, N. H. H., Hansen, B. E., Svejgaard, A., and Fugger, L. H. (1997) Interactions of the human class 11 major histocompatibility complex protein HLD-DR4 with a peptide ligand demonstrated by affinity capillary electrophoresis. *J. Chromatogr. A* **781,** 91–97.
17. Mammen, M., Gomez, F. A., and Whitesides, G. M. (1995) Determination of the binding of ligands containing the N-2,4-dinitrophenyl group to bivalent monoclonal rat anti-dnp antibody using affinity capillary electrophoresis. *Anal. Chem.* **67,** 3526–3535.

18. Shimura, K., and Kasai, K. (1995) Determination of the Affinity Constants of Concanavalin A for Monosaccharides by Fluorescence Affinity Probe Capillary electrophoresis. *Anal. Biochem.* **227,** 186–194.
19. VanderNoot, V. A., Hileman, R. E., Dordick, J. S., and Linhardt, R. J. (1998) Affinity capillary electrophoresis employing immobilized glycosaminoglycan to resolve heparin-binding peptides. *Electrophoresis* **19,** 437–441.
20. Chu, Y.-H., Dunayevskiy, Y. M., Kirby, D. P., Vouros, P., and Karger, B. L. (1996) Affinity capillary electrophoresis-mass spectrometry for screening combinatorial libraries. *J. Am. Chem Soc.* **118,** 7827–7835.
21. Busch, M. H. A., Carels, L. B., Boelens, H. F. M., Kraak, J. C., and Poppe, H. (1997) Comparison of five methods for the study of drug-protein binding in affinity capillary electrophoresis. *J. Chromatogr. A* **777,** 311–328.
22. Lin, S., Hsiao, I.-Y., and Hsu, S.-M. (1997) Determination of the dissociation constant of phosvitin-anti-phosphoserine interaction by affinity capillary electrophoresis. *Anal. Biochem.* **254,** 9–17.
23. Larsen, K. L. and Zimmermann, W. (1999) Analysis and characterisation of cyclodextrins and their inclusion complexes by affinity capillary electrophoresis. *J. Chromatogr. A* **836,** 3–14.
24. Taga, A., Uegaki, K., Yabusako, Y., Kitano, A., and Honda, S. (1999) Simultaneous determination of the association constants of oligosaccharides to a lectin by capillary electrophoresis. *J. Chromatogr. A* **837,** 221–229.
25. Busch, M. H. A., Carels, L. B., Boelens, H. F. M., Kraak, J. C., and Poppe, H. (1997) Comparison of Five Methods for the Study of Drug-Protein Binding in Affinity Capillary Electrophoresis. *J. Chromatogr. A* **777,** 311–328.
26. Amini, A. and Westerlund, D. (1998) Evaluation of association constants between drug enantiomers and human alpha-1-acid glycoprotein by applying a partial-filling technique in affinity capillary electrophoresis. *Anal. Chem.* **70,** 1425–1430.
27. Dunayevskiy, Y. M., Lyubarskaya, Y. V., Chu, Y.-H., Vouros, P., and Karger, B. L. (1998) Simultaneous measurement of nineteen binding constants of peptides to vancomycin using affinity capillary electrophoresis-mass spectrometry. *J. Med. Chem.* **41,** 1201–1204.
28. Chu, Y-.H. and Whitesides, G.M. (1992) Affinity capillary electrophoresis can simultaneously measure binding constants of multiple peptides to vancomycin. *J. Org. Chem.* **57,** 3524–3525.

9

Screening Major Binding Sites on Human Serum Albumin by Affinity Capillary Electrophoresis

Hee Seung Kim, John Austin, and David S. Hage

Summary

A screening method is described for determining whether a drug or small solute has significant interactions at the two major binding sites on human serum albumin (HSA). This method uses affinity capillary electrophoresis (ACE) to perform a mobility shift assay, where the solute of interest is injected in both the presence of pH 7.4, 0.067 *M* phosphate buffer, and the same buffer containing a known concentration of HSA. Dextran is also used in the running buffer to adjust the mobility of HSA. Two types of modified HSA are used in this assay. The first is modified with 2-hydroxy-5-nitrobenzyl bromide (HNB), which selectively blocks HSA's warfarin-azapropazone site. The second type of HSA is modified with tetranitromethane (TNM), which decreases binding at the indole-benzodiazepine site. By comparing the mobility of a solute in the presence of these two modified forms of HSA vs normal HSA, it is possible to detect solute interactions at these binding sites. This approach is illustrated using warfarin and ibuprofen as examples of test solutes.

Key Words

Affinity capillary electrophoresis; drug binding sites; drug screening assay; human serum albumin; 2-hydroxy-5-nitrobenzyl bromide; modified albumin; tetranitromethane.

1. Introduction

The binding of drugs to transport proteins in blood is important in determining the activity, toxicity, excretion, and metabolism of these agents in the body. Human serum albumin (HSA) is an important protein in this process. HSA has a molecular mass of 66.5 kDa and is the most abundant plasma protein. It consists of a single polypeptide chain with 585 amino acids held together through 17 disulfide bonds *(1)*. Many low mass solutes show reversible binding to HSA.

From: *Methods in Molecular Biology, vol. 276: Capillary Electrophoresis of Proteins and Peptides*
Edited by: M. A. Strege and A. L. Lagu © Humana Press Inc., Totowa, NJ

This includes both endogenous and exogenous substances, such as long-chain fatty acids, steroids, salicylate, warfarin, and ibuprofen *(2–4)*. Many of these interactions occur at relatively well-defined regions on HSA, the two most important of which are the warfarin-azapropazone and indole-benzodiazepine sites (i.e., Sudlow sites I and II) *(5)*.

Numerous techniques can be used to examine the binding of solutes to HSA *(5–8)*. One such approach is affinity capillary electrophoresis (ACE) *(9,10)*. This is often performed by examining the mobility of an injected solute in the presence of a running buffer that contains a known concentration of HSA as an additive. Advantages of this technique include its small sample size requirements, its short analysis times, and its relative ease of automation.

In recent work, an ACE method was reported that can quickly identify the location of a drug's binding sites on HSA *(11)*. This is accomplished by using two modified forms of HSA which have been altered at specific amino acids within HSA's warfarin-azapropazone and indole-benzodiazepine sites. A comparison is then made between the mobility of a drug in the presence of these modified proteins vs the drug's mobility in the presence of normal HSA (*see* **Fig. 1**). If a difference is seen in these mobilities, this indicates the drug was interacting at the altered region. This chapter will describe how to perform this assay, using warfarin and ibuprofen as examples of test solutes.

2. Materials

The following materials are needed to prepare the modified HSA, to condition the ACE capillary, and to prepare the running buffers for this assay. A source for the drug or solute of interest is also required, although only small quantities of this agent are needed. In this chapter, racemic warfarin and racemic ibuprofen (Sigma, St. Louis, MO) are used to illustrate the method, but a variety of other compounds can also be tested.

1. HSA (Cohn fraction V, essentially fatty acid and globulin free) (Fluka, Milwakee, WI).
2. Dextran (average mass, 2×10^6 Da; Sigma).
3. Tetranitromethane (TNM; Sigma).
4. 2-Hydroxy-5-nitrobenzyl bromide (HNB; Sigma).
5. Methanol (HPLC grade).
6. 95% Ethanol (prepared from HPLC-grade ethanol).
7. 10 *M* Urea solution.
8. 0.050 *M* Tris-HCl, pH 8.0.
9. 0.067 *M* Potassium phosphate, pH 7.4.
10. 1 *M* NaOH.
11. 1 *M* HCl.
12. γ-Methacryloxypropyltrimethoxysilane (Aldrich, Milwaukee, WI).

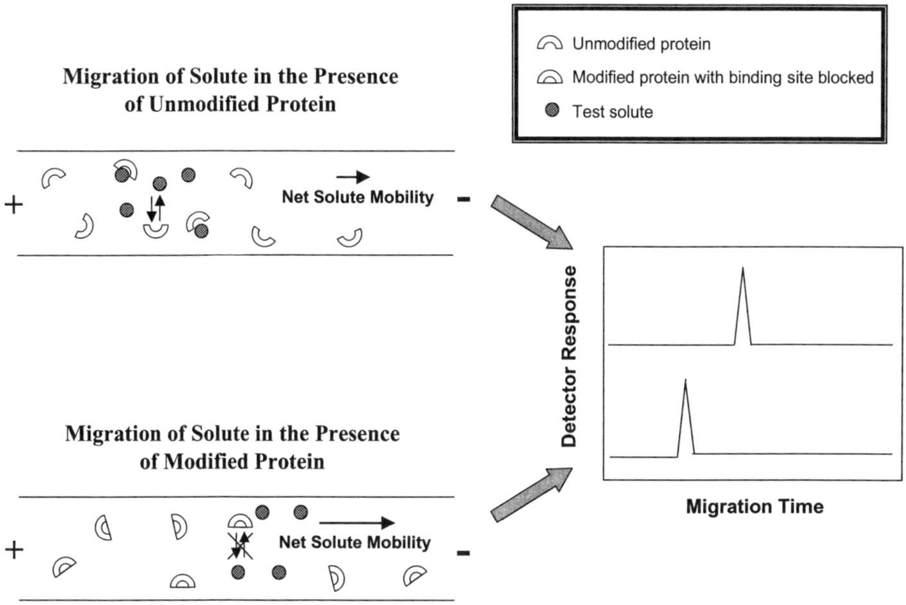

Fig. 1. General illustration of an ACE mobility shift assay using normal and modified proteins. In the top experiment, the test solute is able to bind to a normal protein that is used as a buffer additive; the result is a shift in the solute's apparent mobility away from that expected for the same solute in the absence of any protein. In the bottom experiment, a modified protein is now used that has completely or partially blocked binding sites; the test solute has less binding to this protein, giving rise to an apparent mobility closer to that for the solute in the absence of any protein.

13. Acrylamide (Sigma).
14. Ammonium persulfate.
15. *N,N,N',N'*-Tetramethylethylenediamine (TEMED; Sigma).
16. Sodium nitrate.
17. BCA protein assay kit (Pierce, Rockford, IL).
18. 0.22 μm Cellulose acetate filters (Fisher Scientific, Pittsburgh, PA).
19. Slyde-A-Lyzer dialysis membrane (12 kDa molecular weight cutoff; Pierce).
20. Fused-silica capillary (50 μm id, 365 μm od; Polymicro Technologies, Phoenix, AZ).
21. Microcapillary union (Upchurch Scientific, Oak Harbor, WA).
22. Biofocus 3000 capillary electrophoresis system (Bio-Rad, Hercules, CA) or comparable system.

In addition to these specific materials, other general items that are needed include 1-mL disposable syringes, various micropipets (10–100 and 1000 μL), a nitrogen gas source, a standard UV/vis absorbance spectrometer, an ultrasonicator bath, and a vortex mixer.

In dealing with the listed reagents, it should be noted that tetranitromethane is a known carcinogen and must be handled in a well-ventilated hood. γ-Methacryloxypropyltrimethoxysilane is a moisture-sensitive agent and should be kept in dessicator. Acrylamide is light sensitive and must be kept in dark. Also, ammonium persulfate is a strong oxidizer and should be kept separate from other chemicals, especially those that can act as a reducing agent.

3. Methods

The methods described in this section outline: (1) the modification of HSA, (2) preparation of the fused-silica capillary, (3) reagent and sample preparation, and (4) the ACE assay.

3.1. Modification of HSA

The reactions used to modify HSA for the ACE assay are summarized in **Fig. 2**. The warfarin-azapropazone site of HSA is modified by reacting HSA's trytophan 214 residue (Trp214) with HNB. This is performed according to methods described in the **ref.** ***(12)***. The indole-benzodiazepine site of HSA is modified by reacting tyrosine 411 (Tyr411) of HSA with TNM, based on a previous procedure that uses a 4:1 mol excess of TNM vs HSA ***(12,13)***.

3.1.1. HNB Modification of HSA

HNB is a modification reagent that is highly selective for tryptophan at low pH ***(14,15)***. This adds a hydroxynitrobenzyl group to position 3 of the indole ring in tryptophan, as shown in **Fig. 2A** ***(16)***. In the past, HNB has been used to estimate the number of tryptophans in a protein and to determine if tryptophans are present at a protein's active site ***(14–16)***. HSA has only one tryptophan residue (Trp214), which lies within the warfarin-azapropazone site of HSA. It is known from previous work that the modification of Trp214 with HNB disrupts ligand binding at this site, but does not change binding at the other major site of HSA ***(17)***. The following approach describes how to modify HSA with HNB for use in the ACE assay.

1. Add 0.1744 g HNB to a 10-mL volumetric flask and dilute to the mark with methanol. The HNB should dissolve quickly in the methanol.
2. Weigh 0.5 g of HSA into a beaker and add 10 mL of a 10-*M* urea solution. Allow adequate time for the protein to completely dissolve (*see* **Note 1**).
3. Add all 10 mL of the HNB/methanol solution to the HSA/urea solution. Mix these thoroughly, then cover the container and allow this mixture to react for approx 2 h (*see* **Note 2**).
4. Dialyze the mixture, using dialysis tubing (12,000 molecular weight cutoff) or dialysis cartridge cassettes (such as Slyde-A-Lyzer from Pierce). Perform this dialysis against 1 L of water at 4°C for 60 h. After the first 12 h, place the sample

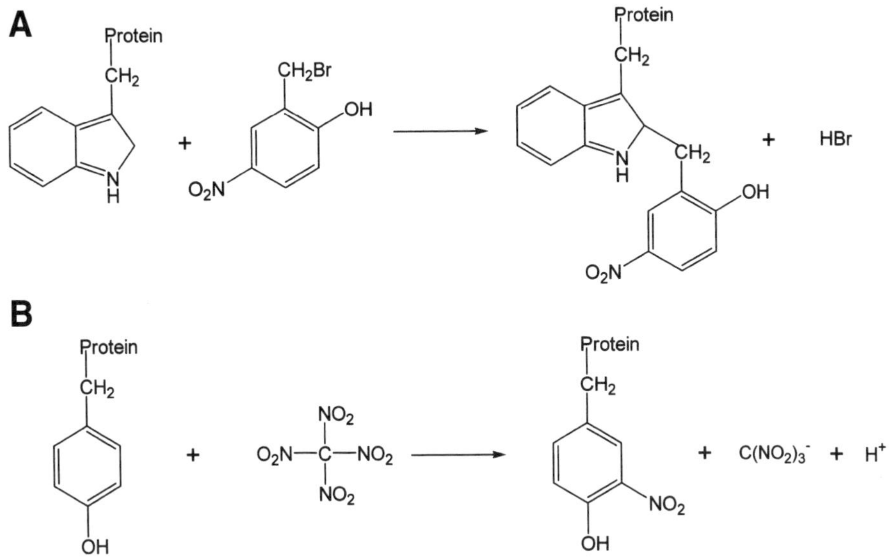

Fig. 2. Reactions for (**A**) the modification of tryptophan residue by HNB, and (**B**) the modification of tyrosine by TNM

in a fresh portion of water and repeat every 4 h for the next 12 h. For the remaining 48 h, change the water every 12 h. A final dialysis step against 1 L of pH 7.4, 0.067 *M* potassium phosphate buffer should be performed at 4°C for 72 h (changing to fresh buffer every 24 h) to remove any residual HNB or side products.

5. Keep the modified HSA solution in a refrigerator (4°C) for short-term storage. For longer-term storage, it is recommended that the TNM-modified HSA be freeze dried and stored at –20°C.

3.1.2. Determining the Extent of HSA Modification by HNB

It is helpful after preparing the HNB-modified HSA to confirm the degree of modification that has occurred with this protein. This can be accomplished by using an assay that makes use of the absorbance of the HNB-tryptophan product at 410 nm and the inherent absorbance of HSA at 280 nm. This approach is described here. An alternative approach is to measure the absorbance at 410 nm to give the amount of HNB-modified residues, with the total amount of HSA determined by a bicinchoninic acid (BCA) protein assay, as described in **Subheading 3.1.4.** for the TNM-modified HSA.

1. First prepare a solution of normal, unmodified HSA to give a concentration of exactly 1.0 mg/mL in pH 7.4, 0.067 *M* potassium phosphate buffer.
2. Dilute the modified HSA with pH 7.4, 0.067 *M* potassium phosphate buffer to give a solution with an approximate HSA concentration of 1–2 mg/mL.

3. Measure the absorbances of the HNB-modified HSA at 280 and 410 nm. Determine the absorbance for normal HSA at 280 nm, using the pH 7.4, 0.067 *M* potassium phosphate buffer as a blank. Some typical spectra that are obtained in these measurements are shown in **Fig. 3**. It can be seen from this figure that there is some decrease in the absorbance of HSA at 280 nm after this protein has been modified with HNB.
4. Use these measured absorbances with the following equation to determine the moles of HNB-modified tryptophan residues that are present per mole of HSA. This gives the extent of modification for the treated HSA.

$$\text{Extent of HSA modification} = \frac{\left(A_{\text{HNB-HSA,410}}\right)\left(A_{\text{HSA,280 nm}}\right)\left(66{,}500 \text{ g/mol}\right)}{\left(18{,}000\right)\left[A_{\text{HNB-HSA,280}} - \left(0.14 \cdot A_{\text{HNB-HSA,410}}\right)\right]}$$

In the above equation, $A_{\text{HNB-HSA,410}}$ is the absorbance of the HNB-modified HSA at 410 nm, $A_{\text{HNB-HSA,280}}$ is the absorbance of the HNB-modified HSA at 280 nm, and $A_{\text{HSA,280}}$ is the absorbance of the unmodified HSA at 280 nm. The denominator of this equation includes a factor that corrects for the decrease in absorbance at 280 nm for HSA when it is modified with HNB. The result that is obtained from this equation should give the moles of HNB-modified tryptophan residues per mole HSA. The expected answer should be approx 1.0, with values in the range of 1.2–0.8 being acceptable for later use in the ACE assay.

3.1.3. TNM Modification of HSA

TNM will react with tyrosine residues in a protein to produce the product 3-nitrotyrosine, as shown in **Fig. 2B** ***(12,13)***. It was originally thought that TNM reacted only with tyrosines close to the surface of a protein ***(17)***, but newer results suggest that TNM can quickly enter the interior of a protein and has a reactivity that depends only on the microenvironments of the individual tyrosine residues. Of the 18 tyrosines in HSA, only nine have been found to react with TNM. Of these 18 residues, only two react when small amounts of TNM are used (i.e., a mole ratio of 4:1 ratio or less for TNM to HSA) ***(12,13)***. The most reactive of these two residues is Tyr411, which is located at or near the indole-benzodiazepine site of HSA ***(12,13,21)***. Furthermore, it has been shown that the use of a low mole ratio of TNM to HSA allows the production of a modified form of HSA in which solute binding is decreased at the indole-benzdodiazepine site but not at the warfarin-azapropazone site ***(21)***. The method given here describes how this modified HSA can be prepared for use in the ACE screening assay.

1. Make a 0.2-*M* stock solution of TNM by adding 24 μL of this reagent to 976 μL of 95% ethanol.

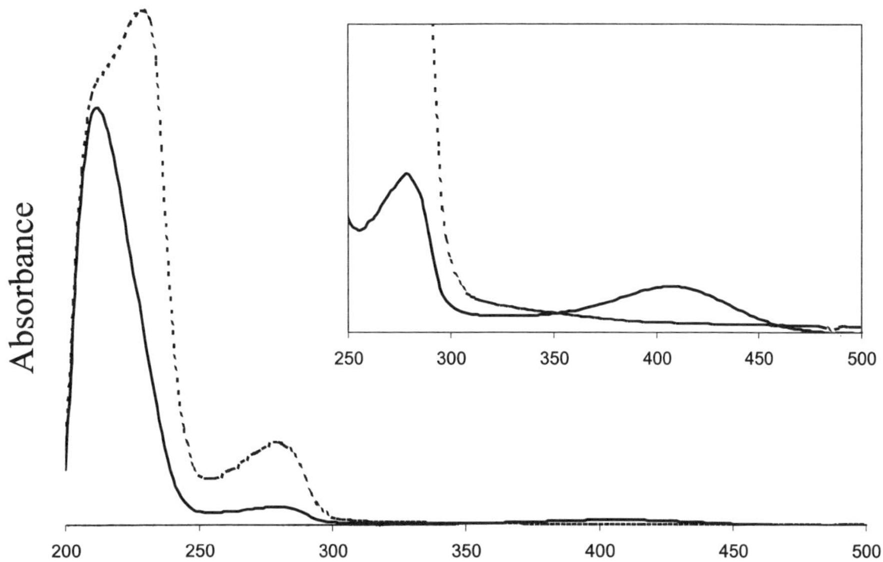

Fig. 3. Typical UV/vis absorption spectra for unmodified HSA (dashed line) and HSA that has been treated with HNB (solid line). These two protein solutions had approximately equal concentrations.

2. Weigh 0.17 g HSA and dissolve in 10 mL of pH 8.0, 0.05 *M* Tris-HCl buffer (*see* **Note 3**).
3. Add 50 µL of 0.2-*M* TNM stock solution to the HSA solution while vortexing or rapidily stirring the TNM solution (*see* **Note 4**).
4. Allow this mixture to react at room temperature for 10 min (*see* **Notes 5** and **6**).
5. Dialyze the mixture, using 12,000 molecular weight cutoff dialysis tubing or dialysis cartridge cassettes, against 1 L of water at 4°C for 60 h, changing to fresh water every 12 h. A final dialysis step against 1 L of pH 7.4, 0.067 *M* potassium phosphate buffer should be performed at 4°C for 72 h (changing to a fresh portion of buffer every 12 h) to remove any residual TNM or side-products (*see* **Note 7**).

3.1.4. Determining the Extent of HSA Modification by TNM

The extent of TNM modification of HSA can be determined by using a colorimetric assay. This makes use of the absorbance of the 3-nitrotyroxine product at 428 nm and the measurement of HSA by a method like the BCA protein assay.

1. Dilute the modified HSA sample with 0.1 *M* NaOH to give a concentration of approx 1–2 mg/mL HSA.
2. Measure the absorbance of this HSA solution at 428 nm, using 0.1 *M* NaOH as the blank. Some typical spectra obtained by this measurement are shown in **Fig. 4**.

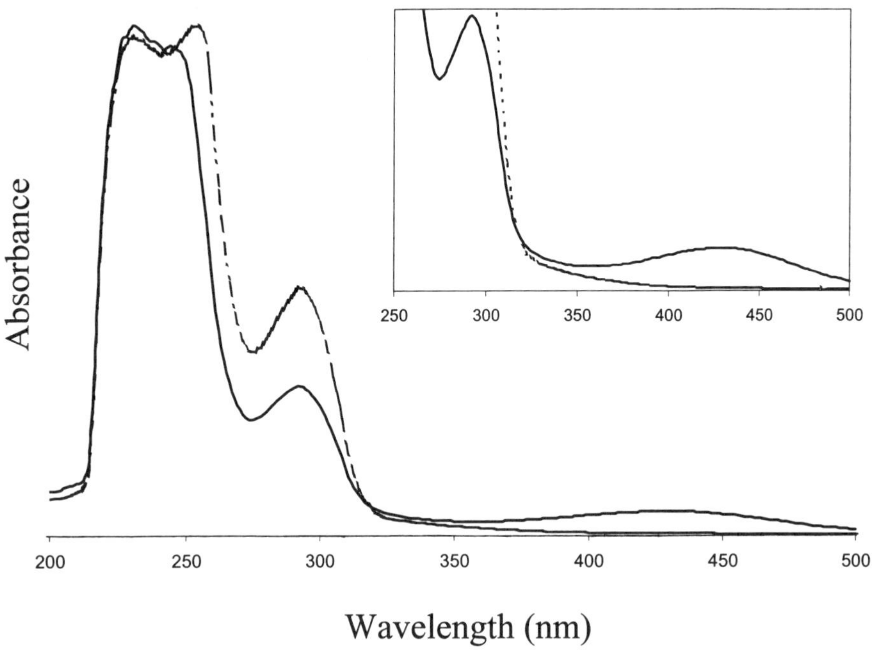

Fig. 4. UV/vis absorption spectra for unmodified HSA (dashed line) and HSA which has been treated with TNM (solid line). These two protein solutions had approximately equal concentrations.

3. Calculate the molar concentration of the 3-nitrotyrosine product in this mixture by using a molar absorptivity for this product of 4100 M^{-1} cm^{-1}.
4. Use a separate aliquot of the modified HSA sample to determine its protein content by the BCA assay, according to instructions provided by the manufacturer of the BCA reagent. Express the results of this assay in terms of the molar concentration of HSA in the sample.
5. Determine the ratio of the concentrations found in **steps 3** and **4**. This gives the rates of mol 3-nitrotyrosine to mol HSA, which represents the extent of HSA modification. A typical ratio that should be obtained when using the modification procedure in **Subheading 3.1.3.** is a mol/mol ratio of 2.3, but slightly higher or lower ratios (e.g., 2.0–2.5) are also acceptable for use in the ACE method.

3.2. Capillary Preparation

For a reproducible ACE assay, several steps must be followed to properly prepare the capillary that will be used in this assay. These steps include: (1) creation of a detection window in the capillary, (2) modification of the interior wall of the capillary with linear polyacrylamide, and (3) conditioning of the capillary before it is used in the ACE assay. Each of these procedures is described in this subheading.

3.2.1. Creation of a Detection Window

The detection window in the capillary must be prepared prior to the capillary's modification with a linear polyacrylamide coating and placement of the capillary into a CE system. This is critical because the linear polyacrylamide coating inside capillary wall can be disrupted during the creation of the window. This detection window is usually prepared by burning away a small amount of the capillary's outside polyimide coating through the use of a match or lighter. During this process, it is recommended that a small piece of aluminum foil be placed at both ends of desired detection window to prevent the window from becoming too wide. After the window has been created, the burned polyimide coating can be removed by washing it away with acetone or methanol (*see* **Note 8**).

3.2.2. Conditioning the Capillary Before Modification With Linear Polyacrylamide

One factor that researchers must consider in CE is the presence of several different types of functional groups on a silica capillary's interior surface. These groups include free silanols, siloxanes and germinal or vicinal silanols. Each of these groups can have a different chemical reactivity towards silanization reactions, as are used in coating the capillary with polyacrylamide, resulting in a heterogeneous polymerization process. In order to produce more homogeneous polymerization, capillary conditioning is required. This procedure results in the maximum number of free silanol groups on the inside capillary wall, thus giving more uniform reactivity. In this work, this is accomplished by passing several conditioning solutions through the capillary, which are applied through the use of a syringe and microcapillary union.

1. Fill the capillary with 1 *M* sodium hydroxide and allow to sit for 1 h.
2. Rinse the inside of the capillary with double-distilled water (ddH_2O).
3. Fill the capillary with 0.1 *M* HCl and allow the capillary to sit for 1 min. Rinse the interior of the capillary with ddH_2O.
4. Dry and purge the inside of the capillary with nitrogen gas for 1 h. This can be accomplished by using a commercial nitrogen gas cylinder with a regulator that is connected to the microcapillary union.

3.2.3. Capillary Modification With Linear Polyacrylamide

Because HSA can have significant binding to the wall of an uncoated silica capillary *(18)*, the use of an uncoated capillary with this protein can give rise to a large variation in the migration time of HSA between runs, with a change as large as 1 min (15–20%) sometimes being observed. To help minimize this problem, the capillary can be coated with a hydrophilic layer to prevent such adsorption from occurring. This accomplished here in this work by placing a linear polyacrylamide coating to the interior of the silica capillary. **Figure 5** shows the reactions involved in this process *(19)*. This makes use of a two step process that involves silanization and reaction with acrylamide. This same type of capillary can be obtained commercially from manufacturers like Bio-Rad and Supelco. As shown in **Table 1**, the use of this coated capillary reduces the run-to-run variations in HSA migration times to less than ±1.2–1.5 % when used in the ACE assay.

1. Prepare a silane solution by mixing three parts of μ-methacryloxypropyl-trimethoxysilane and with 97 parts of 60% acetone (v/v) in water. Typically, 5 mL of this silane solution is enough to prepare several 50-cm-long capillaries.
2. After a detection window has been placed on a water-free fused-silica capillary, fill this capillary with the silane solution and allow it to react for 10 h.
3. Prepare an acrylamide solution by dissolving acrylamide (30 mg/mL) and ammonium persulfate (2 mg/mL, the polymerization initiator) in ddH_2O.
4. Mix the acrylamide solution with TEMED (0.8 μL of TEMED per 1 mL of the acrylamide solution).
5. Immediately inject the acrylamide/TEMED mixture into the fused-silica capillary while also removing the silane solution within the capillary.
6. Seal both ends of acrylamide-filled capillary with clay or rubber septa to prevent evaporation of the acrylamide solution.
7. After 10 h, remove the acrylamide solution from the capillary by rinsing with ddH_2O.
8. When not in use, store the coated capillary by placing both of its ends in water to avoid possible shrinkage of the acrylamide gel.

To see if coating process was successful, the coated capillary can be placed onto a CE system and injected with a small plug of a neutral marker, such as acetone or mestyl oxide. If the coating process has taken place, then the electroosmotic flow should be greatly diminished and no peaks for the neutral markers should be seen even after long periods of time (i.e., 2–3 h) in the presence of an applied electric field.

3.3. Reagent and Sample Preparation

Following modification of the HSA and preparation of the fused-silica capillary, other items that need to be prepared for the assay include: (1) the electrolyte to be used in the capillary, (2) running buffers that contain this electrolyte, as well as dextran as a mobility modifier and normal or modified HSA as binding agents, and (3) samples of the drugs or solutes to be assayed.

3.3.1. Electrolyte Preparation

In this assay, a pH 7.4, 0.067 *M* potassium phosphate buffer is used as the supporting electrolyte. This particular buffer is commonly used in studies of drug and solute binding to HSA because it mimics the pH and ionic strength of

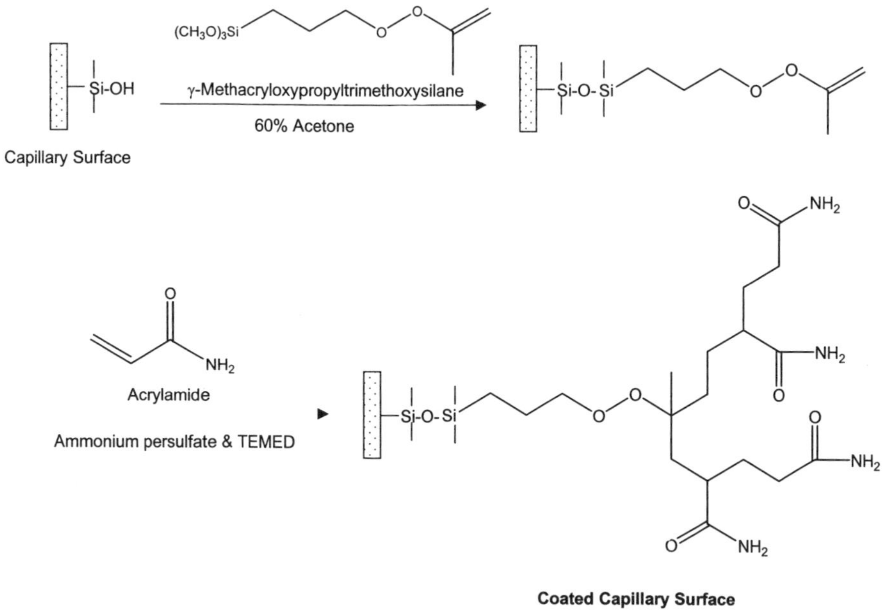

Fig. 5. Reactions for the preparation of a fused-silica capillary that is coated with linear polyacrylamide.

Table 1
The Mobility of HSA and Some Test Solutes in the Presence or Absence of Dextran[a]

	Electrophoretic mobility ($cm^2\ V^{-1}\ s^{-1}$)	
Compounds	Phosphate buffer only	Dextran (10 g/L) as a buffer additive
Normal HSA	1.76 (±0.02) × 10^{-4}	1.37 (±0.03) × 10^{-4}
TNM-modified HAS	1.80 (±0.02) × 10^{-4}	1.36 (±0.02) × 10^{-4}
HNB-modified HAS	1.75 (±0.03) × 10^{-4}	1.38 (±0.03) × 10^{-4}
Warfarin	2.01 (±0.02) × 10^{-4}	1.99 (±0.02) × 10^{-4}
Ibuprofen	2.35 (±0.02) × 10^{-4}	2.35 (±0.02) × 10^{-4}

[a]The numbers in parentheses represent a range of ±1 S.D. for triplicate measurements. The experimental conditions are described in the text. The results shown here are from **ref. *11***.

plasma. This buffer is prepared using deionized or ddH_2O and standard preparations of monobasic and dibasic postassium phosphate salts. After this buffer has been prepared, it is filtered through a 0.22-µm cellulose acetate filter to remove bacteria and particulate matter and is degassed for 30 min in a sonicating bath under vacuum prior to use (*see* **Note 9**). If not used immediately, store this solution in a refrigerator at 4°C. This buffer is used in later steps to prepare the CE running buffer and samples of drugs that will be analyzed by the ACE method.

3.3.2. Preparation of Running Buffer

The electrolyte that was prepared in the previous section requires the addition of two other agents for use in the ACE assay. The first of these additives is dextran, which is used to adjust the mobility of HSA so that it differs from the mobility of the drug or solute of interest (*see* **Table 1**). The other type of additive that is needed is normal or modified HSA, which is used to examine the binding of the solute to this protein and to determine the sites on HSA at which this binding occurs.

1. Dextran is added to the electroloyte by placing approx 10 g/L of this agent in the pH 7.4, 0.067 *M* potassium phosphate buffer from **Subheading 3.3.1.** This dextran solution is then filtered through a 0.22-µm cellulose acetate filter and degassed for 30 min in a sonicator bath (*see* **Note 10**).
2. A running buffer that contains normal, unmodified HSA is prepared by placing 3–5 g/L HSA in the running buffer from **step 1**. This gives a solution that contains pH 7.4, 0.067 *M* potassium phosphate buffer plus 10 g/L dextran.
3. Similar but separate solutions to those prepared in step 2 should be made for the TNM- and HNB-modified HSA. These each should be prepared to contain the same concentration of protein as the running buffer that contains the normal HSA from **step 2** (*see* **Note 11**).
4. Store all of these running buffers in a refrigerator at 4°C. Under these conditions, these solutions should be stable for several months when they are properly prepared. Do not use these solutions if there is any signs of bacterial or microbial growth within their containers.

3.3.3. Sample Preparation

The actual concentration of the test solute that should be used in the ACE assay will vary, but all of these samples should be prepared in the pH 7.4, 0.067 *M* potassium phosphate buffer that is used as the CE electrolyte. The preparation of such a sample will be illustrated here by using racemic warfarin and racemic ibuprofen as examples.

1. Place 10 mg of the warfarin or ibuprofen in a 10-mL volumetric flask and dissolve in the pH 7.4, 0.067 *M* potassium phosphate buffer from **Subheading 3.3.1.** This gives a final concentration for these drugs of 1.0 g/L (*see* **Note 12**).

2. Take 0.5 mL of the solution prepared in **step 1** and mix with 0.5 mL of pH 7.4, 0.067 *M* potassium phosphate buffer containing HSA alone or HSA plus dextran, as prepared in **Subheading 3.3.2.**
3. Take a 0.4-mL aliquot of each sample solution, place it into a CE injection vial and combine it with 0.1 mL of a 0.1 g/L solution of sodium nitrate in the running buffer. The sodium nitrate is used in the ACE assay as a mobility marker to correct for run-to-run variations in electroosmotic flow.
4. Keep unused portions of these samples in enclosed containers and store them at 4°C in a refrigerator. These samples should be stable for more than a month prior to use. However, dispose of any portions of these samples that have been used in the ACE assay. Although the sample volumes that are prepared in **step 3** are sufficient to perform 100 individual ACE measurements, it is not recommended that these solutions be used to perform more 10 assays because of possible contamination of these samples or changes in their concentration during the sample introduction step. For instance, the inlet capillary tip usually contains an unknown amount of solution from the inlet buffer reservoir, which can contaminate and dilute these samples over the course of a large number of injections.

3.4. Affinity Capillary Electrophoresis

The actual ACE assay involves: (1) selection of the appropriate electrophoresis conditions, (2) injection of the test solute in the presence of various types of HSA (i.e., normal HSA, HNB-modified HSA, and TNM-modified HSA), and (3) analysis of the resulting data. Each of these procedures is described here.

3.4.1. Electrophoresis Conditions

The actual conditions used in the ACE assay will differ slightly with the type of CE instrument being employed and the type of drug or solute being examined. The following conditions are those that have been used to perform this assay with a Biofocus 3000 CE using warfarin, ibuprofen and related drugs as test solutes.

1. Obtain a coated fused-silica capillary (as prepared in **Subheading 3.2.**) that has a total length of 30 cm and an effective length of 25 cm to the detection window.
2. Select detection conditions that will allow the test solute to be monitored as it migrates through the CE capillary. This can be accomplished for warfarin and ibuprofen by monitoring their absorbances at 310 and 280 nm, respectively.
3. Set the capillary temperature to 37°C. Although studies can also be performed at other temperatures, 37°C gives the best approximation of the binding behavior that would be expected for drugs in the human body.
4. Use electrokinetic injection (~10 nL) for each test sample (*see* **Note 13**).
5. After the sample has been injected, apply a voltage of approx 12.5 kV to the system (i.e., an electrical field strength of roughly 0.42 kV/cm). This should produce an applied current of approx 40 µA for the pH 7.4, 0.067 *M* potassium

phosphate running buffer and gives an analysis time of roughly 8 min per injection (*see* **Note 14**).

6. After each assay has been finished, rinse the capillary by injecting a fresh portion of the electrolyte and applying a voltage of approx 12.5 kV for 5 min.
7. Once the whole assay has been completed, shut down the CE system according to instructions of the instrument's manufacturer and place both tips of the capillary in pH 7.4, 0.067 *M* potassium phosphate buffer. Store this buffer-filled capillary in a refrigerator at 4°C.

3.4.2. Initial ACE Assay Without HSA As an Additive

For any new test solute, it is recommended that initial CE studies be performed with only the background electrolyte being used as the running buffer (i.e., with no dextran or HSA added). This is done to examine the inherent mobilities of the test solutes and compare these to the mobility that is expected for HSA. This helps determine whether the ACE assay will be useful in examining the interaction of these drugs with HSA, because this assay requires an observable shift in mobility of the drug upon its binding to the protein. This can be quickly determined in the following manner.

1. Place 1 mL of the pH 7.4, 0.067 *M* potassium phosphate buffer in each of the two running buffer reservoirs of the CE system.
2. Perform **step 5** from **Subheading 3.4.1.** until the baseline of the CE system is stable. Typically, the equilibration time for a capillary in the electrolyte is less than 10 min.
3. Inject separate samples of the drug of interest and HSA, with sodium nitrate being added to each as a mobility marker, as discussed in **Subheading 3.3.3., step 3**.
4. Compare the relative migration times vs sodium nitrate for the drug and HSA. If these migration times differ by more than 30%, proceed to the experiments in **Subheading 3.4.3.** that use HSA as a running buffer additive.
5. If less than a 30% difference in mobility is noted between the test solute and HSA, redo the mobility comparison with a running buffer that now contains 10 g/L dextran. If an improvement in the difference in mobility is noted, use this dextran-containing buffer in all further experiments. If desired, the level of dextran in this buffer can be increased or decreased to further adjust the mobility difference.
6. If no improvement in the mobilty difference is noted after dextran has been added to the running buffer, then an alternative method for drug binding studies should be considered.

3.4.3. Screening ACE Assay With HSA As an Additive

After conditions have been identified in **Subheading 3.4.2.** that give a measurable difference in mobility between HSA and the test solute, running buffers should be prepared according to these conditions that also contain a known amount of normal or modified HSA (*see* **Subheading 3.3.**). The migration of

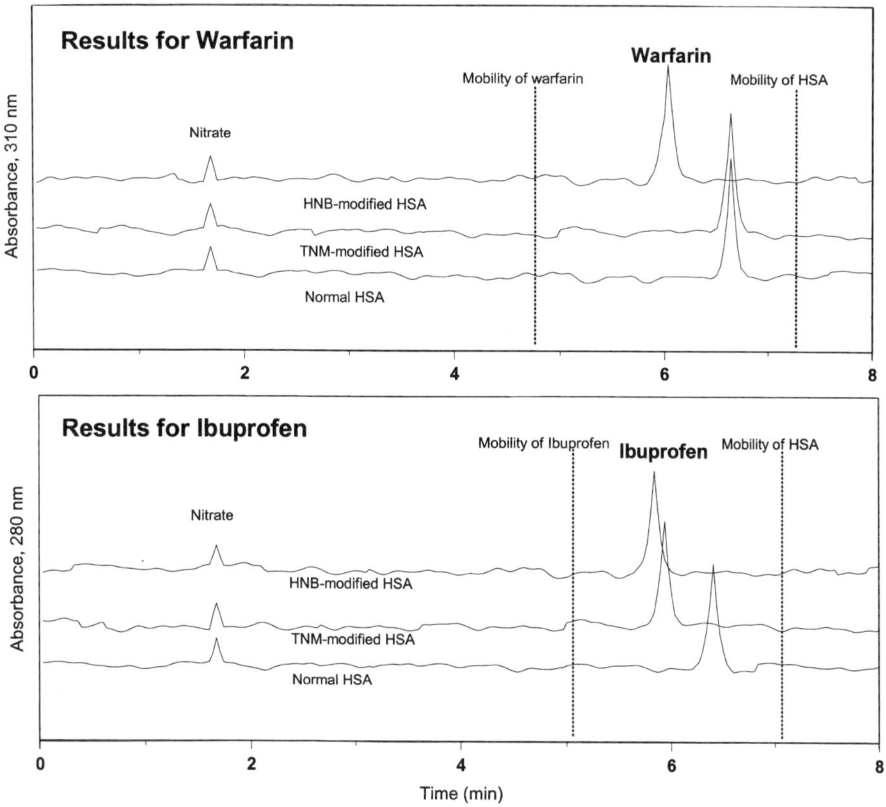

Fig. 6. Typical electropherograms obtained for (top) racemic warfarin and (bottom) racemic ibuprofen.

the solute in the presence of each type of HSA is then compared to determine (1) if binding is taking place between this solute and HSA and (2) if this binding is occurring at one of the two major binding sites on this protein.

1. Place 1 mL of running buffer that contains 5 g/L of normal or modified HSA (plus the required amount of dextran) into the two running buffer reservoirs of the CE system.
2. Perform **step 5** of **Subheading 3.4.1.** until the baseline of the CE system is stable. This typically requires an equilibration time of less than 10 min.
3. Inject the drug or solute sample onto the system and begin the CE run.
4. Repeat this experiment for each type of HSA (*see* **Note 15**).

Figure 6 shows some typical results that have been obtained with this assay for racemic warfarin and racemic ibuprofen in the presence of normal or modified HSA and 10 g/L dextran. The results that were calculated from these as-

Table 2
Mobilities of Various Solutes in the Presence of Normal or Modified HSA[a]

	Electrophoretic mobility ($cm^2 V^{-1} s^{-1}$)		
Injected compounds	Normal HAS	HNB-modified HSA	TNM-modified HSA
Warfarin	$1.52 (\pm 0.02) \times 10^{-4}$	$1.66 (\pm 0.01) \times 10^{-4}$	$1.52 (\pm 0.01) \times 10^{-4}$
Ibuprofen	$1.57 (\pm 0.01) \times 10^{-4}$	$1.72 (\pm 0.03) \times 10^{-4}$	$1.67 (\pm 0.01) \times 10^{-4}$

[a]The numbers in parentheses represent a range of ±1 S.D. for triplicate measurements. The results shown here were obtained from **ref. *11***.

says are shown in **Table 2**. From this data, it can be seen that the modification of HSA with TNM gave essentially the same mobility for warfarin as was obtained with unmodified HSA. This was expected because TNM modifies the indole-benzodiazepine site of HSA where warfarin has no interactions. In contrast to this, the modification of HSA with HNB (which alters the warfarin-azapropazone site of HSA) did have a measurable effect on the mobility of warfarin, causing a shift in its migration time from 6.7 min to roughly 6.0 min. This shift toward the mobility of the nonbound warfarin represents weaker binding between this drug and the modified HSA. This agrees with earlier measurements performed between *R*-warfarin and HNB-modified HSA, in which a twofold decrease in binding strength was seen for this protein vs normal HSA ***(20,21)***. Furthermore, this shift in mobility confirms that warfarin has interactions at or near the warfarin-azapropazone site.

Another illustration of the ACE assay is given by using ibuprofen as an example. This drug is known to interact with the indole-benzodiazepine site of HSA ***(5)***. When compared to warfarin, this drug has a similar migration time and overall binding affinity for normal HSA, but it has different changes in this mobility when combined with HNB- or TNM-modified HSA. For example, ibuprofen gave a shift in mobility when either of these modified forms of HSA were added to the running buffer. This agrees with an earlier finding that ibuprofen can increase the free fraction of warfarin in serum and cause displacement of warfarin from serum albumin ***(22)***, and a previous report in which two binding sites on HSA were found for the *S*-enantiomer of ibuprofen ***(23)***. This suggests that both of the major sites of HSA are involved in binding to ibuprofen.

4. Notes

1. Urea is necessary to promote unfolding of HSA during the HNB reaction, because Trp214 is located in the interior of this protein and normally does not have much exposure to solvent.

2. The HNB stock solution can be adjusted to other concentrations, with typical ratios of 100–1000 mol HNB to mol HSA being used. The conditions given here result in a 100-fold mol excess of HNB vs HSA.
3. The reaction of TNM with tyrosine is pH dependent. At pH values below 6.0–7.0, this reaction does not occur and pHs above 9.0 promote unwanted side reactions. The pH value of 8.0 that is used here is the best choice for speed and specificity in this reaction.
4. The volumes and concentrations given here for the HSA and TNM stock solutions are designed to give a final mole ratio for TNM vs HSA of 4:1. This ratio can be changed by adding different amounts of the TNM stock solution to the HSA solution. Lower mole ratios will produce a lower degree of HSA modification by TNM, whereas higher ratios will give unwanted side reactions that involve additional tyrosine residues or the formation of 3,5-dinitrotyrosine adducts.
5. The TNM reaction can occur on the order of seconds for some tyrosine residues. Thus, fast and efficient mixing is needed when combining the TNM and HSA solutions to avoid a large local excess of TNM and the possibile reaction of undesired tyrosine residues.
6. Longer reaction times are not detrimental to this step.
7. Both TNM and the nitroformate anion, $C(NO_2)_3^-$, which is formed during the TNM reaction, have an affinity for the nonpolar interior of HSA. Extensive dialysis is needed to completely remove these agents from the final HSA solution.
8. Once the outside polyimide coating has been removed from a fused-silica capillary, this part of the capillary becomes quite fragile. It is always recommended that the detection window be made as short as possible to prevent this breaking from occurring.
9. It is recommended that this buffer be stored in a refrigerator to prevent microbial growth and minimize bubble formation. Under these conditions, this buffer should be stable for at least several months. However, a degassing procedure (i.e., sonication under vacuum) should be performed for this buffer about once a week when it is used over long periods of time.
10. In other work, it has been shown that the addition of dextran will alter the migration of HSA and bovine serum albumin (BSA) but has only a slight effect on the mobility of smaller solutes ***(24,25)***. In this ACE assay, as the dextran concentration is increased from 0 to 10 g/L an approximately linear change in mobility is observed for each type of HSA. The use of a larger amount of dextran is discouraged because of its effects on the viscosity and conductivity of the running buffer.
11. In this step, the starting concentration of each HSA solution can be determined by a BCA protein assay. The HSA solutions are then diluted to the same final concentration (10 g/L) by using adjustable micropipets.
12. For drugs or solutes with low solubilities, sonication is often helpful in placing enough of these subtances in solution for analysis. The actual concentration of solute that is required will depend on the detection properties of this agent. The sample concentrations that are used here for warfarin and ibuprofen are typical of those that would be selected for drugs that have good UV/vis absorbance. Lower

sample concentrations may be possible when using fluorescent solutes and other detection modes to monitor the migration of these solutes.

13. It is highly recommended that electrokinetic injection be used, rather than pressure injection in the ACE assay, to prevent leaking of the polyacrylamide coating from inside the CE capillary.
14. Lower applied voltages can be used to reduce the operating current, but this will also lengthen the time of the assay and give slightly broader solute peaks.
15. The apparent mobility of a solute (μ_{app}) is actually the sum of the inherent mobility of the solute (μ_{ep}) and the mobility caused by electroosmotic flow (μ_{eo}). However, in this ACE method, electroosmotic flow within the coated capillary is very small. This means that the apparent and inherent mobilities of the solute are approximately equal. Under these conditions, the apparent mobility (as well as the inherent mobility) of the solute can be determined from the equation $\mu_{ep} \approx \mu_{app} = L_d\ L_t/t_s V$, where L_d is the effective capillary length, L_t is the total capillary length, t_s is the migration time of solute, and V is the applied voltage. It is also helpful in such work to divide each calculated mobility by the mobility of the internal marker during the same run. This corrects for any run-to-run variations in the solute mobility owing to fluctuations in the current, applied voltage, or viscosity of the running buffer.

Acknowledgments

This work was supported by the National Institutes of Health by Grant RO1 GM44931.

References

1. Ho, J. X. and Carter, D. C. (1994) Structure of serum albumin. *Adv. Protein Chem.* **45,** 153–203.
2. He, X., Munson, S. H., Twigg, P. D., et al. (1989) Three-dimensional structure of human serum albumin. *Science* **244,** 1195–1198.
3. Peters, T., Jr. (1996) *All About Albumin*. Academic, New York.
4. Brodersen, R. (1979) Bilirubin. Solubility and interaction with albumin and phospholipid. *J. Biol. Chem.* **254,** 2364–2369.
5. Birkett, D. J., Wade, D. N., and Sudlow, G. (1976) Further characterization of specific drug binding sites on human serum albumin. *Mol. Pharmacol.* **12,** 1052–1061.
6. Tillement, J. P.,Zini, R., D'athis, P., and Vassent, G. (1974) Binding of certain acidic drugs to human albumin: theoretical and practical estimation of fundamental parameters. *Eur. J. Clin. Pharmacol.* **7,** 307–313.
7. Kragh-Hansen, U., Brennan, O. S., Galliano, M., and Sugita, O. (1990) Binding of warfarin, salicylate, and diazepam to genetic variants of human serum albumin with known mutations. *Mol. Pharmacol.* **37,** 238–242.
8. Hage, D. S. and Tweed, S. A. (1997) Recent advances in chromatographic and electrophoretic methods for the study of drug-protein interactions. *J. Chromatogr. B* **699,** 499–525.
9. Heegaard, N. H. H., Nilsson, S., and Guzman, N. A. (1998) Affinity capillary electrophoresis: important application areas and some recent developments. *J. Chromatogr. B* **715,** 29–54.

10. Hage, D. S. (2001) Chromatographic and electrophoretic studies of protein binding to chiral solutes. *J. Chromatogr. A* **906,** 459–481.
11. Kim, H. S., Austin, J., and Hage, D. S.(2002) Identification of drug-binding sites on human serum albumin using affinity capillary electrophoresis and chemically modified proteins as buffer additive. *Electrophoresis* **23,** 956–963.
12. Lundblad, R. L. (1991) *Chemical Reagents for Protein Modification.* CRC, Boca Raton, FL.
13. Fehske, K., Muller, W., and Wollert, U. (1979) A highly reactive tyrosine residue as part of the indole and benzodiazepine binding site of human serum albumin. *Biochem. Biophy. Acta* **577,** 346–359.
14. Fehske, K., Muller, W., and Wollert, U. (1978) The modification of the lone tryptophan residue in human serum albumin by 2-hydroxy-5-nitrobenzyl bromide. Characterization of the modified protein and the binding of L-tryptophan and benzodiazepines to the tryptophan-modified albumin. *Hoppe-seyler's Physil. Chem.* **359,** 709–717.
15. Karkhanis, Y. D., Carlo, D., and Zeltner, J. (1975) Simplified procedure to determine tryptophan residues in proteins. *Anal. Biochem.* **69,** 55–60.
16. Loudon, G. M. and Koshland, D. E., Jr. (1970) Chemistry of a reporter group: 2-Hydroxy-5-nitrobenzyl bromide. *J. Biol. Chem.* **245,** 2247–2254.
17. Moravec, L., Saber, M., and Meloun, B. (1979) Steric accessibility of tyrosine residues in human serum albumin. *Collect. Czech. Chem. Commun.* **44,** 1657–1670.
18. Yang, J. and Hage, D. S. (1994) Chiral separations in capillary electrophoresis using human serum albumin as a buffer additive. *Anal. Chem.* **66,** 2719–2725.
19. Mazzeo, J. R. and Krull, I. S. (1994) Modification of capillaries and buffers for enhanced separations, in *Handbook of Capillary Electrophoresis* (Landers, J. P., ed.). CRC, Boca Raton, FL, pp. 495–512.
20. Chattopadhyay, A., Tian, T., Kortum, L., and Hage, D. S. (1998) Development of tryptophan-modified human serum albumin columns for site-specific studies of drug-protein interactions by high-performance affinity chromatography. *J. Chromatogr. B* **715,** 183–190.
21. Tian, T. (1998) Application of protein modification to drug binding studies and chiral separations in high-performance affinity chromatography. Master's thesis. Univ Nebraska, Lincoln.
22. Kober, A. and Sjöholm, I. The binding sites on human serum albumin for some nonsteroidal antiinflammatory drugs. *Mol. Pharmacol.* **18,** 421–426.
23. Hage, D. S., Noctor, T. A., and Wainer, I. W. (1995) Characterization of the protein binding of chiral drugs by high-performance affinity chromatography. Interactions of R- and S-ibuprofen with human serum albumin. *J. Chromatogr. A* **693,** 23–32.
24. Sun, P., Wu, N., Barker, G., and Hartwick, R. A. (1993) Chiral separations using dextran and bovine serum albumin as run buffer additives in affinity capillary electrophoresis. *J. Chromatogr.* **648,** 475–480.
25. Li, S., Ryan, P., and Lloyd, D. K. (1995) Protein chiral selector in free-solution capillary electrophoresis and packed-capillary eletrochromatograophy. *J. Chromatogr. A* **694,** 285–296.

10

Using Charge Ladders and Capillary Electrophoresis to Measure the Charge, Size, and Electrostatic Interactions of Proteins

Upma Sharma and Jeffrey D. Carbeck

Summary

This chapter provides an overview of protein charge ladders—collections of protein derivatives that differ in charge—and capillary electrophoresis (CE). The combination of charge ladders and CE is a useful biophysical tool for measuring the net charge of proteins and the role of electrostatics in biochemical processes involving proteins. Methods to synthesize and analyze charge ladders by CE are described. Applications of charge ladders and CE to the simultaneous measurement of net charge and hydrodynamic radius of proteins are presented. Techniques for using charge ladders and CE to measure the role of interactions between charged groups on protein stability and ligand binding are also given. The power of this approach lies in the ability to isolate protein charge as an independent and measurable variable in the study of protein stability and function.

Key Words

Acetylation; affinity capillary electrophoresis; denaturation; electrostatic interactions; hydrodynamic radius; ligand binding; molecular recognition; protein charge ladders; stability.

1. Introduction

Challenges arising in proteomics require methods for the separation, identification, and classification of the proteins expressed and modified by cells. Proteomics also requires techniques for measuring protein function: for example, protein stability, protein–protein interactions, and protein–ligand binding *(1)*. The separation and classification of proteins is typically done on the basis of size (molecular weight) and charge, usually by 2D gel electrophoresis. In this technique, the first dimension involves separation of proteins on the basis of

From: *Methods in Molecular Biology, vol. 276: Capillary Electrophoresis of Proteins and Peptides*
Edited by: M. A. Strege and A. L. Lagu © Humana Press Inc., Totowa, NJ

charge using isoelectric focusing; the second dimension involves separation of proteins on the basis of size using gel electrophoresis. Because this second separation is done under conditions that denature proteins (i.e., in the presence of charged detergent molecules), 2D gel electrophoresis is incompatible with the analysis of protein function.

In this chapter, we describe the combination of capillary electrophoresis (CE) and protein charge ladders—collections of protein derivatives that differ incrementally in number of charged groups *(2–5)*; (*see* **Fig. 1**)—as a biophysical tool that can measure the properties and interactions of native proteins in free solution. CE measures the electrophoretic mobility of proteins in solution (μ_{elec}, $m^2\ V^{-1}\ s^{-1}$), defined as the steady-state velocity per unit of applied electric field. The value of μ_{elec} depends on both the hydrodynamic size and net charge of a protein. The combination of CE and charge ladders can measure both of these properties in a single electrophoresis experiment under conditions that maintain the folded, native state *(6)*. CE is also a useful tool for monitoring protein folding *(7)* and binding of ligands *(8)*. The combination of CE and charge ladders allows protein folding and ligand binding to be measured as a function of the net charge of the protein *(9–11)*. CE and charge ladders, therefore, provide separation and classification of proteins on the basis of charge and size, as well as quantification of stability or binding affinity for ligands as a function of protein net charge. All of this information is obtained in a single set of electrophoresis experiments.

In this chapter, we first describe how protein charge ladders are synthesized and analyzed using CE. Second, we describe how electrophoretic mobilities of charge ladders are used to determine values of net charge and hydrodynamic size of proteins. Third, we describe how CE is used to measure protein–ligand binding and protein folding. Finally, we show how protein charge ladders and CE are used to measure the role of electrostatic interactions in these molecular recognition events.

2. Synthesis of Protein Charge Ladders

2.1. Overview

Charge ladders are synthesized by the partial, random modification of charged groups on a protein (*see* **Fig. 2**), most commonly via acetylation of Lys ε-amino groups with acetic anhydride *(4)*. Because each acetylation converts a positively charged NH_3^+ group into a neutral $NHCOCH_3$ group, charge ladders contain protein derivatives that differ incrementally in net charge. The hydrodynamic size of these derivatives is effectively unchanged by the acetylation (*see* **Note 1**). The change in charge resulting from the modification of a charged group, ΔZ, depends on the pH of the solution and the pK_a of the group that is modified: acetylation of Lys ε-amino groups results in a value of ΔZ of

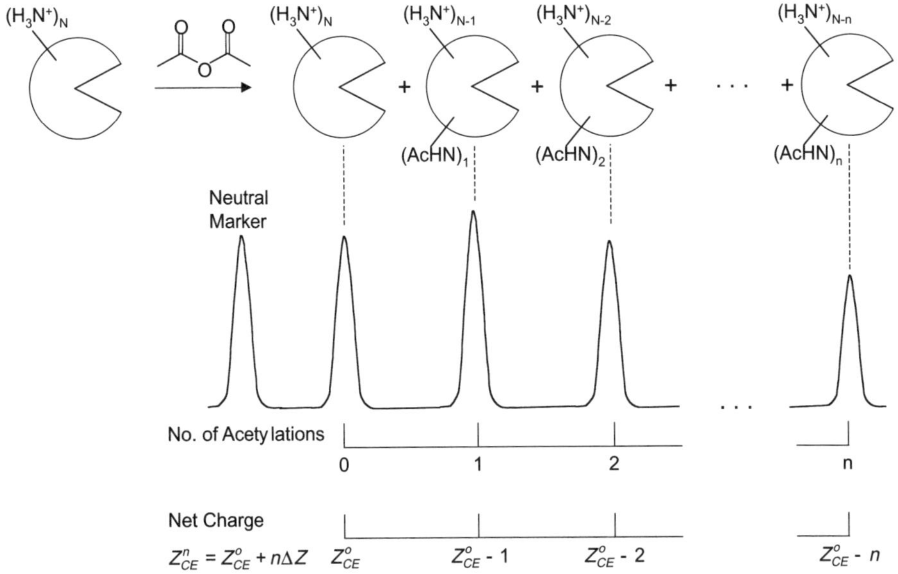

Fig. 1. This scheme illustrates the formation of a charge ladder by the random partial acetylation of Lys ε-amino groups with acetic anhydride, and the separation of the protein derivatives into individual rungs by capillary electrophoresis. The unmodified protein has N NH_3^+ groups; the nth rung of the ladder is composed of proteins that have n neutral $NHCOCH_3$ derivatives (AcHN). The protein derivatives that make up the nth rung of the charge ladder have approximately the same net charge, $Z^n_{CE} = Z^o_{CE} + n\Delta Z$: Z^o_{CE} is the net charge of the unmodified protein; ΔZ is the change in charge caused by the conversion of an NH_3^+ group to its neutral $NHCOCH_3$ derivative. A neutral marker is used to measure the rate of electroosmotic flow. (Reproduced with permission from **ref. *11***. Copyright 2002 Am. Chem. Soc.)

approx –1 for values of pH < 9.0 (*see* **Note 2**). Charge ladders have been produced using reagents other than acetic anhydride to introduced additional charge groups to the proteins (e.g., succinic anhydride, **ref. *4***), and by the amidation of carboxyl groups ***(12)***. Doing so has produced charge ladders with values of ΔZ ranging from –6 to +1.

2.2. Materials

1. Model proteins: lysozyme (chicken egg white), myoglobin (horse heart), ovalbumin (chicken egg), carbonic anhydrase II (human or bovine erythrocytes), α-lactalbumin (bovine milk) (Sigma Chemical Company, St. Louis, MO; *see* **Note 3**).
2. 0.1 N NaOH (Acros, NJ).
3. For acetylation reactions: 1 vol% acetic anhydride, succinic anhydride or 1,2,4-benzenetricarboxylic anhydride in dioxane (Aldrich, Milwaukee, WI).

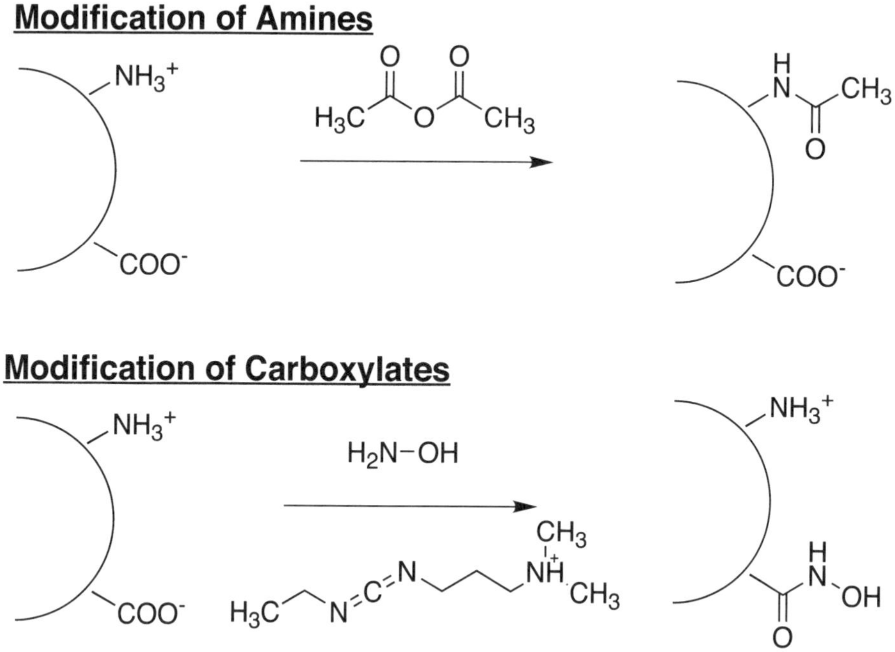

Fig. 2. Formation of charge ladders through modification of different functional groups on a protein. Primary amines can be modified by anhydrides (acetic anhydride is shown) or NHS-esters. Carboxylate groups on proteins can be modified via carbodiimide-mediated coupling of amines.

4. For amidation reactions:
   a. 100 m*M* 1-Ethyl-3- (3-dimethylaminopropyl) carbodiimide hydrochloride (EDAC) (Sigma) in water.
   b. 500 m*M* Hydroxylamine hydrochloride or 500 m*M* methylamine hydrochloride (Sigma) in water.
5. 18 MΩ Ultrapure water (Hydro Service and Supplies, Levittown, PA).
6. pH paper.

2.3. Methods

2.3.1. Synthesis of Charge Ladders Via Acylation of Amino Groups

Protein charge ladders with values of $\Delta Z < 0$ are typically produced by the acylation of Lys ε-amino groups ***(4)***; the *N*-terminal amino group can also be modified.

1. Proteins are dissolved in Ultrapure water to the desired concentration (0.1–0.5 m*M*).
2. The pH of the solution is adjusted to approx 12 by the addition of 20 vol% of 0.1 *N* NaOH; the pH is checked using a drop of the solution and pH paper.

3. An aliquot of acetic ($\Delta Z = -1$), succinic ($\Delta Z = -2$) OR 1,2,4-benzenetricarboxylic anhydride ($\Delta Z = -3$) anhydride solution (typically containing 5–60 equivalents of anhydride per mole of protein) is added to the protein solution.
4. The solution is gently mixed (*see* **Note 4**) and allowed to react for at least 2 min before analysis (the consumption of the anhydride by the combined effects of hydrolysis and aminolysis is typically complete by this time).
5. A single reaction does not usually produce a full charge ladder. CE is used to measure the distribution of products produced with different equivalents of anhydride. (See the following section for a detailed protocol.)
6. The products of several reactions can be mixed to provide the full ladder in a single sample.

2.3.2. Synthesis of Charge Ladders Via Amidation of Carboxyl Groups

Protein charge ladders with values of $\Delta Z > 0$ are produced by the amidation of the carboxyl groups of aspartic and glutamic acid ***(12)***; the C-terminal carboxyl group can also be modified.

1. Proteins are dissolved at a concentration of 0.1 m*M* in deionized water.
2. 50 µL of either the hydroxylamine or methylamine hydrochloride (**Subheading 2.2., item 4b**) are added.
3. Reactions are allowed to proceed for 20 min before analysis by CE (*see* **Note 5**).

3. Analysis of Charge Ladders by CE

3.1. Overview

CE separates the collections of derivatives of a protein charge ladder in free solution into the individual peaks or "rungs" of the ladder. Each peak is composed of a family of protein derivatives that have the same number of modifications and approximately the same net charge (*see* **Fig. 1**). The electrophoretic mobility of the proteins that make up the *n*th rung of the charge ladder μ^n_{elec} where *n* is the number of modified charged groups, is determined experimentally using **Eq. 1**, where *V* is the applied voltage, t^n_x is the migration time for the proteins in the *n*th rung of the ladder, t_{nm} is the migration of a neutral marker, and ℓ_{tot} and ℓ_{det} are the total capillary length and the length to the detector window, respectively. By convention, the sign of μ_{elec} (+ or –) is assigned to be the same as the net charge of the protein.

$$\mu^n_{\text{elec}} = \frac{\ell_{\text{tot}}\,\ell_{\text{det}}}{V}\left(\frac{1}{t^n_x} - \frac{1}{t_{nm}}\right) \quad (1)$$

The neutral marker is added to the sample to measure the rate of electroosmotic flow: that is, the rate of motion of the solution inside of the capillary that arises owing to the presence of fixed charges on the wall of the capillary ***(13)***.

Negatively charged proteins are separated on uncoated capillaries with "normal polarity:" that is, with the cathode of the high-voltage power supply at the outlet. In this way, the direction of electroosmotic flow is toward the outlet.

To analyze positively charged proteins (p*I* > pH of the electrophoresis buffer), it is necessary to minimize adsorption of positively charged proteins onto the negatively charged capillary wall (which present siloxide groups with values of pK_a ~2.2). A variety of chemistries are possible including chemical modification and physical adsorption of polymers. The adsorption of cationic polymers is a convenient means to modify the surface of capillaries, as it requires fewer steps than covalent modification. In particular, the adsorption of the polyamine, Polybrene, onto the bare capillary surface prevents adsorption of many positively charged proteins to the wall of the capillary ***(14)***. In order to maintain the direction of the electroosmotic flow towards the outlet on the positively charged capillary, a "reverse polarity" configuration is used: that is, with the anode of the high-voltage power supply at the outlet.

3.2. Materials

1. Fused-silica capillaries with an external polyimide coating (id = 50 μm, od = 360 μm), (Polymicro Technologies, Phoenix, AZ).
2. Electrophoresis buffer: 25 m*M* tris-hydroxy(methyl)aminomethane (Tris-HCl) and 192 m*M* glycine (Gly) (Sigma) in deionized (DI) water (*see* **Note 6**).
3. Para-methoxybenzyl alcohol (PMBA) (Aldrich).
4. 7 wt% Solution of polybrene (hexadimethrine bromide, Aldrich) in electrophoresis buffer.
5. 0.1 *N* NaOH (Acros).
6. 18 MΩ ultrapure water (Hydro).
7. RBS (Pierce, Rockford, IL).
8. Kimwipe® (Kimberly-Clark, Roswell, GA).

3.3. Methods

3.3.1. Preparation of Capillaries for CE

1. Capillaries are cut to total lengths that vary between 27 and 107 cm.
2. A small window (~5 mm) is created to allow for ultraviolet (UV) detection by burning away the polyimide coating. This region of the capillary is washed with ethanol and wiped carefully with a Kimwipe to remove the charred polyimide, leaving bare glass.

3.3.2. Pretreatment of Capillaries for CE

1. Rinse the capillary for 15 min with RBS at 20 psi.
2. Rinse the capillary for 15 min with 0.1 *N* NaOH at 20 psi.
3. Rinse the capillary for 15 min with DI water at 20 psi.
4. Rinse the capillary for 15 min with electrophoresis buffer at 20 psi.

3.3.3. Coating of Capillaries for CE

1. Rinse the capillary for 15 min with RBS at 20 psi.
2. Rinse the capillary for 15 min with 0.1 *N* NaOH at 20 psi.
3. Rinse the capillary for 15 min with the 7 wt% polybrene solution at 20 psi.
4. Rinse the capillary for 15 min with electrophoresis buffer at 20 psi (*see* **Note 7**).

3.3.4. CE

1. CE experiments are performed either on a Beckman P/ACE MDQ or on a Beckman P/ACE 5500. Analysis is typically done at a fixed temperature (either 25 or 37°C) using the internal cooling system of the instrument to dissipate Joule heating. For thermal denaturation studies, the capillary is maintained at a constant temperature between 25 and 95°C using an external water bath connected to the internal cooling system.
2. An injection sample is typically prepared by diluting the charge ladder, as synthesized, tenfold in electrophoresis buffer.
3. PMBA is added to the injection sample at approx 0.02 vol% to mark the rate of electroosmotic flow. The total volume of the injection sample is usually 20–100 μL.
4. After conditioning the capillary, it is filled with electrophoresis buffer and a plug of sample is injected at 0.5 psi for 5 s (*see* **Note 7**).
5. The ends of the capillary are transferred to vials containing the electrophoresis buffer, and separation proceeds at voltages ranging from 5 to 30 kV (*see* **Note 8**).
6. The sample is separated until all peaks are eluted; typical separation times are between 5 and 15 min.
7. Between separations, uncoated capillaries are rinsed for 1 min with 0.1 *N* NaOH, and 2 min with electrophoresis buffer. Coated capillaries are rinsed for 2 min with electrophoresis buffer.
8. Peak detection is by UV absorbance at 214 nm (*see* **Note 9**). **Figure 3** shows an electropherogram for the charge ladder of human carbonic anhydrase II produced by the partial acetylation of Lys ε-amino groups.

4. Determination of Net Charge and Hydrodynamic Size of Proteins

4.1. Overview

A single measurement of electrophoretic mobility is insufficient to deter mine both the charge and the size of a protein. CE and charge ladders, in contrast, measure the mobility of many protein derivatives (a total 2^N, where N is the total number of charged groups available for modification) in a single CE experiment. Because the change in charge caused by chemical modification of the protein is (or is assumed to be) known, the combination of CE and charge ladders allows the hydrodynamic size and net charge of proteins to be determined independently *(6)*.

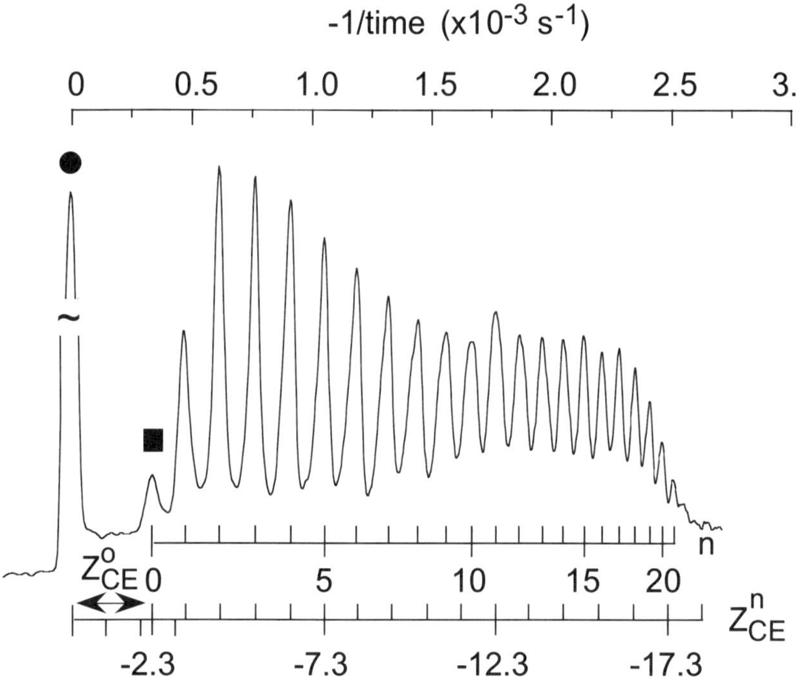

Fig. 3. The charge ladder of human carbonic anhydrase II (HCA II) produced by the partial acetylation of Lys ε-NH_3^+ groups; each acetylation results in an assumed increment of charge, Δ*Z*, owing to the conversion of a Lys ε-NH_3^+ group to its neutral ε-$NHCOCH_3$ derivative (ΔZ ~ –1 at pH = 8.4). The number of ε-$NHCOCH_3$ groups (*n*) and the net charge of the rungs of the charge ladder estimated by CE, Z^n_{CE} are indicated below the electropherogram. The –1/time scale is directly proportional to the electrophoretic mobility. The peak marked with (●) is an electrically neutral marker, and the peak marked with (■) is the native protein. Separations were performed at 25°C on a 47-cm silica capillary (40 cm from inlet to detector; id 50 μm) using an electrophoreis buffer of 25 m*M* Tris-HCl-192 m*M* Gly (pH 8.4). Detection was by direct UV absorbance at 214 nm.

During electrophoresis, a molecule undergoes two counteracting forces. The electrostatic force (F_{elec}, *N*) acting on a molecule is equal to the product of its charge (*Z*, *C*) and the magnitude of the applied electric field (*E*, *V* m^{-1}; *see* **Eq. 2**). The hydrodynamic force (F_{hydro}, *N*) acting on a molecule is equal to the product of its coefficient of friction (*f*, *N* s m^{-1}) and its velocity (*v*, m s^{-1}) relative to the surrounding solvent (*see* **Eq. 3**). The electrophoretic mobility (i.e., the steady-state velocity per unit applied field) is the result of the balance of electrostatic and hydrodynamic forces acting on the charged molecule (*see* **Eq. 4**).

$$F_{\text{elec}} = ZE \tag{2}$$

$$F_{\text{hydro}} = f\,v \tag{3}$$

$$F_{\text{elec}} = F_{\text{hydro}} \Rightarrow \frac{Z}{f} = \frac{v}{E} = \mu \tag{4}$$

We define Z_{CE} as the net charge that gives rise to electrophoretic motion of the protein. It differs from the net charge measured by proton titration Z_{H^+} in that Z_{CE} also includes the effects of other charged species associated with the protein; these species include ions bound specifically or nonspecifically to the protein. Z_{CE} may also include effects related to the environment specific to the CE experiment: for example, an induced asymmetry in the distribution of counter ions surrounding the protein that reduces the effective electric field acting on the protein (*see* **Note 10**).

For each rung of the charge ladder, the charge of that rung $Z^n{}_{\text{CE}}$ can be related to its electrophoretic mobility μ^n_{elec} using Eq. 5. $Z^n{}_{\text{CE}}$ differs from the charge of the native protein, $Z^0{}_{\text{CE}}$, by the net change in charge owing to the modifications. To simplify analysis, we assume this total change in charge is equal to the change in charge per modification, ΔZ (assumed to be a constant across the rungs of the ladder), multiplied by the number of modifications n. The value of ΔZ depends on the pK_a of the charged group that is modified, and the pH of the electrophoresis buffer. At values of pH < 9.0, the value of ΔZ is approx –1 for the modification of a Lys ε-NH_3^+ group (pK_a ~10.5); at values of pH > 5.5, ΔZ is approx +1 for the modification of a Glu- or Asp-CO_2^- group (pK_a ~4.0) (*see* **Notes 2** and **11**).

$$\mu^n_{\text{elec}} = \frac{Z^n_{\text{CE}}}{f} = \frac{1}{f}\left(Z^o_{\text{CE}} + n\Delta Z\right) \tag{5}$$

Equation 5 also relates the mobility of the rungs of the ladder to the coefficient of friction f of the protein. The coefficient of friction measured using CE and charge ladders, in conjunction with colloidal models of electrophoretic mobility, can be used to estimate hydrodynamic radius of the protein R_H. The simplest model for the effective hydrodynamic radius of a solute R_H is Stokes' model for flow of a fluid with viscosity η around a spherical particle (**Eq. 6**).

$$f = 6\pi\eta R_H \tag{6}$$

In an ionic solution, a diffuse double layer of ions surrounds a charged particle and increases the effective frictional drag on the particle. The Debye-Hückel model describes these effects by incorporating a term that accounts for the screening of charges on the solute by small ions in the solution (**Eq. 7**). In this equation, κ (m^{-1}) is the inverse Debye length (κ^{-1} is a measure of the thickness of the diffuse double layer); κ is calculated using **Eq. 8** where F is Faraday's constant, $c_{i\infty}$ is the bulk concentration and z_i is the valence of ion i; ε is the dielectric constant of the buffer. The Debye length is a function of the ionic strength of the solution; the higher the ionic strength, the smaller the diffuse double layer. **Equation 7** reduces to **Eq. 6** when the ionic strength goes to zero (because κ also approaches zero).

$$f = 6\pi\eta R_h\left(1 + \kappa R_h\right) \tag{7}$$

$$\kappa = \left(\frac{F^2 \sum c_{i\infty} z_i^2}{RT\,\varepsilon}\right)^{1/2} \tag{8}$$

Equation 7 can be further modified to account for the distortion of the electric field because of the presence of a nonconducting sphere. Doing so results in Henry's model of electrophoresis ***(15)***, as described by **Eqs. 9** and ***10***. The function g_I describes effects of the protein on the local electric field; $g_I = 1$ when $\kappa R_H < 1$ (the Hückel limit) and $g_I = 3/2$ when $\kappa R_H > 10$ (the Helmholtz-Smoluchowski limit). In between these two limiting cases, g_1 is calculated from **Eq.10**. **Equations 9** and **10** are the basis for determining values of R_H once the coefficient of friction has been measured.

$$f = \frac{6\pi\eta R_h\left(1 + \kappa R_h\right)}{g_1\left(\kappa R_h\right)} \tag{9}$$

$$g_1\left(\kappa R_H\right) = \left(1 + \frac{\kappa^2 R_H^2}{16} - \frac{5\kappa^3 R_H^3}{48} - \frac{\kappa^4 R_H^4}{96} + \frac{\kappa^5 R_H^5}{96} - \frac{11}{96} e^{\kappa R_H} \int_{\infty}^{\kappa R_H} \frac{e^{-r}}{r}\, dr\right) \tag{10}$$

4.2. Methods

4.2.1. Determination of Net Charge of Proteins

1. Measure the migration time of each rung of the ladder and use **Eq. 1** to determine its mobility μ^n_{elec}.
2. Plot μ^n_{elec} vs $n\Delta Z$; in Tris-Gly buffer we assume that the value of ΔZ is approx –1 for the modification of a Lys ε-NH_3^+ group and +1 for the modification of a Glu- or Asp-CO_2^- (*see* **Note 2**).

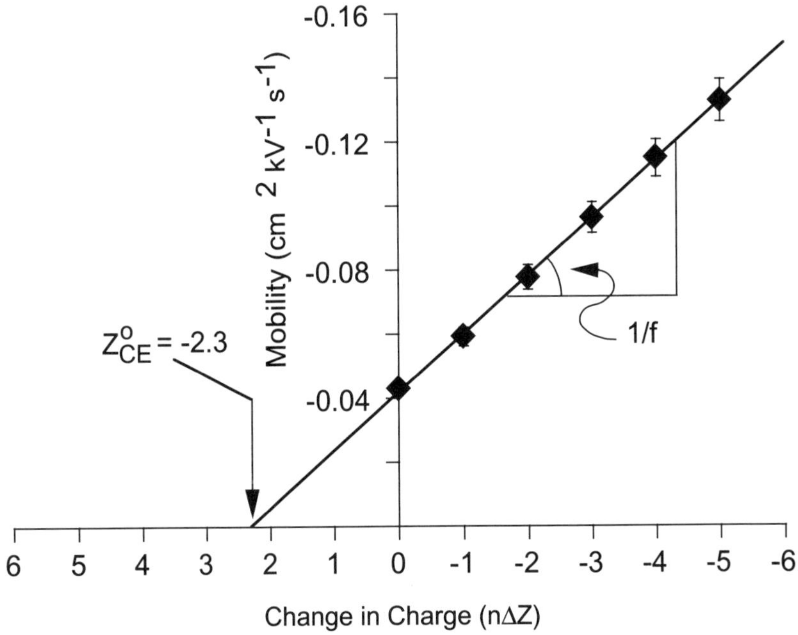

Fig. 4. Values of the electrophoretic mobility (μ_{elec}) (from **Fig. 3**) of the first six rungs of the charge ladders of human carbonic anhydrase II (HCAII), produced by the partial acetylation of Lys ε-NH_3^+ groups, plotted as a function of change in charge $n\Delta Z$ where each acetylation results in an assumed increment of charge, ΔZ, owing to the conversion of a Lys ε-NH_3^+ group to its neutral ε-$NHCOCH_3$ derivative ($\Delta Z \sim -1$ at pH = 8.4). The solid line is a fit of the data to **Eq. 5** using linear least-squares analysis: the x-intercept gives the net charge of the unmodified protein $Z^o{}_{CE}$; the slope of this line give the reciprocal of the effective hydrodynamic drag of the protein $1/f$.

3. Using Excel (Microsoft) or Kaleidagraph (Synergy Software) fit a line to the data using a linear least squares analysis. The line is fit by including the maximum number of points that give a R^2 value greater than 0.999.
4. According to **Eq. 5**, the charge of the native protein $Z^o{}_{CE}$ can be measured by extrapolating the mobility versus $n\Delta Z$ plot to the x-intercept. The values of μ^n_{elec} of the first six rungs of the charge ladder of human carbonic anhydrase correlate approximately linearly with $n\Delta Z$; the charge of the native protein was found to be –2.3 (*see* **Fig. 4**).
5. The slope of the best-fit line is equal to $1/f$. The charge of any rung of the ladder $Z^n{}_{CE}$ is determined using **Eq. 5** by assuming that the coefficient of friction is constant across the rungs of the ladder. This assumption is valid when there is a linear correlation between μ^n_{elec} and $n\Delta Z$ (*see* **Notes 1** and **10**).

4.2.2. Determination of Hydrodynamic Size of Proteins

1. As described in **Subheading 4.2.1.**, we use the slope of plots of μ^{n}_{elec} vs $n\Delta Z$ to get an experimental measurement of the coefficient of friction f.
2. We assume g_I =1 and calculate R_H from **Eq. 9**.
3. We use this value to solve **Eq. 10** for an improved estimate of g_I.
4. Using this new value of g_I, we again use **Eq. 9** to determine a revised estimate of R_H.
5. We repeat this process until R_H converges to a single value.
6. Typically, we use the Solver tool in Excel to automate this procedure. In doing so, we express **Eq.10** using Simpson's Rule to approximate the integral. Solver calculates the value of R_H that minimizes the root-mean-squared (rms) difference between the measured and calculated coefficients of friction.
7. The value of R_H that gives the smallest error is reported as the hydrodynamic radius of the protein to the nearest 0.1 Å. The radius of a particular protein in a buffer was found to vary by ±0.1 Å from run-to-run and by less then ±0.2 Å from capillary-to-capillary. Using the value of f determined from the data in **Fig. 4**, we determined a value of R_H for human carbonic anhydrase of 27 Å ***(6)***. This value is the same as that measured by dynamic light scattering.

4.2.3. Application: Measuring the Charge and Size of a Protein As It Denatures

Although net charge and hydrodynamic size are convenient properties for classifying proteins, these properties also reflect the functional state of the protein when measured under conditions found in real biological systems. The net charge reflects the association equilibrium of protons and other ions with proteins. Binding of these ions plays an important role in protein function: examples include the binding of metal ions to form active, holo, and the thermodynamic linkage of proton binding and protein folding, which leads to the pH dependence of protein stability ***(11)***. The hydrodynamic size is a function of both the molecular weight of the protein and its conformation in solution. Changes in hydrodynamic size can reflect the association between proteins and other molecules (e.g., dimerization), or changes in conformation of the proteins, such as those that accompany protein folding.

As an example, the techniques described in this section have been used to measure the net charge and size of α-lactalbumin (α-LA) as it unfolds by analyzing samples of an α-LA charge ladder at different temperatures. **Figure 5** shows the electropherograms of the charge ladder measured at different temperatures ***(11)***. The response of the UV detector is plotted as a function of electrophoretic mobility, relative to the mobility of a neutral marker (*see* **Note 12**).

The melting temperature of α-LA is 56°C under the conditions of these separations (pH 8.4 Tris-Gly buffer containing 25 m*M* NaCl and 35 μ*M* $CaCl_2$). Individual rungs of the charge ladder are resolved at all temperatures as the protein unfolds. At 75°C, we assume that the protein is fully denatured. A surprising feature of these data is that the mobility of the proteins increases upon unfolding. We expect the configuration of the protein to become more extended and, therefore, the hydrodynamic drag coefficient f to increase as the protein unfolds. Because the mobility is the ratio of charge to effective frictional drag, the observed increase in mobility must be attributable to an increase in magnitude of the net charge of the protein. Changes in Z_{CE} outweigh effects of changes in R_H on the mobility of α-LA as it unfolds.

To quantify these effects, we analyzed the data in **Fig. 5** to determine R_H and Z_{CE} of α-LA as it unfolds. **Figure 6** shows plots of mobility vs $n\Delta Z$ for the charge ladder of α-LA measured in the folded and unfolded states (i.e., at 25 and 75°C, respectively; **ref. *11***). Using the linear least squares analysis of these data and **Eq. 5** we determined values of Z_{CE} and f for the unmodified protein at each of these temperatures. Using **Eqs. 9** and **10**, we determined values of R_H. α-LA unfolds to a compact denatured state, which corresponds to an increase in R_H of approx 2 Å or 11%, relative to the native state. The formation of a compact denatured state (often referred to as a molten globule) by α-LA is well known. This relatively small change in size is accompanied by a large change in magnitude of net charge. The value of Z_{CE} nearly doubles from –5.6 in the native state to –9.6 in the denatured state. **Figure 7** shows that values of Z_{CE} and R_H of the unmodified protein can be measured throughout the transition from the folded to the denatured states. These properties can, therefore, be used to monitor the thermal unfolding of proteins. We have also done similar experiments using chemical denaturants such as urea ***(11)***.

5. Using CE to Measure Molecular Recognition Events and the Role of Electrostatics

5.1. Overview

Molecular recognition events involving proteins, such as ligand binding or folding, in general result in changes in the net charge and hydrodynamic size of proteins, as described in **Subheading 4**. CE easily detects these changes and can, therefore, be used to monitor these events. Because protein charge ladders allow the net charge of proteins to be isolated as an independent variable, biophysical studies using charge ladders and CE can also provide measurements of the role of electrostatic interactions in molecular recognition events involving proteins ***(5,8–11)***. Charge ladders of proteins and CE have been used in similar ways to measure the role of electrostatic interactions in bioprocessing of proteins (i.e., in ultrafiltration; **ref. *16)***. In this subheading, we describe how CE and charge ladders are used in this way to measure the role of electrostatic interactions in protein folding ***(11)*** and receptor–ligand binding ***(10)***.

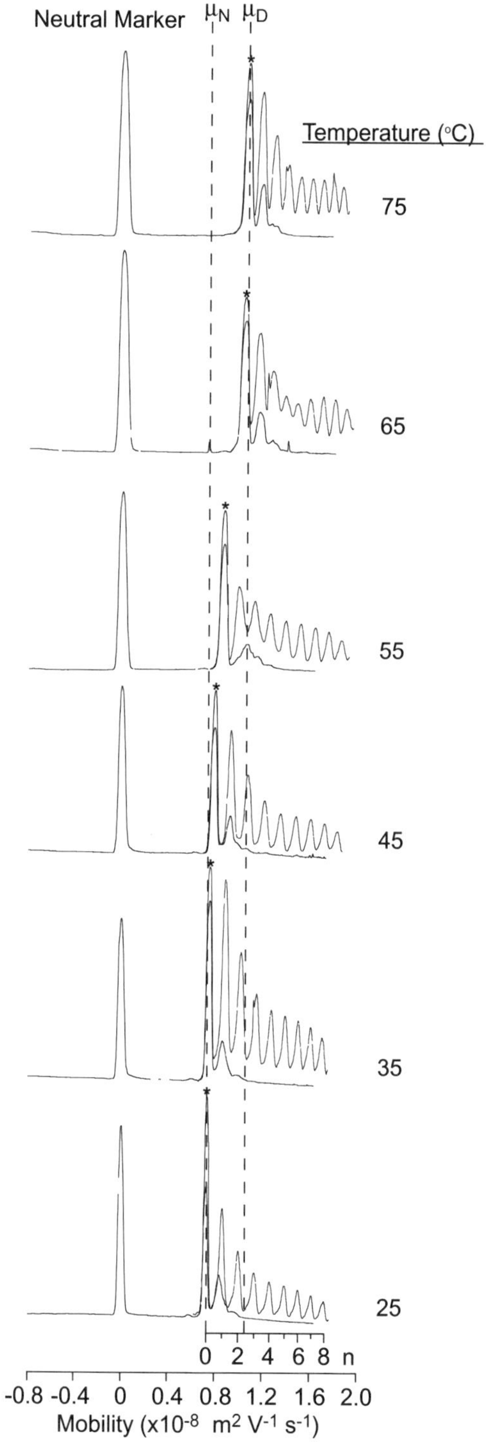
Neutral Marker
μN
μD
Temperature (°C)
75
65
55
45
35
25
0 2 4 6 8 n
-0.8 -0.4 0 0.4 0.8 1.2 1.6 2.0
Mobility (x10-8 m2 V-1 s-1)

To measure molecular recognition events by CE, we describe them in terms of a reversible chemical equation: A↔B. For protein folding, A and B refer to the native and denatured states of the protein, respectively (*see* **Fig. 8**). For ligand binding, A and B refer to the free and the bound states of the protein, respectively.

For each of these states, we measure an electrophoretic mobility μ_A and μ_B. For recognition events that equilibrate rapidly relative to the time scale of separations (typically 1–10 min), capillary electrophoresis measures the average mobility of the protein in the two states. That is, the observed mobility reflects the distribution of protein between the two states (bound and free, or folded and unfolded) at a particular concentration of ligand (C_L, M) or temperature (T, K), as expressed by **Eq. 11**, where θ is the fractional occupancy of the B state, as defined by **Eq. 12** *(8)*.

$$\mu_{\text{elec}}\left(T \text{ or } C_L\right) = \mu_A\left(1-\theta\right) + \mu_B\theta \tag{11}$$

$$\theta\left(T \text{ or } C_L\right) \equiv \frac{C_B}{C_A + C_B} = \frac{\mu_{\text{elec}}\left(T \text{ or } C_L\right) - \mu_A}{\mu_B - \mu_A} \tag{12}$$

The use of CE to monitor folding or ligand binding, therefore, involves the measurement of electrophoretic mobilities of proteins at different concentrations of ligand in the electrophoresis buffer (for analysis of ligand binding) or at different temperatures of the electrophoresis buffer (for analysis of protein

Fig. 5. Electropherograms of the thermal denaturation of the charge ladder of holo α-LA (in gray) superimposed on those of the unmodified protein (in black). The ladder was produced by the partial acetylation of Lys ε-amino groups using acetic anhydride. Separations were done in a buffer composed of 25 m*M* Tris-HCl, 192 m*M* Gly (pH 8.4), 20 m*M* NaCl, and 35 μ*M* $CaCl_2$. The UV absorbance at 214 nm is plotted as a function of electrophoretic mobility. The peak marked with (*) corresponds to unmodified protein; the number of acetylated Lys ε-NH_3^+ groups, n, is indicated below the corresponding peak. The dashed lines indicate the electrophoretic mobility of unmodified α-LA in the native (μ_N) and denatured (μ_D) states. The rungs of the charge ladder, as well as the peak that corresponds to the unmodified protein, show broadening near the middle of the transition (i.e., near the melting temperature of holo α-LA, ~56°C). This broadening may reflect the finite rate of conversion between the folded and denatured states. Broadening of the rungs of the charge ladder may also reflect some heterogeneity in the free energies of unfolding of different derivatives of α-LA that have the same number of acetylated Lys ε-NH_3^+ groups. (Reproduced with permission from **ref.** ***11***. Copyright 2002 Am. Chem. Soc.)

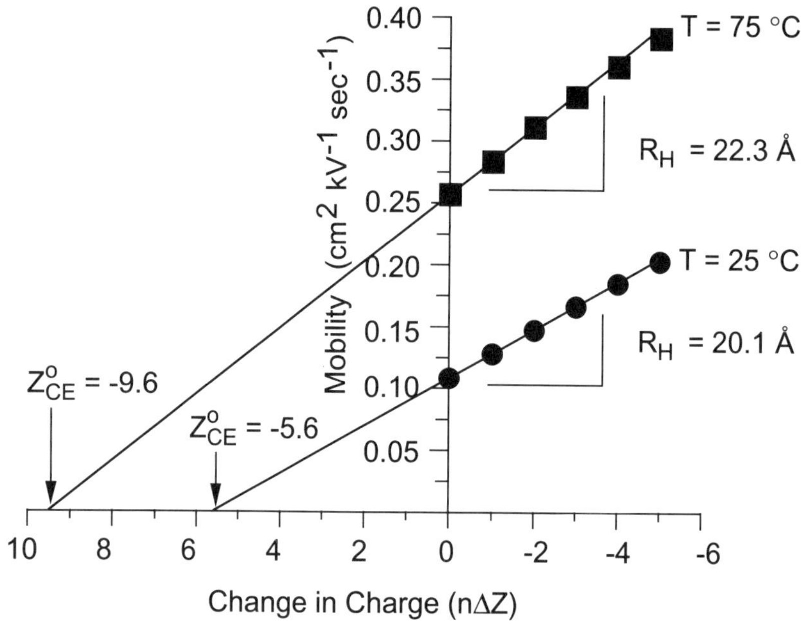

Fig. 6. Determination of the charge $Z^o{}_{CE}$ and hydrodynamic radius R_H of proteins in the folded and denatured state. Values of electrophoretic mobility of the rungs of the charge ladder of holo α-lactalbumin from **Fig. 5** are plotted as a function of $n\Delta Z$. Circles (●) represent the native protein at 25°C, and squares (■) represent the compact denatured state at 75°C. The data are fit to **Eqs. 5** and **9** using linear least-squares analysis: the *x*-intercept gives the net charge of the unmodified protein $Z^o{}_{CE}$ the hydrodynamic radius R_H is determined from the slope of the line. (Reproduced with permission from **ref.** ***11***. Copyright 2002 Am. Chem. Soc.)

folding) (*see* **Fig. 8**). This approach can also be used to monitor chemical denaturation of proteins (e.g., by the addition of urea to the electrophoresis buffer) ***(11)***.

5.2. Methods

5.2.1. Protein–Ligand Binding (Affinity Capillary Electrophoresis [8])

1. The mobility of the protein or the rungs of the protein charge ladder is measured in the electrophoresis buffer to obtain μ_A, the mobility of the receptor in the absence of ligand (*see* **Note 13**).
2. A known concentration of ligand C_L is added to the electrophoresis buffer, and the mobility of the protein sample is measured in this new buffer.
3. Buffers with increasing concentration of ligand are used until the addition of ligand no longer causes the mobility of the protein to change. The mobility in the presence of this saturating concentration of ligand is μ_B.

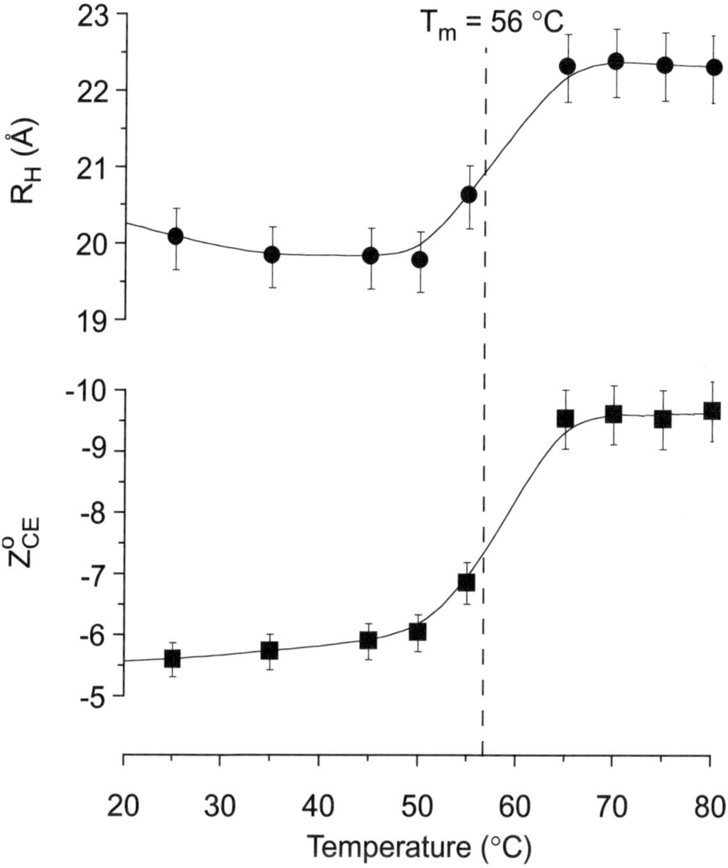

Fig. 7. Values of hydrodynamic radius R_H and net charge $Z^o{}_{\mathrm{CE}}$ of holo α-lactalbumin determined at different temperatures from analysis of the data in **Fig. 5**. The dashed line indicates the melting temperature of this protein 56°C. (Reproduced with permission from **ref. *11***. Copyright 2002 Am. Chem. Soc.)

4. From the mobility of the sample at each concentration of ligand, the fractional occupancy in the bound state θ is determined using **Eq. 12**.
5. Protein–ligand binding is characterized by an equilibrium constant K_b (in this case, expressed as the equilibrium dissociation constant) that relates the relative concentration of the different states of the protein.

$$K_b = \frac{C_A C_L}{C_B} \tag{13}$$

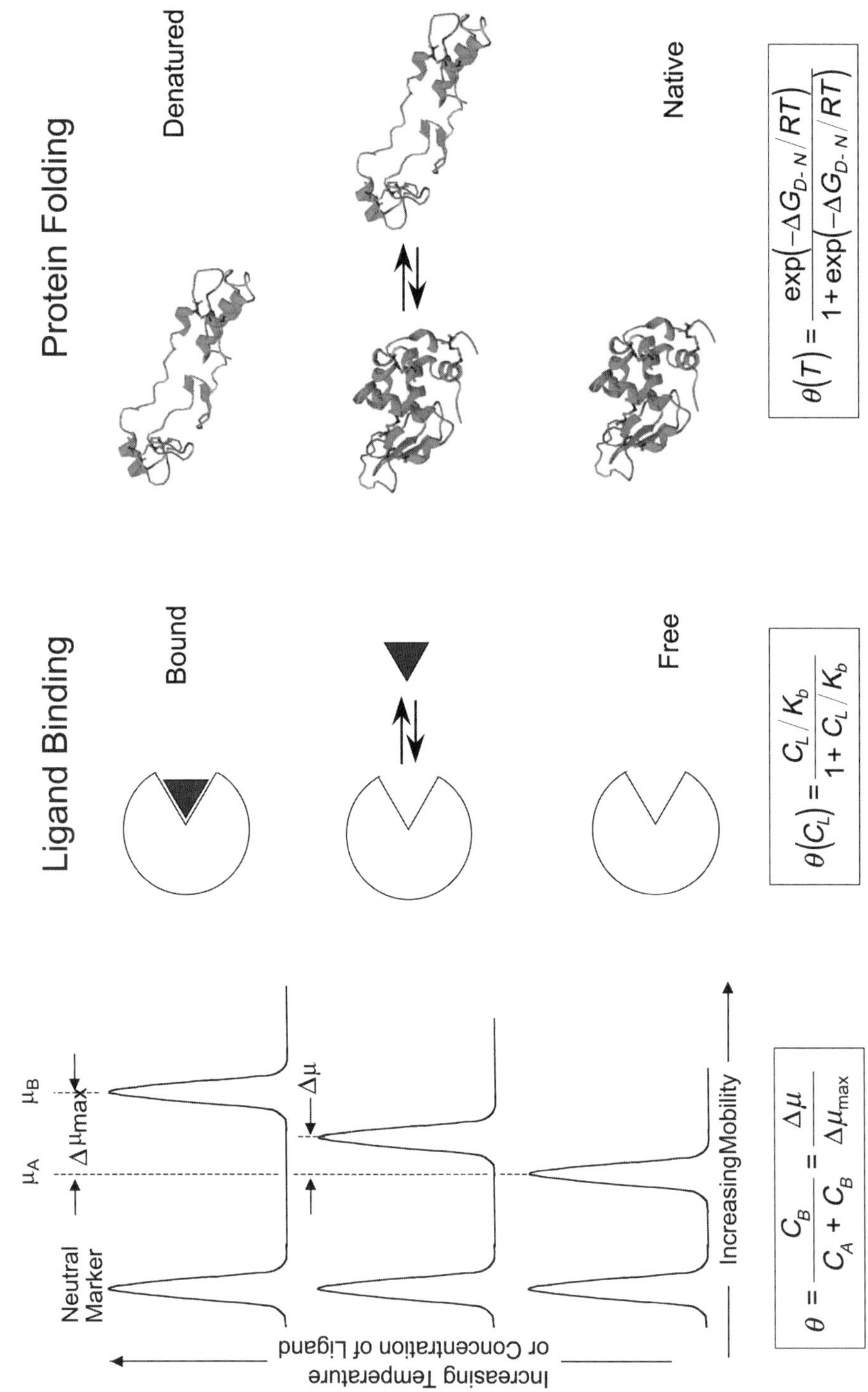
Protein Folding
Denatured
Native
$\theta(T) = \frac{\exp(-\Delta G_{D-N}/RT)}{1+\exp(-\Delta G_{D-N}/RT)}$
Ligand Binding
Bound
Free
$\theta(C_L) = \frac{C_L/K_b}{1+C_L/K_b}$
Neutral Marker
μ_A
μ_B
$\Delta\mu_{max}$
$\Delta\mu$
Increasing Mobility
Increasing Temperature or Concentration of Ligand
$\theta = \frac{C_B}{C_A + C_B} = \frac{\Delta\mu}{\Delta\mu_{max}}$

6. Combination of **Eqs. 12** and **13** gives **Eq. 14**, which can be used to determine values of K_b by fitting the nonlinear binding isotherm (θ vs C_L) to **Eq.14** (e.g., using Kalediagraph from Synergy Software) with K_b as the only adjustable parameter (*see* **Note 14**). The isotherm can also be used to obtain the value of K_b graphically; it is the concentration of ligand for which $\theta = 1/2$.

$$\theta = \frac{C_B/C_A}{1 + C_B/C_A} = \frac{C_L/K_b}{1 + C_L/K_b} \tag{14}$$

7. From the dissociation constant, a standard state free energy of binding ΔG_b^o is determined using **Eq.15**.

$$\Delta G_b^o = -RT \ln(K_b) \tag{15}$$

5.3. Application: Electrostatic Interactions in Ligand Binding

Shifts in mobility are observed upon the binding of ligand to the rungs of a charge ladder (*9,10*). **Figure 9** shows a set of electropherograms of the charge ladder of bovine carbonic anhydrase II (BCAII) measured with different concentrations of the benzene sulfonamide inhibitor **4** (shown in **Fig. 10**) in the electrophoresis buffer *(9)*. As the concentration of this ligand increases, the fraction of the proteins bound to the ligand also increases. Because this ligand carries a net negative charge, the effect of binding is a shift in positions of the peaks in the electropherograms to the right (i.e., to increasing values of mobility; the protein is also negatively charged at pH 8.4, the conditions of these experiments).

The data in **Fig. 9** were analyzed to determine the binding affinity of the rungs of the charge ladder of BCAII for a charged ligand *(9)*. The mobilities of the rungs of the charge ladder measured at 190 μ*M* of the ligand **4** represent the saturated or fully bound state of the proteins and provide values of μ_B for each of the rungs; mobilities measured without ligand provide values of μ_A. Values of mobility measured at intermediate concentrations of ligand are used, together

Fig. 8. (*opposite page*) This cartoon illustrates how CE measures the fractional binding of a ligand or unfolding of a protein. The protein is assumed to equilibrate rapidly between the two states, relative to the time scale of the CE separation (typically 5–10 min). In this way, we assume a two-state model and the value of mobility at some temperature T or concentration of ligand C_L represents the concentration-averaged mobility of the two states (bound and free, or native and denatured). Values of mobility measured at different concentrations of ligand or temperatures are used to determine equilibrium constants for ligand binding or folding, respectively.

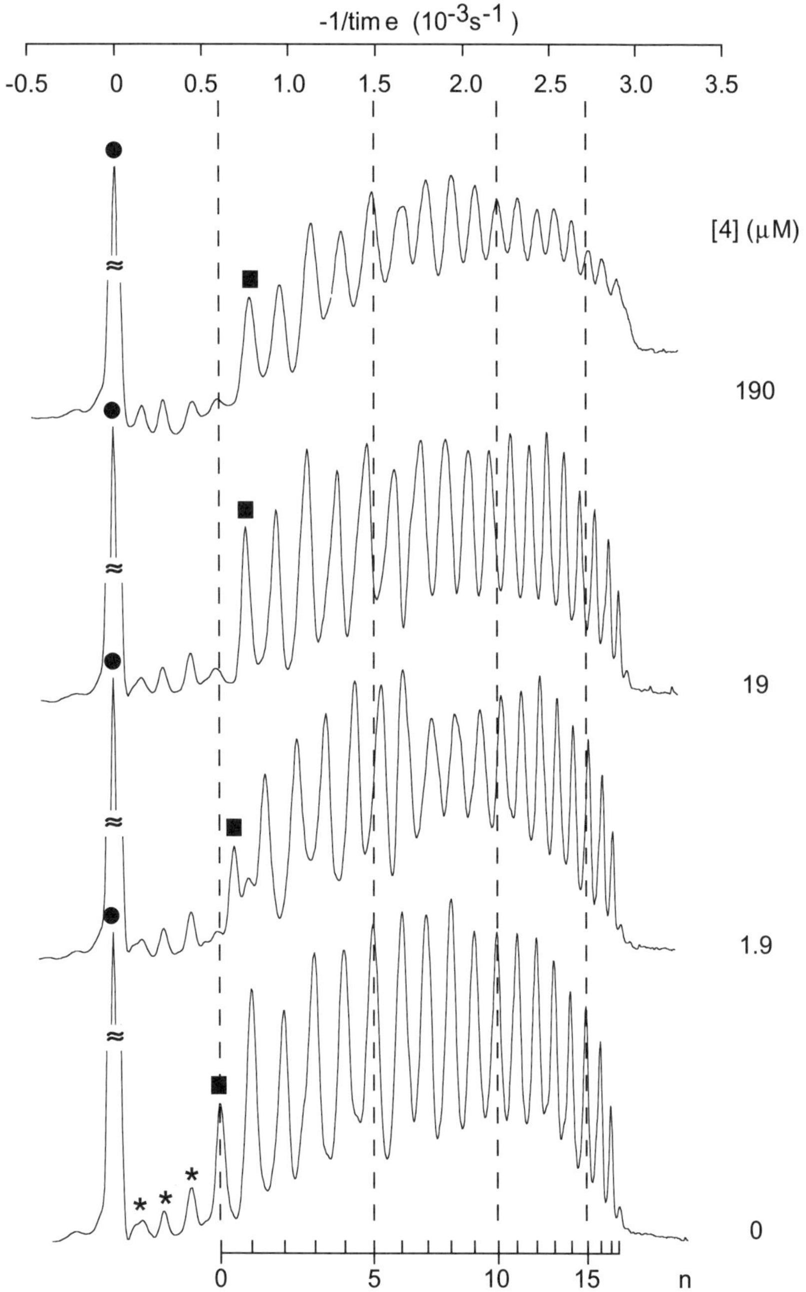
-1/time ($10^{-3}s^{-1}$)
-0.5
0
0.5
1.0
1.5
2.0
2.5
3.0
3.5
[4] (μM)
190
19
1.9
0
*
*
*
0
5
10
15
n

with **Eq. 12**, to determine values of θ vs C_L for each of the rungs of the ladder. Nonlinear least squares analyses of these data using **Eq. 15** provide values of K_b for the binding of this ligand to each of the rungs of the charge ladder.

Figure 10 shows the results of these and similar experiments done with four different ligands–all of them benzene sulfonamides substituted in the paraposition with differently charged or neutral groups–that bind to BCAII *(9)*. Values of the K_b have been converted to standard state free energies of binding (ΔG_b^o) and are plotted as a function of the net charge of the rungs of the charge ladder. These data allow comparison of the binding affinities of unmodified BCAII for each of these different ligands; they also provide a direct measure of the effects of long range electrostatic interactions on the binding affinities. Values of ΔG_b^o for the neutral ligands are approximately independent of the net charge on the protein. These results imply that the modifications to amino acids that produced the charge ladders had little effect on the structure of the active site of this enzyme. For the charged ligands, a linear relationship between the free energy of binding and the net charge of the protein was measured: increasing the net negative charge of the protein resulted in more favorable binding of the positively charged inhibitor **1**, and less favorable binding of the negatively charged inhibitor, **4**.

5.4. Protein Folding

The free energy of protein unfolding by thermal denaturation can also be measured using CE *(7,11)*. The procedure is similar to that for protein–ligand binding. Rather than using different concentrations of ligand, values of μ_{elec} are measured at different temperatures of the electrophoresis buffer.

1. The mobility of the protein or the rungs of the protein charge ladder is measured at 25°C to obtain μ_A, the mobility of the folded protein.
2. The coolant temperature is increased (either using the internal system of the CE machine or an external water bath attached to the internal system) and the mobility is measured; this measured mobility has to be corrected for changes in viscosity (*see* **Note 12**).

Fig. 9. (*opposite page*) Electropherograms demonstrating the changes in the electrophoretic mobility of the rungs of the charge ladder of bovine carbonic anhydrase II caused by the binding of ligand **4** (*see* **Fig. 10**). The peaks marked with a filled circle (●) are neutral marker and the peaks marked with a filled square (■) are the native protein. The peaks marked with an asterisk (*) are impurities. The number of acetylated Lys ε-amino groups (*n*) is indicated below the electropherogram. The-1/time scale shown on top of the figure applies to all the electropherograms. (Reproduced with permission from **ref.** *5*. Copyright 1998 Am. Chem. Soc.)

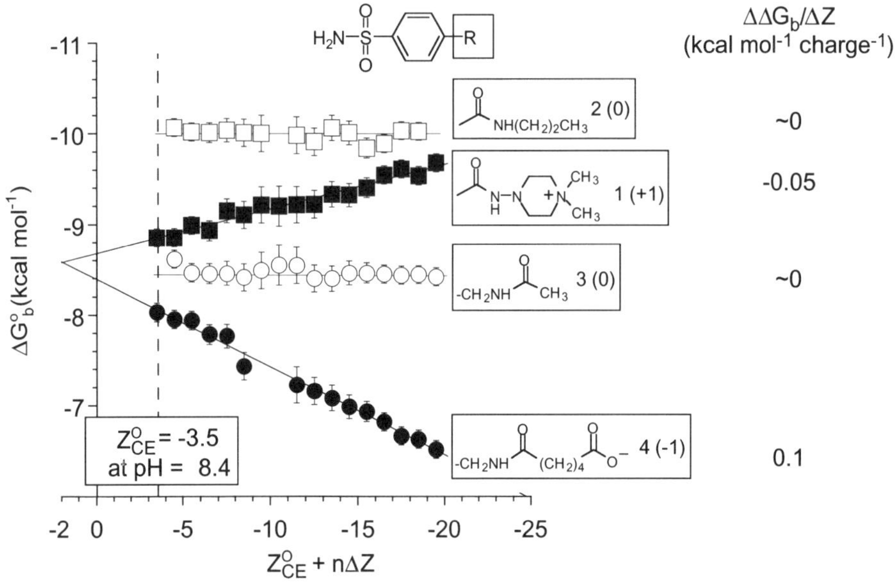

Fig. 10. Dependence of the standard state free energy of binding ($\Delta G^o{}_b$) on the net charge ($Z^n{}_{CE}$) of the rungs of the charge ladder of bovine carbonic anhydrase II, and on the charge on the ligands. The binding affinity of each rung of the charge ladder to ligands **1–3** was measured by CE in 25 mM Tris-HCl-192 m*M* Gly (pH 8.4). The slopes ($\Delta\Delta G^o{}_b/\Delta Z^n{}_{CE}$) from the linear regression analyses of ΔG_b vs $Z^n{}_{CE}$ yielded the magnitudes of influence of charges on CAII-ligand interactions. (Reproduced with permission from **ref. 5**. Copyright 1998 Am. Chem. Soc.)

3. The temperature is increased until further increases no longer change the mobility; this mobility μ_B, is the mobility of the denatured state.
4. The appropriate equilibrium constant for folding is given by **Eq. 16**, where A and B refer to the folded and denatured protein, respectively, (*see* **Fig. 5**).

$$K_{D-N}(T) \equiv \frac{C_B}{C_A} \tag{16}$$

5. Values of electrophoretic mobility measured at different temperatures provide values of θ as a function of **T**, which can be related to the free energy of unfolding $\Delta G_{D\text{-}N}$ using **Eq. 17**.

$$\theta(T) = \frac{\exp\left[-\Delta G^o_{D-N}(T)/RT\right]}{1 + \exp\left[-G^o_{D-N}(T)/RT\right]} \tag{17}$$

Equation 17 cannot be used directly to determine the free energy of protein folding because it is itself a function of temperature as shown in **Eq. 18**. The thermodynamic parameters that are determined are the melting temperature T_m defined as the temperature at which $\Delta G^o_{D\text{–}N} = 0$, and the enthalpy of unfolding at T_m, $\Delta H_{D\text{-}N}(T_m)$ (*see* **Eq. 19**).

$$\Delta G^o_{D\text{–}N} = \Delta H_{D\text{–}N} - T\Delta S_{D\text{–}N} \tag{18}$$

$$\Delta G^o_{D\text{–}N}(T_m) = 0 = \Delta H_{D\text{–}N}(T_m) - T_m\,\Delta S_{D\text{–}N}(T_m) \Rightarrow \Delta S_{D\text{–}N}(T_m) = \frac{\Delta H_{D\text{–}N}(T_m)}{T_m} \tag{19}$$

Protein folding is often characterized by a change in heat capacity ΔC_p which, in turn, gives rise to a temperature dependence of $\Delta H_{D\text{-}N}$ and $\Delta S_{D\text{-}N}$, as shown in **Eqs. 20** and **21**.

$$\Delta H_{D\text{–}N}(T) = \Delta H_{D\text{–}N}(T_m) + \Delta C_p\,(T_m/T) \tag{20}$$

$$\Delta S_{D\text{–}N}(T) = \Delta S_{D\text{–}N}(T_m) + \Delta C_p \ln(T/T_m) = \frac{\Delta H_{D\text{–}N}(T_m)}{T_m} + \Delta C_p\,\ln\,(T/T_m) \tag{21}$$

Combining these expressions with **Eqs. 17** and **18** gives **Eq. 22**, an expression that can be used to determine the thermodynamics of protein folding.

$$\theta(T) = \frac{\exp\left[\Delta H^o_{D\text{–}N}(T_m)\left(1 - T/T_m\right) - T\Delta C_p\left(1 - T_m/T + \ln T/T_m\right)\right]}{1 + \exp\left[\Delta H_{D\text{–}N}(T_m)\left(1 - T/T_m\right) - T\Delta C_p\left(1 - T_m/T + \ln T/T_m\right)\right]} \tag{22}$$

6. Fitting **Eq. 22** to values of θ vs T using a nonlinear least squares analysis provides the van't Hoff enthalpy $\Delta H_{D\text{-}N}(T_m)$, the melting temperature T_m and the change in heat capacity ΔC_p of the protein upon unfolding.
7. **Equations 18**, **20**, and **21** are used to determine the free energy of unfolding at a specific temperature, typically at 25°C.

5.5. Application: Electrostatics in Protein Folding

To quantify the effects of long-range electrostatics on the energetics of folding, we determined $\Delta G^o_{D\text{-}N}$ as a function of charge for α-lactalbumin. From the data in **Fig. 5**, we determined the fraction of unfolded protein as a function of temperature for members of charge ladders of α-LA. Analysis of these data using **Eq. 22** yielded values of T_M and $\Delta G^o_{D\text{–}N}$ for the members of the charge ladders ***(11)***. Values of ΔC_p for this protein are small and were assumed equal to zero in fitting **Eq. 22** to the data.

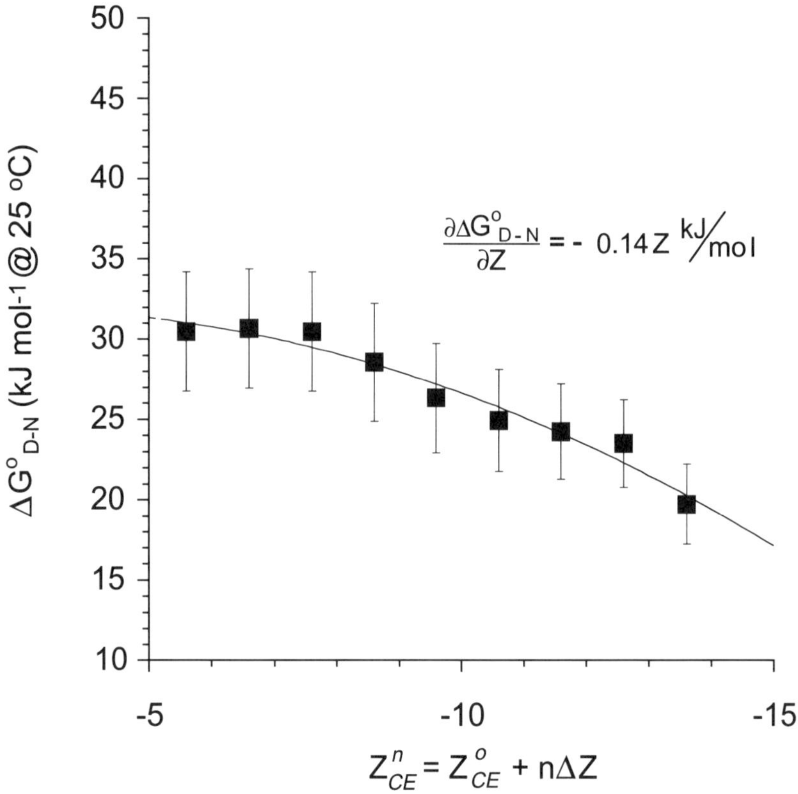

Fig. 11. Dependence of free energy of unfolding ($\Delta G^o_{D\text{-}N}$) at 25°C on net charge (Z^n_{CE}) of the rungs of charge ladders of α-lactalbumin. The curve is a fit of the data to a quadratic equation; the first derivative of this equation gives the differential dependence of $\Delta G^o_{D\text{-}N}$ on net charge: $\partial \Delta G^o_{D\text{-}N}/\partial Z = 0.14Z$ kJ/mol. (Adapted from **ref. *11*.**)

Figure 11 compares values of ΔG^o_{D-N} for the members of ladders of holo α-LA as a function of charge of the folded state ***(11)***. The line is a fit of a quadratic equation to the data; this fit shows that ΔG^o_{D-N} correlates with the square of net charge of the members of the charge ladder. This quadratic dependence of ΔG^o_{D-N} on charge is consistent with electrostatic interactions that act between charges on the protein. A simple estimate of contributions of electrostatic interactions to the energetics of protein folding is the energy required to distribute the net charge evenly on the surface of a sphere; this charging free energy is proportional to the square of net charge, consistent with the data for holo α-LA. The differential dependence of ΔG^o_{D-N} on net charge for holo α-LA,

$\partial\Delta G^{o}_{D\text{-}N}/\partial Z = -0.14Z$ kJ/mol per unit of charge, was determined from the first derivative of curves fit to $\Delta G^{o}_{D\text{-}N}$ vs Z. This differential provides a direct measure of the average effects of long-range electrostatic interactions to the free energy of folding of α-LA.

6. Conclusions

The combination of CE and protein charge ladders is a convenient tool for measuring the properties and interactions of proteins in solution under conditions that are similar to those found in real biological systems. Charge ladders can be used to measure the charge and hydrodynamic size of a protein in a single experiment under nondenaturing condition. Because CE is sensitive to the size and charge of a protein, it can also be used to measure molecular recognition events such as protein–ligand binding and protein folding. Charge ladders and CE allow for the isolation of charge as the independent variable in a system, and thereby allow the direct measure of the effect of charge on the energetics of molecular recognition events involving proteins. CE and charge ladders therefore provide separation and classification of proteins on the basis of charge and size, as well as quantification of stability or binding affinity of proteins as a function of net charge. All this information is obtained in a single set of electrophoresis experiments.

7. Notes

1. The assumption that the hydrodynamic drag or coefficient of friction of proteins is constant across the rungs of a ladder is justified when using modifying agents of low molecular weight, relative to that of the protein. As the molecular weight of the modifying agent increases, this assumption becomes less valid ***(4)***.
2. We assume that the value of ΔZ of a protein modified by acetylation of Lys ε-amino groups is –1 in Tris-Gly buffer. The cooperative nature of proton binding to amino acids on proteins can result in values of ΔZ other than –1. This effect has been estimated using a mean field theory ***(17)***. By measuring the pK_a of the *N*-terminal amino acid of lysozyme across the rungs of the ladder, we used this group as a thermodynamic "reporter" of the effects of acetylation on the binding affinity of other amino acids for protons. The value of pK_a varied from 7.8 to 7.5 from the native to the fully acetylated state, owing to the cooperativity amongst ionizable groups on this protein. We found that the addition of 100 m*M* NaCl caused the variations in values of pK_a across the rungs of the ladder to disappear, indicating that the addition of salt attenuates the effects of cooperativity in ionization on values of ΔZ ***(18)***.
3. Proteins are used as received; because CE naturally separates different molecular species, the presence of impurities does not usually have a significant impact on the analysis.

4. Excessive mixing can cause proteins to denature at the air-water interface. Typically, samples less than 1 mL in volume are mixed by repeated pipeting.
5. Protein charge ladders produced by acylation are typically analyzed directly by CE without further purification. The amidation of carboxyl groups produces byproducts that can interfere with the analysis of charge ladders. Furthermore, it may be desirable to exchange the protein charge ladder into a buffered solution. We have found Sephadex gel purification columns (NICK spin columns, Pharmacia) to be useful for purification and buffer exchange, following the protocol provided with the columns. This purification can cause a two- to tenfold dilution of the product.
6. This buffer (which has a value of pH of 8.4 at 25°C) contains approx 10 m*M* of the Tris-HCl as a cation; most of the Gly is, therefore, present as a zwitterion. This buffer is desirable for CE separations because it has a low conductivity, which results in minimal Joule heating at field strengths as high as 1 kV/cm. Additionally, zwitterions can reduce protein adsorption to the capillary wall and unwanted protein–protein interactions ***(19)***.
7. Spurious peaks and variations in baseline UV intensity may be observed during the first one or two separations performed on a capillary coated with polybrene, but should then disappear. It is suggested that these first runs be made with an injection that contains only the neutral marker. Doing so provides a measure of the stability of coating; a migration time that does not vary from run-to-run reflects a stable coating.
8. The product of the applied voltage and current should be monitored to determine the total power input to the capillary. The amount of Joule heating, which results in an increase in the internal temperature of the capillary, is directly proportional to this power.
9. Charge ladders that have been labeled fluorescently (e.g., with fluorescein) have also been separated by CE and detected by laser-induced fluorescence. Doing so allows increased sensitivity of detection at lower concentration of protein, relative to detection by UV absorbance.
10. The effects of ion polarization and relaxation on the electrophoretic mobility of proteins have been quantified using a combination of experimental data and electrokinetic models of colloids ***(6)***. We concluded that such effects are only important when a nonlinear correlation between mobility and charge is observed; such a correlation corresponds to electrostatic potentials at the surface of proteins that exceed 25 mV.
11. Uncertainty in the value of ΔZ results in similar uncertainties in Z^o_{CE} and R_H (i.e., a 10% error in the value of ΔZ results in ~10% error in Z^o_{CE} and R_H) ***(18)***.
12. Because values of electrophoretic mobility are sensitive to changes in the viscosity of the solution, we have corrected these data for changes in viscosity caused by increasing temperature by monitoring the change in current with temperature *(20)*. In this way, shifts in the position of peaks reflect only changes in the charge and size of the protein with increasing temperature.

13. The definition of receptor and ligand is arbitrary; either member of an interacting pair of molecules can be defined as the receptor. A smaller quantity of the receptor is required, relative to the ligand, and its concentration need not be known for analysis by ACE. These considerations dictate that the molecule that is more precious be used as the receptor and the less-precious molecule used as the ligand.
14. Scatchard analysis can also be performed to determine the dissociation constant K_b; however, because this analysis does not weight values of θ at different concentrations equally, it is susceptible to error in the determination of values of K_b.

Acknowledgment

The authors thank the National Science Foundation (Contract CTS-0134429), and the Camille and Henry Dreyfus Foundation for financial support.

References

1. Collins, F. S., Patrinos, A., Jordan, E., Chakravarti, A., Gesteland, R., and Walters, L. (1998) New Goals for the U.S. Human Genome Project: 1998-2003. *Science* **282,** 682–689.
2. Creighton, T. E. (1980) Counting integral numbers of amino acid residues per polypeptide chain. *Nature* **284,** 487–489.
3. Gao, J., Gomez, F. A., Härter, R., and Whitesides, G. M. (1994) Determination of the effective charge of a protein in solution by capillary electrophoresis. *Proc. Natl. Acad. Sci. USA* **91,** 12,027–12,030.
4. Colton, I. J., Anderson, J. R., Gao, J. M., Chapman, R. G., Isaacs, L., and Whitesides, G. M. (1997) Formation of protein charge ladders by acylation of amino groups on proteins. *J. Amer. Chem. Soc.* **119,** 12,701–12,709.
5. Carbeck, J. D., Colton, I. J., Gao, J. M., and Whitesides, G. M. (1998) Protein charge ladders, capillary electrophoresis, and the role of electrostatics in biomolecular recognition. *Acc. Chem. Res.* **31,** 343–350.
6. Carbeck, J. D. and Negin, R. S. (2001) Measuring the size and charge of proteins using protein charge ladders, capillary electrophoresis, and electrokinetic models of colloids. *J. Amer. Chem. Soc.* **123,** 1252–1253.
7. Hilser, V. J. and Freire, E. (1995) Quantitative analysis of conformational equilibrium using capillary electrophoresis: applications to protein folding. *Anal. Biochem.* **224,** 465–485.
8. Colton, I. J., Carbeck, J. D., Rao, J., and Whitesides, G. M. (1998) Affinity capillary electrophoresis. A physical-organic tool for studying interactions in biomolecular recognition. *Electrophoresis* **19,** 367–382.
9. Gao, J. M., Mammen, M., and Whitesides, G. M. (1996) Evaluating electrostatic contributions to binding with the use of protein charge ladders. *Science* **272,** 535–537.
10. Caravella, J. A., Carbeck, J. D., Duffy, D. C., Whitesides, G. M., and Tidor, B. (1999) Long-range electrostatic contributions to protein-ligand binding estimated using protein charge ladders, affinity capillary electrophoresis, and continuum electrostatic theory. *J. Amer. Chem. Soc.* **121,** 4340–4347.

11. Negin, R. S. and Carbeck, J. D. (2002) Measurement of electrostatic interactions in protein folding with the use of protein charge ladders. *J. Amer. Chem. Soc.* **124,** 2911–2916.
12. Carbeck, J. D., Colton, I. J., Anderson, J. R., Deutch, J. M., and Whitesides, G. M. (1999) Correlations between the charge of proteins and the number of ionizable groups they incorporate: studies using protein charge ladders, capillary electrophoresis, and Debye-Hückel theory. *J. Amer. Chem. Soc.* **121,** 10,671–10,679.
13. Grossman, P. D. (1992) in *Capillary Electrophoresis: Theory and Practice* (Colburn, J. C., ed.), Academic, San Diego, CA.
14. Cordova, E., Gao, J. M., and Whitesides, G. M. (1997) Noncovalent polycationic coatings for capillaries in capillary electrophoresis of proteins. *Anal. Chem.* **69,** 1370–1379.
15. Henry, D. C. (1931) The cataphoresis of suspended particles. I. The equation of cataphoresis. *Proc. Royal Soc. London* **123,** 106–129.
16. Menon, M. K. and Zydney, A. L. (2001) Protein charge ladders: a new technique for studying electrostatic interactions in ultrafiltration systems. *J. Membr. Sci.* **181,** 179–184.
17. Menon, M. K. and Zydney, A. L. (2000) Determination of effective protein charge by capillary electrophoresis: effects of charge regulation in the analysis of charge ladders. *Anal. Chem.* **72,** 5714–5717.
18. Sharma, U., Negin, R. S., and Carbeck, J. D. (2003) Effects of cooperativity in proton binding on the net charge of proteins in charge ladders. *J. Phys. Chem. B* **107**, 4653–4666.
19. Basak, S. K., Velayudhan, A., and Ladisch, M. R. (1994) Characterization of buffers for electrokinetic separations. *Appl. Biochem. Biotechnol.* **44,** 243–261.
20. Ishihama, Y., Oda, Y., Asakawa, N., and Iwakura, M. (1997) Nano-scale monitoring of the thermally-induced unfolding of proteins using capillary electrophoresis with in-column incubation. *Anal. Sciences* **13,** 931–938.

11

Frontal Analysis Continuous Capillary Electrophoresis for Protein–Polyelectrolyte Binding Studies

Emek Seyrek, Toshiaki Hattori, and Paul L. Dubin

Summary

A novel technique, frontal analysis continuous capillary electrophoresis (FACCE), has been described as an effective way to study protein–polyelectrolyte binding. FACCE involves continuous sampling, integrating sample injection and separation into one process that provides advantages over conventional frontal chromatography. The method provides rapid and precise determination of binding isotherms, and allows for quantitative binding analysis in terms of binding constant and the binding-site size by considering the protein as the ligand and allowing the polyelectrolyte to bind to a number of proteins with variable levels of cooperativity. FACCE is particularly suitable for binding systems involving rapid binding kinetics because it allows for the determination of the concentrations of free or bound ligands under conditions that avoid perturbation of the binding equilibrium. This chapter focuses on studies of the binding of bovine serum albumin (BSA) to heparin using FACCE. These investigations are demonstrated within the context of this chapter as representative of a model protein–polyelectrolyte system from which extensions to other systems can be made.

Key Words

Binding constant; binding isotherm; binding site size; bovine serum albumin; capillary electrophoresis; heparin; patch binding; protein–polyelectrolyte.

1. Introduction

Capillary electrophoresis (CE) has been used increasingly for protein–ligand binding studies, such as the association of proteins with drugs *(1–4)*, inorganic ions *(5,6)*, sugars *(7)* and micelles *(8,9)*. The CE techniques applied involve either the measurement of mobility or the determination of the concentrations of free or bound ligand *(1–3,10)*. The main problems observed in binding studies based on those CE techniques are: (1) the difficulty in calibration for quan-

From: *Methods in Molecular Biology, vol. 276: Capillary Electrophoresis of Proteins and Peptides*
Edited by: M. A. Strege and A. L. Lagu © Humana Press Inc., Totowa, NJ

titative studies, and (2) the perturbation of the binding equilibria caused by the dynamic behavior of the systems under study. A novel method that addresses these problems, frontal analysis continuous capillary electrophoresis (FACCE) *(11,12)*, has been developed for the study of protein–polyelectrolyte (PE) binding.

Because both biological and synthetic polyelectrolytes have contour lengths that are large compared to the polyelectrolyte-binding sites on the protein surface, it is necessary to view the protein in these cases as the ligand, one polyelectrolyte capable of binding a number of proteins. There is no fundamental problem in discriminating between bound and free proteins, and there is a sound theoretical framework for the analysis of such macromolecular binding *(13)*. Experimentally, the binding of proteins to linear macromolecules has been studied by turbidimetry, light scattering, electrophoretic mobility, viscometry, fluorescence, potentiometric titration, and dialysis equilibrium *(14)*. However, only the last directly yields binding isotherms, and it is unacceptably slow. Thus, FACCE addresses the need for rapid and precise determination of such binding isotherms.

In contrast to conventional frontal chromatography (CFC), FACCE combines continuous sampling, integrating sample injection and separation into one process as illustrated in **Fig. 1**. The electroosmotic flow that transports all components toward the cathode arises from the negatively charged wall of the fused-silica capillary. CFC employs regular chromatography methods but with a sample volume relatively large compared to the capacity of the separation column. Careful selection of the amount of sample injected, the flow rate of the mobile phase, and the length of the separation column, leads to continuous but distinct plateaus, followed by resumption of detector baseline response as shown in the electropherograms in **Fig. 1A**. In FACCE, the capillary is filled and equilibrated with the run buffer prior to sample introduction. The inlet end of the capillary is then immersed in the sample vial and a voltage is applied across the capillary to initiate the sample introduction and separation process. Species separated by electrophoresis appear as discrete and progressive plateaus in the electropherograms as shown in **Fig. 1B**. The separation profile of FACCE is particularly suitable for the study of binding in systems with rapid binding kinetics, inasmuch as it allows for the determination of the concentrations of free or bound ligands without complete separation avoiding perturbation of the binding equilibrium.

FACCE is particularly effective in multiple complexation equilibria where more than one protein binds to a single ligand; measurement of the concentration of free ligand is determined not from mobility but from the peak height which directly indicates the free protein concentration. The stoichiometric relationship between bound protein and the protein–ligand complex can then be fit to appropriate binding isotherms to yield binding constants and the bind-

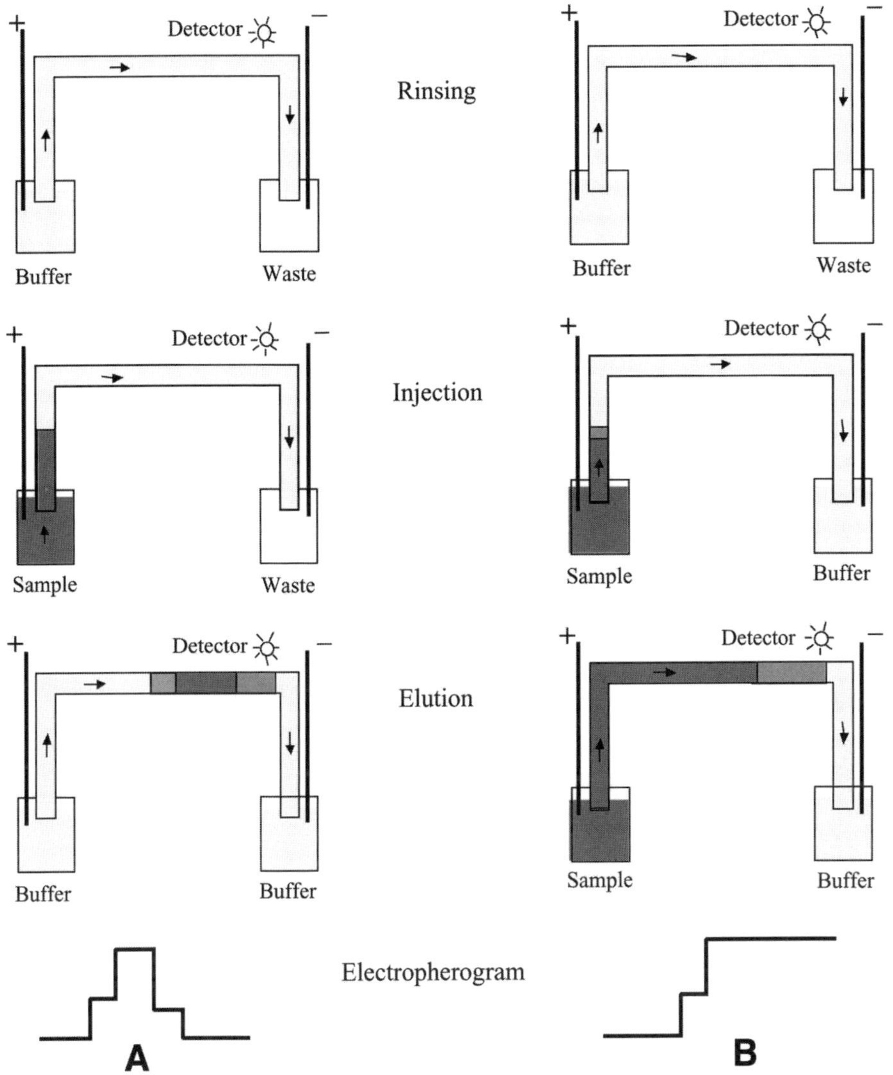

Fig. 1. Schematics of CFC (**A**) and FACCE (**B**). Reprinted with permission from **ref.** ***12***.

ing-site size, both of which are parameters essential to the elucidation of the binding mechanism. Binding of proteins with both synthetic and biological polyelectrolytes is an example of multiple complexation and has, therefore, been studied with FACCE ***(11–25)***. In this paper, analysis by FACCE of the binding of bovine serum albumin (BSA) and heparin will be discussed as an example. Because heparin is a negatively charged polyelectrolyte, it interacts with the positive charges on the protein. The positive charges of the protein increase with decreasing pH, and the binding force increases. However, binding often occurs at pH above the isolecetric point (p*I*) of the protein where the protein bears the same net charge as the polyelectrolyte. For example, the p*I* of BSA is 4.9 and binding at $I = 0.01$ *M* occurs at pH 7.0, where both the protein and the polymer bear net-negative charges. This is an indication of "patch binding" in which the electrostatic attraction between polyanion and a local protein positive region overcomes the repulsion between polyanion and the global protein charge ***(15)***.

Many functions have been ascribed to the interaction of heparin with various proteins ***(26)***. Although no specific function has been identified with the heparin–BSA interaction, this system has been used to develop the methodology that would enable application of the technique to, for example, heparin–protein cognate pairs, DNA-binding proteins, and other biological polyelectrolyte–protein systems. The extended use of FACCE to such protein–polyelectrolyte systems will facilitate a better understanding of many biological phenomena.

2. Materials

1. Heparin (sodium salt, porcine intestinal mucosa, Calbiochem, La Jolla, CA, nominal M_r 13,500–15,000).
2. BSA (fatty acid free, M_r 68,000, Boehringer Mannheim, Indianapolis, IN).
3. For turbidimetric titrations: Brinkmann PC800 probe colorimeter detecting at 420 nm, equipped with a 1-cm path-length fiber optics probe, and a pH meter.
4. FACCE rinsing and run buffer: phosphate buffer prepared at desired pH and ionic strength.
5. For FACCE rinsing: 1 *N* NaOH before each use of capillary column, 0.1 *N* NaOH for rinsing between consecutive runs.
6. FACCE instrument: P/ACE 5500 CE (Beckman, Fullerton, CA).
7. Fused-silica capillary of dimensions 50 μm × 27 cm (Polymicro Technologies Inc, Phoenix, AZ).
8. Milli-Q water for all buffer and solution preparations (Millipore, Milford, MA).
9. All protein and PE solutions should be prepared fresh, and complete solubilization should be achieved prior to experiments.

3. Methods

3.1. Identification of Soluble Complex Region for Protein–Polyelectrolyte System

Binding equilibria in any protein–PE system can only be studied under conditions corresponding to the formation of the complex without second-order reactions, such as aggregation or phase separation. Any protein–PE system can be identified in terms of one of these three states. For a polyanion–protein system at fixed ionic strength, the system progresses, upon decrease in pH, from: (a) noninteracting solution, to (b) soluble complex phase, and finally to (c) complex aggregate or biphasic state ***(27–29)***. Turbidity is sufficiently sensitive to detect the two relevant transitions; therefore, these three regions can be effectively identified by pH-turbidimetric titration. Titrations should be done at various ionic strengths to enable construction of a phase boundary: a plot of transition pH vs ionic strength, which shows the three regions corresponding to (a), (b), and (c) states. This will allow for determination of the working conditions for FACCE experiments, where the pH and ionic strength of the solution should lie in the complex formation region (b).

3.1.1. Preparation of Protein–Polyelectrolyte Solutions for pH-Turbidimetric Titration

1. The concentrations of protein and PE solutions to be prepared should be determined such that protein is in excess when they are mixed. A weight ratio of protein to PE of 10 would be appropriate for a typical titration.
2. Dissolve the protein and PE separately in desired salt solution.
3. Mix appropriate amounts of protein and PE solutions to achieve the mixture with the desired weight ratio.
4. Prepare a blank protein solution for blank titration that has the protein concentration in the mixture.
5. Add 1 *M* NaOH gradually to the mixture to adjust the pH to 10.0.
6. As an example, 4 mg/mL of BSA and 0.4 mg/mL of heparin solutions were prepared in 0.01 *M* NaCl solution, and 10 mL of both were combined to obtain a final mixture with weight ratio of 10, making a total volume of 20 mL. (The required amount of sample volume depends on the size of the probe and the container.) A "blank" protein solution was 4 mg/mL BSA in the same NaCl solution.

3.1.2. pH-Turbidimetric Titration

The soluble complex region lies between pH at the initial point of increasing turbidity (pHc) and pH at the point of the abrupt increment of turbidity (pHϕ) ***(14,15)***.

1. Titrate the protein–PE solution with 1 *M* HCl using, e.g., Gilmont microburet, while monitoring simultaneously pH and % transmittance (%T) at 420 nm.
2. Plot 100-%T, proportional to turbidity vs pH, and identify pHc and pHϕ.
3. A sample plot for heparin and BSA at $I = 0.01$ *M* is shown in **Fig. 2**. The soluble complex exists in region 2 which lies between pHc (7.1) and pHϕ (5.0).

3.2. CE

3.2.1. Equipment

CE is performed using a Beckman P/ACE 5500 CE with programmatic autosampling. Operating temperature is at 25°C. The dimensions of the fused-silica capillary are 50 μm × 27 cm with an effective length (the distance from the inlet end to the UV detector) of 20 cm.

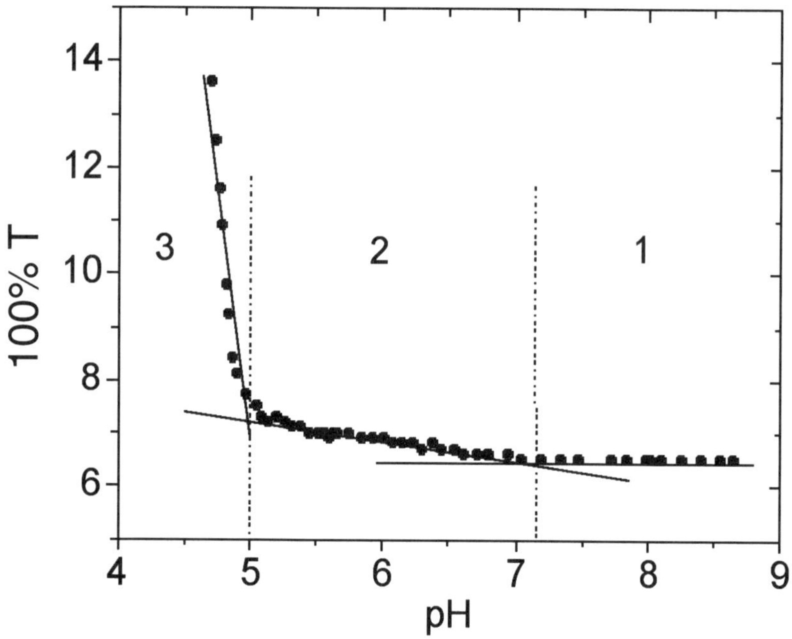

Fig. 2. Turbidimetric titration curve for 1 g/L BSA and 1 g/L heparin at I=0.01 M, Region 1, no complex formation as turbidity does not change. Region 2, complex formation as turbidity starts to increase gradually. Region 3, phase separation as there is an abrupt change in turbidity. The dashed lines represent the defined pHc and pHϕ values dividing regions 1, 2, and 3. Reprinted with permission from **ref. *19***.

3.2.2. Procedure for FACCE

Each measurement follows the procedure given below.

1. Wash the capillary with 0.1 *M* NaOH solution for 5 min. (This rinsing is necessary between each run to remove any adsorbed protein on the capillary surface. *See* **Note 1**.)
2. Rinse the capillary with water for 5 min.
3. Run phosphate buffer solution of desired pH and ionic strength through the capillary for 5 min.
4. Immerse the negative pole tip of the capillary into the sample and the positive pole tip into the buffer solution.
5. Apply a constant voltage and monitor UV absorbed spectra at a certain wavelength. The operating voltage and the wavelength should be determined so that the best resolution is obtained for the desired components to be measured. The typical applied voltage range is 5–15 kV. For most proteins, 200 or 214 nm would be the best detection wavelength.
6. As an example, the electropherograms shown in **Fig. 3** was obtained by applying a voltage of 10 kV at 214-nm wavelength.

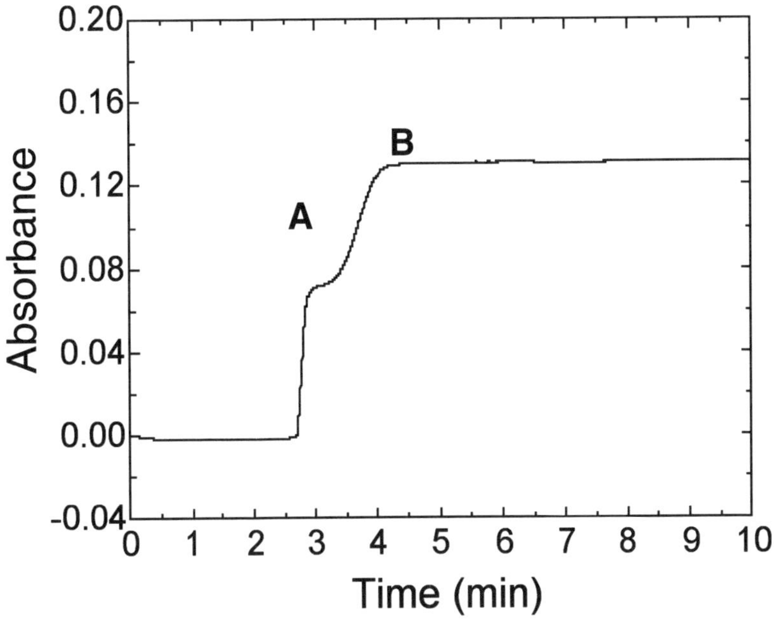

Fig. 3. Electropherogram of a sample FACCE experiment with BSA (4 g/L) and heparin (0.2 g/L) mixture at pH=6.8, *I*=0.01 *M* with plateaus representing (a) free BSA, (b) BSA–heparin complex.

3.3. Determination of Free–Protein Concentration

The electropherogram of FACCE shows multiple plateaus, the number of which depends on the number of components in the mixture. In a protein–PE mixture, two plateaus are typically observed (*see* **Fig. 3**), the first plateau corresponding to free protein and the second to protein–PE complex. The concentration of free protein can be determined by the height of the first plateau using a calibration curve ***(12)***.

3.3.1. Calibration Curve for Free Protein

1. Solutions of protein from 0.1 to 1.5 mg/mL are prepared in NaH_2PO_4–Na_2HPO_4 buffer solution at the desired pH and ionic strength. Measure each sample solution as explained in **Subheading 3.2.2.**
2. Measure the height of the single plateau in the electropherogram at each concentration.
3. Plot absorbance (plateau height) vs concentration of protein.

3.3.2. Determination of Free Protein in a Protein–PE System

1. Make up a series of solutions containing protein and PE dissolved in the appropriate buffer solution. The concentrations needed for the analysis depend on the strength of UV signal. For example, for BSA and heparin, the protein concentration was 0.4–4.0 g/L with heparin at 0.2 g/L.
2. Measure each sample solution as explained in **Subheading 3.2.2**.
3. Determine the concentration of free protein by absorbance of first plateau height using the calibration curve (*see* **Note 2**).

3.4. Data Analysis

There are several procedures for fitting binding isotherm data *(30,31)*. Particularly appropriate in the present case is the one based on a binding theory of large ligands to a 1D homogeneous lattice given by McGhee and von-Hippel *(32,33)*. This is appropriate for specific or nonspecific binding involving cooperative or noncooperative interaction between binding sites.

3.4.1. Constructing the Binding Isotherm

The binding isotherms were obtained by plotting the concentration of free protein, L_{free}, which is calculated as explained above vs the average number of bound protein per unit charged group on heparin, ν (*see* **Note 3**) and (**Fig. 4**).

3.4.2. Determination of Binding Parameters Via McGhee and von-Hippel Equation

General equation for McGhee and von Hippel model *(31)* is given in **Eq. 1**:

$$\frac{\nu}{L_{free}} = K_b(1-n\nu)\left[\frac{(2w+1)(1-n\nu)+\nu-R}{2(w-1)(1-n\nu)}\right]^{n-1} \times \left[\frac{1-(n+1)\nu+R}{2(1-n\nu)}\right]^2 \quad (1)$$

with

$$R=\sqrt{[1-(n+1)\nu]^2+4w\nu(1-n\nu)} \quad (2)$$

where ν represents the binding density (in units of moles of bound ligand per mole of total lattice residue), L_{free} is the free-ligand concentration, K_b is the binding constant, n is the number of binding sites, and w is the cooperativity parameter.

For $w=1$, which is the case for noncooperative binding, **Eq. 1** reduces to **Eq. 3** for noninteracting ligands. In the case of BSA–heparin, **Eq. 3** was used for analysis of binding parameter as no additional cooperativity term was needed to obtain a good fit to the binding isotherms.

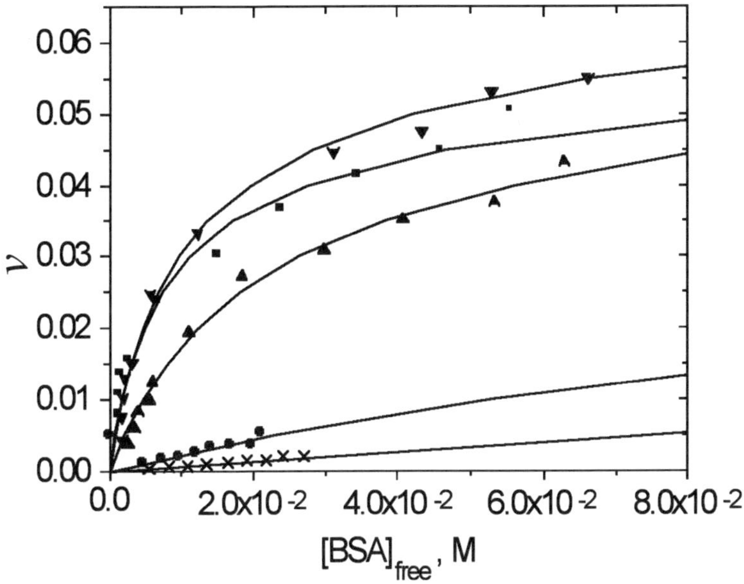

Fig. 4. Binding isotherms for BSA and heparin at ionic strengths of (▼) 2 m*M*, (■) 7 m*M*, (▲) 10 m*M*, (•) 30 m*M*, (×) 50 m*M* in phosphate buffer at pH=6.8; the solid lines are fits to the McGhee von Hippel equation (from **ref. *19***).

$$\frac{\nu}{L_{\text{free}}} = K_b\left(1 - n\nu\right)\left[\frac{1-n\nu}{1-\left(n-1\right)\nu}\right]^{n-1} \tag{3}$$

Equation 3 yields two parameters, the intrinsic binding constant K_b and the binding-site size *n*. The binding isotherms are fitted to **Eq. 3** where ν is the number of bound BSA per ionic site of heparin, L_{free} is the concentration of free BSA, K_{obs} is the observed binding constant, and *n* is the binding site size in number of heparin charge groups (*see* **Note 3**). The nonlinear curve fitting can be carried out by a software such as Origin (Microcal Software, Inc.) to obtain the parameters K_{obs} and *n*. The fitted curves are also shown in **Fig. 4** for BSA and heparin at various ionic strengths. The calculated parameters are given in **Table 1**.

4. Notes

1. The adsorption of protein on the capillary wall is a significant problem. In case of BSA, strong adsorption of protein was observed at pH < 6.5; therefore, the measurements were made above this pH. The adsorption conditions which may differ for each protein may require the use of a coated capillary.

Table 1
Binding Constants and Binding Site Size for BSA–Heparin Interaction Calculated by Nonlinear Curve Fitting of Fig. 4 Data to Eq. 3

Ionic strength (*M*)	log *K*	n (±0.8)
0.05	1.88	12.7
0.03	2.39	13
0.01	3.44	11.4
0.007	3.89	12.9
0.002	3.85	10.8

2. A typical FACCE electropherogram shows two steps; however, more have also been observed ***(28)***. In addition, in some cases, the electropherogram had a spike peak before the first plateau ***(34)***, which may arise from adsorption of protein onto the capillary wall. Although this behavior did not significantly affect determination of free BSA here, the use of a coated capillary might be necessary to reduce adsorption when the effect becomes more significant.
3. Binding density ν is calculated by dividing the bound protein concentration (in mol/L) by the charge of heparin (in eq.mol/L). Because heparin has an equivalent weight of 200 g/mol, a 0.2 g/L heparin solution contains 1.0×10^{-3} eq. mol/L. Binding density ν is defined in this manner because the binding site on heparin does not correspond to a well-defined portion of molecule, which is the typical case for nonspecific binding arising from long range electrostatic forces. The size of the apparent binding site *n* is defined for the same reason in terms of the number of charges it encompasses. An estimate of the number of disaccharide units involved in binding can easily be obtained by dividing *n* by the average number of charges of one disaccharide, 3.7 for heparin ***(15)***. The length of the binding site can then be calculated from the length of a disaccharide unit, 11 Å.

Acknowledgment

Paul L. Dubin acknowledges the support of NSF grant CHE 9987891.

References

1. Kraak, J. C., Busch, S., and Poppe, H. (1992) Study of protein-drug binding using capillary elecrophoresis. *J. Chromatogr.* **608,** 257–264.
2. Ohara, T., Shibukawa, A., and Nakagawa, T. (1992) Capillary electrophoresis/frontal analysis for microanalysis of enantioselective protein binding of a basic drug. *Anal. Chem.* **67,** 3520–3525.
3. Oravcova, J., Sojkova, D., and Linder, W. (1996) Comparison of the Hummel-Dreyer method in high-performance liquid chromatography and capillary electrophoresis conditions for study of the interaction of (RS)-, (R)- and (S)-carvedilol with isolated plasma proteins. *J. Chromatogr. B.* **682,** 349–357.

4. Chu, Y. H. and Whitesides, G. M. (1992) Affinity capillary electrophoresis can simultaneously measure binding constants of multiple peptides to vancomycin. *J. Org. Chem.* **57,** 3524–3526.
5. Heegard, N. H. H. and Robey, F. A. (1993) Capillary electrophoresis-based assay for the binding of Ca^{2+} and phosphorylcholine to human C-reactive protein. *J. Immun. Met.* **166,** 103–110.
6. Kajiwara, H., Hirano, H., and Oono, K. (1991) Binding shift assay of parvalbumin, calmodulin and carbonic anhydrase by high-performance capillary electrophoresis. *J. Biochem. Biophys. Meth.* **22,** 263–268.
7. Honda, H., Taga, A., Suzuki, K, Suzuki, and S., Kakehi, K. (1992) Determination of the association constant of monovalent mode protein-sugar interaction by capillary zone electrophoresis. *J. Chromatogr.* **597,** 377–382.
8. Takeda, K., Sasaoka, H., Sasa, K., Hirai, H., Hachiya, K., and Moriyama, Y. (1992) Size and mobility of sodium dodecyl sulfate-bovine serum albumin complex as studied by dynamic light scattering and electrophoretic light scattering. *J. Coll. Int. Sci.* **154,** 385–392.
9. Karim, M.R., Shinagawa, and S., Takagi, T. (1994) Electrophoretic mobilities of the complexes between sodium dodecyl sulfate and various peptides or proteins determined by free solution electrophoresis using coated capillaries. *Ectrophoresis.* **15,** 1141–1146.
10. Chu, Y. H., Lees, W. J., Stassinopoulos, A, and Walsh C. T. (1994) Using affinity capillary electrophoresis to determine binding stoichiometries of protein–ligand interactions. *Biochem.* **33,** 10,616–10,621.
11. Gao, J. Y. (1998) Binding of proteins to polyelectrolytes studied by capillary electrophoresis. Ph.D. thesis, Chemistry, Indiana University–Purdue University, Indianapolis.
12. Gao, J. Y., Dubin, P. L., and Muhoberac, B. B. (1997) Measurement of the binding of proteins to polyelectrolytes by frontal analysis continuous capillary electrophoresis. *Anal. Chem.* **69,** 2945–2951.
13. Clore, G. M., Gronenborn, A. M., and Davies, R.W. (1982) Theoretical aspects of specific and non-specific equilibrium binding of proteins to DNA as studied by the nitrocellulose filter binding assay. Co-operative and non-co-operative binding to a one-dimensional lattice. *J. Mol. Biol.* **155,** 447–466.
14. Xia, J. and Dubin, P. L. (1994) Protein–polyelectrolyte complexes, in *Macromolecular Complexes in Chemistry and Biology*, (Dubin, P. L., et al., eds.) Springer-Verlag, New York, NY, pp. 247–271.
15. Hattori, T., Hallberg, R. K., and Dubin, P. L. (2000) Roles of electrostatic interaction and polymer structure in the binding of β-lactoglobulin to anionic polyelectrolytes: Measurement of binding constants by frontal analysis continuous capillary electrophoresis. *Langmuir* **16,** 9738–9743.
16. Kimura, K., Seyrek, E., and Dubin, P. L. (2001) Binding of bovine serum albumin to heparin determined by turbidimetric titration and frontal analysis continuous capillary electrophoresis. *Anal. Chem.* **295,** 158–167.

17. Hattori, T., Kimura, K., Seyrek, E., and Dubin, P. L. (2001) The use of frontal analysis continuous capillary electrophoresis to compare protein binding by natural and synthetic polyelectrolyte. *Analyt. Sci. (Japan)* **17,** 93–95.
18. Gao, J. Y. and Dubin, P. L. (1999) Binding of proteins to copolymers of varying hydrophobicity. *Biopolymers* **49,** 185–193.
19. Seyrek E., Dubin, P. L., Tribet, C., and Gamble, E.A. (2003) Ionic strength dependence of protein–polyelectrolyte interactions. *Biomacromolecules* **4,** 273–282.
20. Hallberg, R. K. and Dubin, P. L. (1998) Effect of pH on the binding of b-lactoglobulin to sodium polystyrenesulfonate. *J. Phys. Chem. B* **102,** 8629–8633.
21. Porcar, I., Cottet, H., Gareil, P., and Tribet, C. (1999), Association between protein particles and long amphiphilic polymers: effect of the polymer hydrophobicity on binding isotherms. *Macromolecules* **32,** 3922–3929.
22. Porcar, I., Gareil, P., and Tribet, C. (1998) Formation of complexes between protein particles and long amphiphilic polymers: binding isotherms versus size and surface of the particles. *J.Phys. Chem. B.* **102,** 7906–7909.
23. Ladaviere, C., Lorenzo, C., Elaissari, A., Mandrand, B., and Delair, T. (2000) Electrostatically driven immobilization of peptides onto (maleic anhydride-alt-methyl vinyl ether) copolymers in aqueous media. *Bioconjugate Chem.* **11,** 146–152.
24. Tanaka, Y. and Terabe, S. (2002) Estimation of binding constants by capillary electrophoresis. *J. Chromatog. B.* **768,** 81–92.
25. Rundlett, K. L. and Armstrong, D. W. Methods for the determination of binding constants by capillary electrophoresis. *Electrophoresis.* **22,** 1419–1427.
26. Conrad, H. E. (1998) *Heparin-Binding Proteins.* Academic, San Diego, CA.
27. Mattison, K. W., Yingfan, W., Grymonpré, K., and Dubin, P. L. (1999) Micro- and macro-phase behavior in protein–polyelectrolyte complexes. *Macromol. Symp.* **140,** 53–76.
28. Park, J. M., Muhoberac, B. B., Dubin, P. L., and Xia, J. (1992) Effect of protein charge heterogeneity in protein–polyelectrolyte complexation. *Macromolecules* **25,** 290–295.
29. Mattison, K. W., Brittain, I. J., and Dubin, P. L. (1995) Protein–polyelectrolyte phase boundaries. *Biotech. Prog.* **11,** 632–637.
30. Scatchard, G. (1949) The attraction of proteins for small molecules and ions. *Ann. N.Y. Acad. Sci.* **51,** 660–672.
31. Heck, H. D. (1971) Statistical theory of cooperative binding to proteins, the Hill equation and the binding potential. *J. Am. Chem. Soc.* **93,** 23–29.
32. McGhee J. D. and von Hippel P. H. (1974) Theoretical aspects on DNA-protein interactions: co-operative and non-co-operative binding of large ligands to a one-dimensional homogeneous lattice. *J. Mol. Biol.* **86,** 469–489.
33. Olson, S. T., Halvorson, H. R., and Björk I. (1991) Quantitative characterization of the thrombin–heparin interaction. *J. Biol. Chem.* **266,** 6342–6352.
34. Gao, J. Y., Dubin, P. L., and Muhoberac, B. B. (1998) Capillary electrophoresis and dynamic light scattering studies of structure and binding characteristics of protein–polyelectrolyte complexes. *J. Phys. Chem. B.* **102,** 5529–5535.

12

Analysis of Proteins by CE, CIEF, and Microfluidic Devices With Whole-Column-Imaging Detection

Jiaqi Wu, Xing-Zheng Wu, Tiemin Huang, and Janusz Pawliszyn

Summary

The recently developed whole-column-imaging detection technique for capillary electrophoresis (CE) and capillary isoelectric focusing (CIEF), a commercial whole-column-imaged CIEF instrument and its standard operation protocol are introduced. Furthermore, new developments and applications of whole-column-imaging detection in protein–protein interaction study, in protein separation using microfluidic devices and CIEF methods without carrier ampholytes, as well as in 2D separation techniques are reviewed. Miniaturization of whole-column-imaging CIEF and axially illuminated fluorescence whole-column-imaging CIEF are also discussed.

Key Words

Capillary electrophoresis; capillary isoelectric focusing; conjugation reaction; microfluidic device; protein; whole-column-imaging detection.

1. Introduction

In the past two decades, capillary electrophoresis (CE) has grown as a powerful separation technique with high-separation efficiency, and low-sample consumption. Various CE modes such as capillary zone electrophoresis (CZE), capillary gel electrophoresis (CGE), micellar electrokinetic capillary chromatography (MEKC), capillary electrochromatography (CEC), capillary isoelectric focusing (CIEF), and capillary isotachophoresis (CITP) have been developed *(1)*. Conventional CE is carried out in a long capillary (from ~10 to 100 cm). Recently, many CE experiments have been done in a microchannel fabricated in a chip *(2–5)* or a short capillary *(6,7)* with a length of several centimeters. Because CE separation efficiency depends on applied electric field, high-separation efficiency may still be achieved and the separation speed

From: *Methods in Molecular Biology, vol. 276: Capillary Electrophoresis of Proteins and Peptides*
Edited by: M. A. Strege and A. L. Lagu © Humana Press Inc., Totowa, NJ

is usually greatly accelerated in the short capillary or microfabricated device with a high-electric field. This meets the growing demands for high-speed, high-throughput DNA diagnostic, and screening applications *(7,8)*.

In conventional CE experiments, sample solution is injected from one end of the capillary or from a cross channel near one end of the microfabricated chip. After being separated along the electromigration distance, sample zones are detected at a detection point (window) near the other end of the capillary or the chip. This is the single-point detection. Using the single-point detection, even though all samples are completely separated before they reach the detection point, one still has to wait until all peaks pass through the detection point to complete the whole separation. Especially in CIEF *(9)*, the formation of pH gradient and focusing of sample zones usually take about 5 min while the following mobilization step requires 10–40 min in order to detect all focused sample zone at the detection point. The diffusion of the sample zone in the migration process also broadens the peak and decreases the peak height, thus decreasing the separation efficiency and detection sensitivity. Also, it is difficult to obtain detailed insight of the separation process and interaction between sample and capillary wall in the single-point detection, although they can be speculated from the electropherogram. Therefore, an ideal detection method is whole-column-imaging detection that allows real-time monitoring and direct visualization of the separation process.

The concept of whole-column-imaging detection was first proposed for CIEF *(10–17)*. Recently, it was further extended to other CE modes such as CZE *(6,18)*. In the whole-column-imaging detection, a short capillary (3–6 cm) is used as a separation capillary. Three types of whole-column-optical-imaging detection: refractive index gradient *(10–12)*, fluorescence *(15)*, and absorbance *(16)* imaging detection have been established. Recently, whole-column absorbance imaging detection for CIEF has been commercialized *(17,19,20)*. Experience and practice in CIEF experiments have proved that the whole-column-imaging detection permits the direct observation of the isoelectric focusing dynamics and greatly accelerates the analysis speed.

Proteins can exhibit microheterogeneity because of their molecular modifications, such as glycosylation, oxidation, deamidation, phosphorylation, and amino terminal modifications. In most instances, the modification of the proteins causes a change in their charges. For protein samples in biotechnology laboratories, the microheterogeneity may impact its biological function. Thus, identifying the microheterogeneity is essential for identification and characterization of these proteins. The high resolution of isoelectric focusing (IEF) makes it the most effective technique in assessing charge heterogeneity in protein molecules. Slab gel IEF is known to suffer from low speed, poor reproducibility, and nonquantitation. On the other hand, IEF performed in capillary format (i.e., CIEF), was believed to be able to provide quantitative IEF analysis with fast speed. However, the extra mobilization process of the conventional CIEF introduces problems to CIEF, such as uneven separation resolution, poor reproducibility and increased analysis time, and method development time. These problems do not exist in whole-column-imaging CIEF, because the detection finishes at the same time as the IEF process. Therefore, whole-column-imaging CIEF can offer fast speed gel-like IEF separation combined with quantitative on-column detection and automatic sample injection; thus, it is an ideal tool for assessing charge microheterogeneity of proteins

Here, construction of the commercial whole-column-imaging detection CIEF instrument, and its standard operation procedure are described by demonstrating CIEF rapid characterization of monoclonal antibody and monitoring stability of a humanized monoclonal immunoglobulins (IgG). Furthermore, its new developments and applications are also introduced.

2. Materials

1. 1% Methyl cellulose solution (prepared from SIGMA P/N: M0387 methylcellulose powder).
2. Pharmalyte, pH 3.0–10.0 (Sigma +1-314-771-5765 P/N: P1522).
3. Ampholyte, pH 4.0–7.0 (Sigma P/N: A9203).
4. Phosphoric acid, 87%, reagent grade.
5. Sodium hydroxide, 50% (w/w) reagent grade.
6. p*I* Markers 5.3, 6.6, 7.4, and 8.6 (Bio-Rad +1-510-724-7000, P/N: 148-2100).
7. Monoclonal antibody, anti-α-chorionic gonadotropin (α-hCG; 4.44 mg/mL, Calbiochem-Novabiochem +1-619-450-9600, P/N: 230744).
8. Humanized monoclonal IgG (generously donated by a biotechnology company).
9. Deionized water.
10. Commercial imaged CIEF instrument, the *i*CE280 IEF Analyzer: **Fig. 1** shows a block diagram of the whole-column-imaging detection CIEF system, *i*CE280 IEF analyzer (Convergent Bioscience, Toronto, Canada). The separation column of the instrument is a 50-mm-long, 100 μm id, 200-μm od silica capillary. Its outside polyimide coating is removed for whole-column detection. Its inner wall is coated with fluorocarbon to substantially reduce electroosmotic flow (EOF). The light source of the imaging detector is a xenon lamp. The light beam from the lamp is focused onto the separation column by a bundle of optical fibers and a set of lenses. Monochromatic light is obtained by placing a 280-nm-bandpass optical filter before the lamp. The whole column ultraviolet (UV) absorption image is captured by a camera, which includes an imaging lens and a charge-coupled device (CCD) sensor.

    The flow path of the *i*CE280 analyzer is also shown in **Figs. 1** and **2**. The column cartridge's inlet capillary is connected by a press-fit nut to a two-position, eight-port PEEK switch valve. The pressure needed for the sample introduction is provided by a low-pressure syringe pump that is connected to the switch valve and operates continuously.

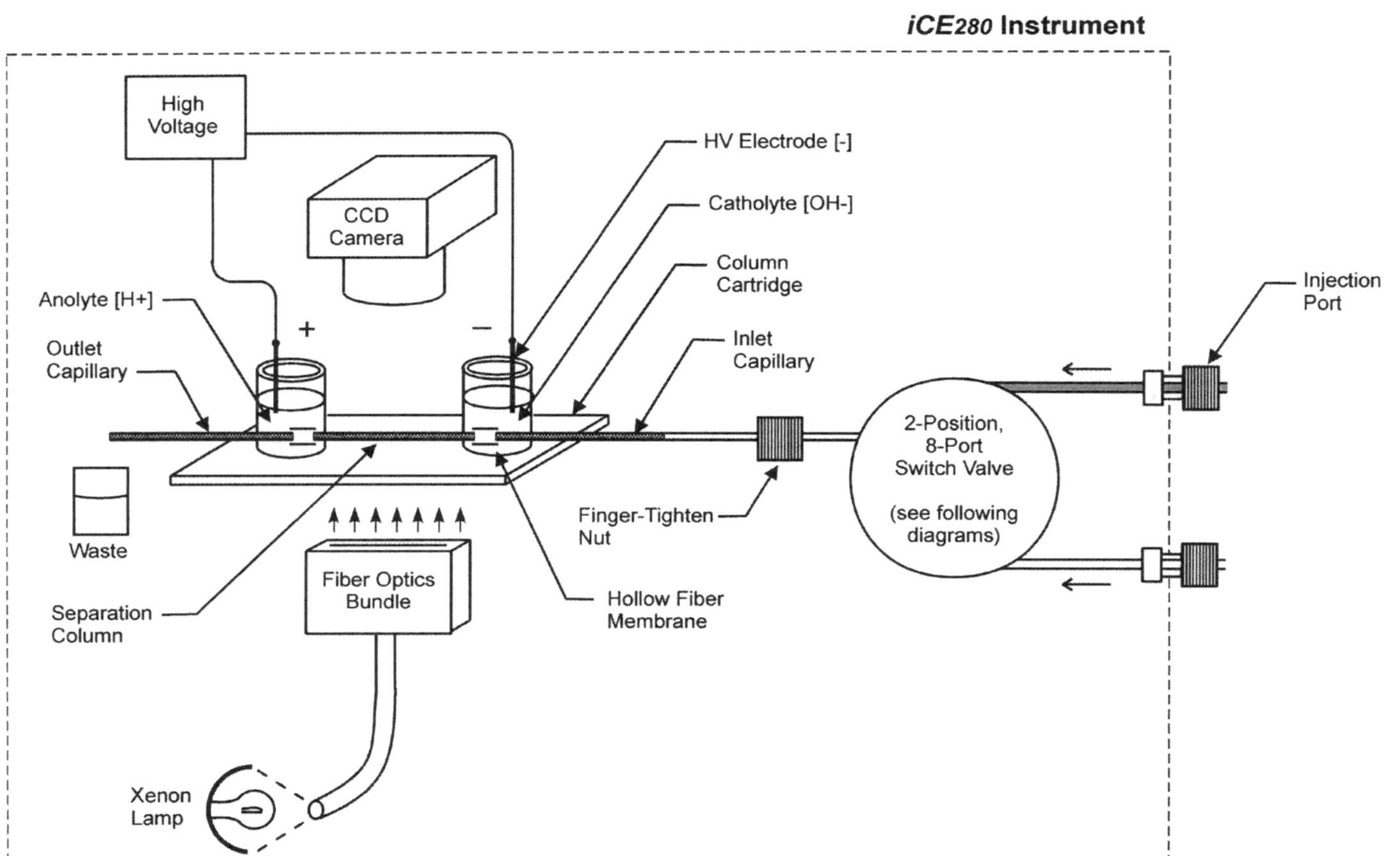

Fig. 1. The block diagram of the iCE280 instrument.

Operation of the *i*CE280 Analyzer: A column cartridge is first installed into the cartridge holder inside the instrument and connected to the switch valve. The two electrolyte tanks are filled with anolyte (usually 80 m*M* H_3PO_4) and catholyte (usually 100 m*M* NaOH). The syringe pump is filled and turned on.

The sample introduction procedure for the manual mode is similar to a conventional liquid chromatography (LC) instrument. As shown in **Fig. 2**, the sample loop is filled from the injection port when the switch valve is in position 1 (load position). The switch valve then rotates to position 2 (inject position) and the sample stored in the loop is pushed into the column cartridge by the syringe pump. Once the column is filled with the sample, the switch valve returns to the position 1 (load position). A 3-kV DC voltage is applied to the two electrolyte tanks to start isoelectric focusing. The focusing process usually lasts 5–7 min. During focusing, the process can be monitored by having the CCD camera take a picture of the whole column and display the UV absorption image of it every 30 s. At the end of the focusing process, the voltage is turned off and the sample and separation columns are rinsed for a few seconds by the syringe pump. The instrument then is ready for the next sample. The sample introduction procedure can be automated using a LC autosampler. In the automatic mode, the sample throughput is up to 8 samples/h.

The *i*CE280 analyzer includes quantitation software for rapid batch reprocessing of electropherograms.

11. Capillary Cartridge: As shown in **Fig. 1**, the column is packaged into a cartridge by sandwiching it between two glass slides. The two ends of the column are connected to the inlet and outlet capillaries by two sections of porous hollow fiber membranes. The two sections of the hollow fiber membranes isolate protein sample and carrier ampholytes within the column from external electrolytes in the two electrolyte tanks. Two connection capillaries for sample introduction are glued to the other ends of hollow fiber membranes. (*See* **Notes 1** and **2** for comments regarding cartridge conditioning and careful handling.)

3. Methods

*3.1. Rough Determination of Isoelectric Point (p*I*) of a Protein*

The general procedure for roughly determining p*I* of a protein is as follows.

1. Dissolve protein sample into carrier ampholyte solution (its final concentration is usually about 4%) containing methylcellulose (its final concentration is usually 0.5%). (For more details regarding the concentration of the sample and the selection of carrier ampholyte, *see* **Notes 3** and **4**, respectively.)
   Sample solution can be prepared in a 1.5-mL centrifuge tube. In the tube, add 80 μL deionized water, 100 μL 1% methylcellulose, 10 μL Pharmalyte pH 3.0–10.0 (or Ampholyte pH 4.0–7.0), and 10 μL sample. This makes the final volume of 200 μL. The concentration of the carrier ampholytes is 5%, and methylcellulose concentration is 0.5%. Degas the sample using a centrifuge for 3 min.
2. Install a conditioned column cartridge (*see* **Notes 1** and **2**) in the CIEF instrument.

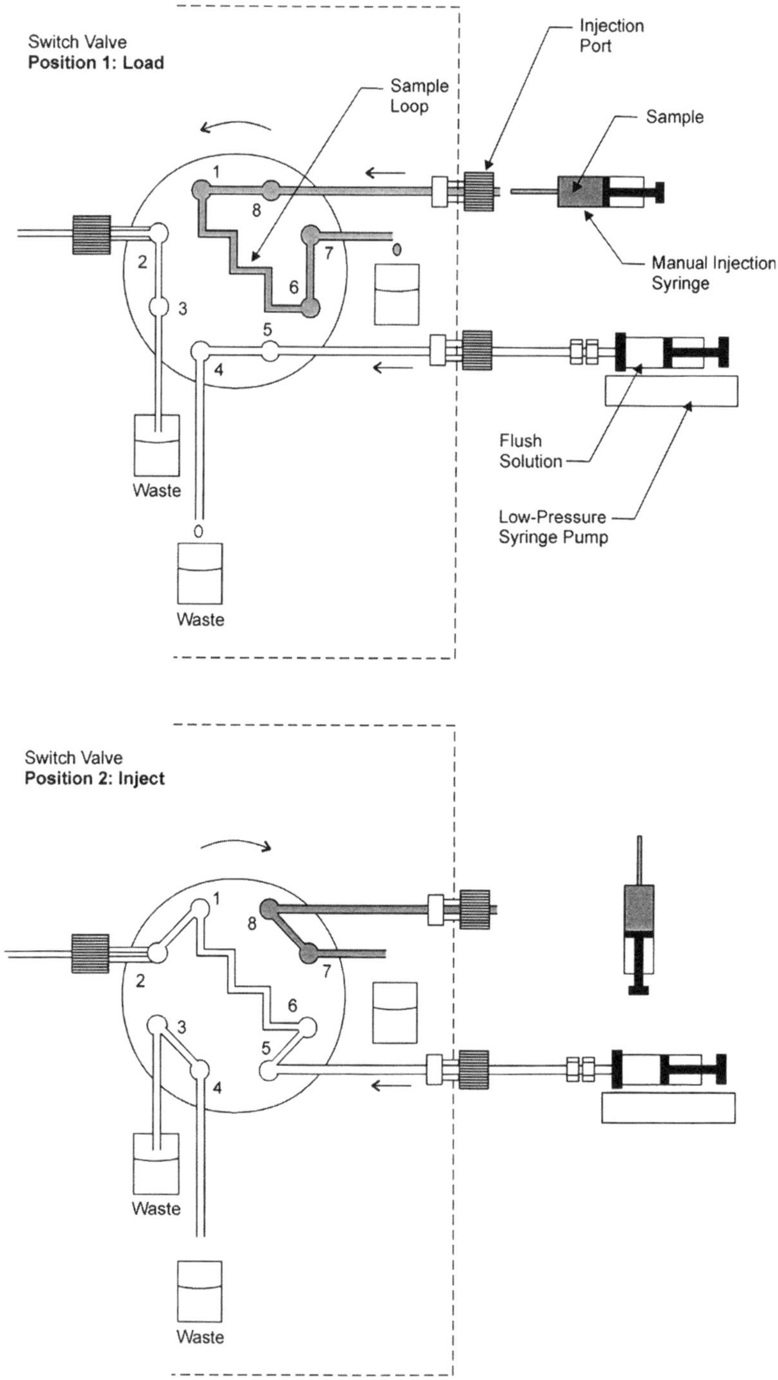

Fig. 2. Flow path in the iCE280 instrument.

3. Fill one electrolyte tank on the column cartridge with a catholyte (typically NaOH) and the other tank with an anolyte (typically H_3PO_4).
   The anolyte, 0.08 *M* phosphoric acid, is prepared by diluting 0.8 mL 1 *M* phosphoric acid into 9.2 mL 0.1% methylcellulose. 1 *M* phosphoric acid is prepared from 87% phosphoric acid. The catholyte, 0.1 *M* sodium hydroxide, is prepared by adding 53 μL of 50% sodium hydroxide into 10 mL of 0.1% methylcellulose.
4. Inject 100 μL sample into the sample loop of the *i*CE280 Analyzer.
5. Switch the valve of the *i*CE280 Analyzer to the "Inject" position so that the sample in the sample loop is transferred into the separation column.
6. Apply a high DC voltage (1500 or 3000 V) across the electrolyte tanks on the column cartridge to start the IEF process.
7. After a few minutes under high-voltage conditions, the carrier ampholytes establish a linear pH gradient and the sample proteins focus in bands, or zones, along the pH gradient at their p*I*. For the initial run conditions for an unknown sample, start from 6 min focusing time.
8. At the end of the IEF process, which normally takes about 5 min, the CCD camera (i.e., whole-column-imaging detector) records all protein zones within the capillary column.
9. The DC voltage is turned off and the process is complete.
10. Introduce a wash solvent into the loop and injecting the solvent into the column for a few seconds so that the next run can be carried out (*see* **Note 5**).

For the α-hCG sample, the p*I* value of the sample is roughly estimated based on peak position in electropherogram relative to the pH range of the carrier ampholyte (pH of the anodic and cathodic ends are 3.0 and 7.0, respectively). The p*I* value of the α-hCG sample is roughly estimated to be between 5 and 7.

3.2. Accurate Determination of pI *of a Protein*

The operation procedure for accurate determination of p*I* of a protein is basically same as in **Subheading 3.1.** except that two p*I* makers are added into the sample solution and the calculation of p*I* is made from the two p*I* makers. (*See* **Note 6** for more details regarding the selection of p*I* markers.)

1. Dissolve protein sample into carrier ampholyte solution containing methylcellulose and pI markers.
   In a 1.5-mL-centrifuge tube, add 80 μL deionized water, 100 μL 1% methylcellulose, 8 μL Pharmalyte pH 3.0–10.0 (or Ampholyte pH 4.0–7.0), 1 μL p*I* marker 5.3, 1 μL p*I* marker 7.4 (6.6 or 8.6), and 10 μL sample. This makes the final volume of 200 μL. The concentration of the carrier ampholytes is 4%, and methylcellulose concentration is 0.5%. Degas the sample using a centrifuge for 3 min.

2.–9. Same as **Subheading 3.1.**

10. Calculate p*I* value of the sample from the two p*I* markers by the CIEF analyzer.
11. Introduce a wash solvent into the loop and inject the solvent into the column for a few seconds so that the next run can be carried out.

The result for α-hCG sample with BioMarkers 5.3 and 7.4 is shown in **Fig. 3**. Five major peaks are resolved under these conditions. The p*I* values of the peaks are calculated to be from 5.8 to 6.2 using the p*I* calibration routine in the *i*CE280 analyzer software.

The separation resolution can be enhanced by using narrow pH range carrier ampholytes. Because the p*I* value of the sample is determined to be in the 5.8–6.2 range, Ampholyte, pH 4.0–7.0, is selected to run the sample to increase the resolution. The result is shown in **Fig. 4**. Compared to **Fig. 3**, the resolution is much improved. All peaks have baseline resolution. Again, using the *i*CE280 analyzer software, p*I* values of all peaks, as well as their peak area percentages, are calculated and labelled on each peak (the first number is p*I* value, the second number is percentage).

The standard deviation in peak identification is less than 0.2% and in peak area percentage it is less than 5%. Considering the good precision of the method, this result can be used as a fingerprint to identify and characterize this antibody. The method can monitor changes in relative peak area and degradation products. Thus, it is useful in stability studies for proteins. Stability studies are routinely conducted in biotechnology laboratories to monitor a variety of degradation products that can occur when proteins are stored in solution for a period of time.

Following is an example of application of *i*CE280 analyzer to stability study of a humanized monoclonal IgG. First, the initial sample is run with two p*I* markers, p*I* 5.3 and 8.6. By the two p*I* markers, as shown in **Fig. 5**, all sample peaks are identified by their p*I* values. Three peaks are observed: major peak at p*I* 7.8, a double peak at p*I* 7.5, and a minor peak at p*I* 8.0. Next, aged samples (incubation at 40C° for 0–4 wk) are run using the same conditions for the initial sample. The results, shown in **Fig. 6**, clearly show the trend of the sample changes with the aging process.

For some protein samples with small dissolubility, protein precipitation will need to be addressed (*see* **Note 7**). Also, for sample with high concentration salt, sample dilution or desalting should be considered (*see* **Note 8**). Details of cartridge stability are given in **Note 9**.

3.3. Recent Topics of the Whole-Column-Imaging Detection of Protein

3.3.1. Protein–Protein Interaction Studied by Whole-Column-Imaging CIEF

If two proteins have strong interaction, change in their p*I* is expected. Especially when the interaction is so strong that a protein–protein complex or conjugate is formed, a new peak with completely different p*I* from the two reactant proteins is expected in the whole-column-imaging CIEF result. This has been successfully demonstrated with model proteins bovine serum albumin (BSA),

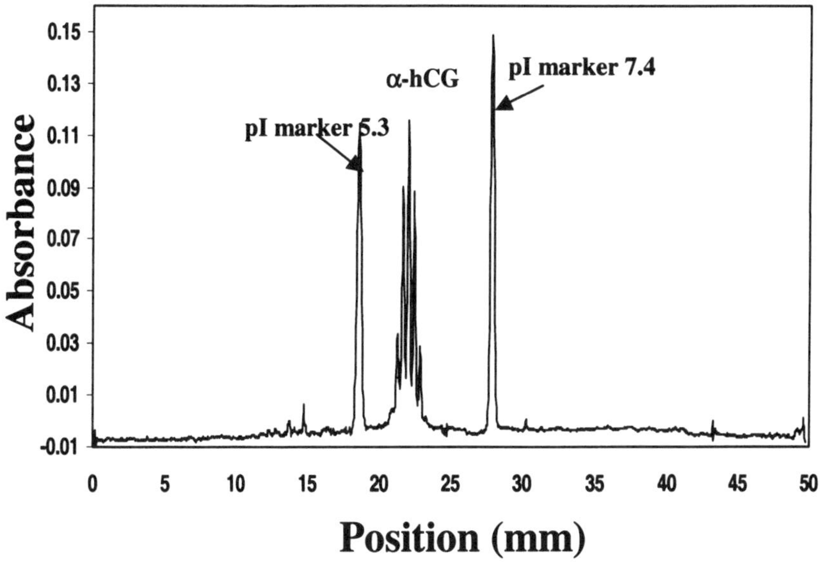

Fig. 3. Whole-column-imaging CIEF results of α-hCG sample with two BioMarkers. Carrier ampholyte, pH 3.0–10.0.

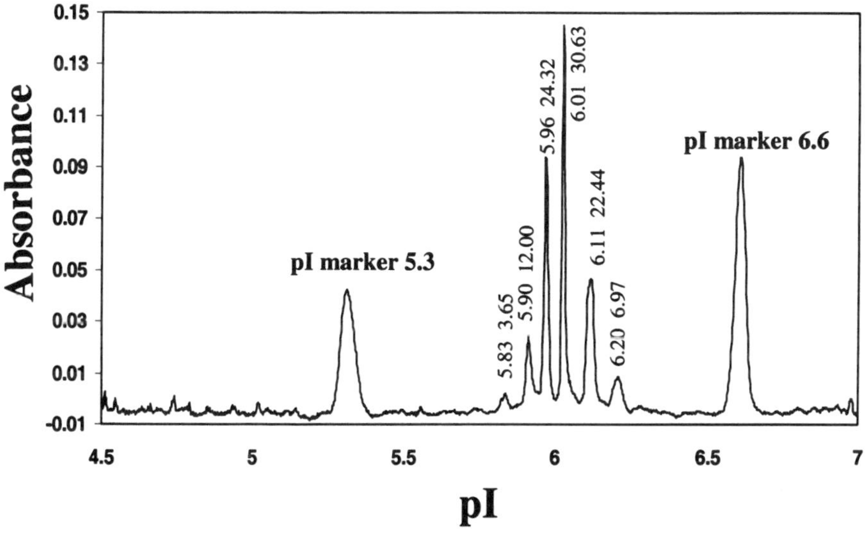

Fig. 4. Whole-column-imaging CIEF results of α-hCG sample with two BioMarkers. Carrier ampholyte, pH 4.0–7.0.

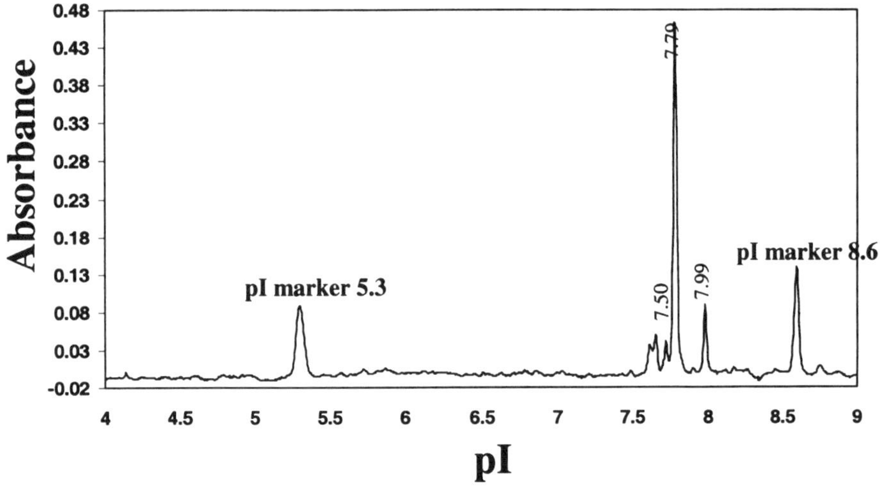

Fig. 5. Whole-column-imaging CIEF results of humanized monoclonal IgG sample with two BioMarkers. Carrier ampholyte, pH 3.0–10.0.

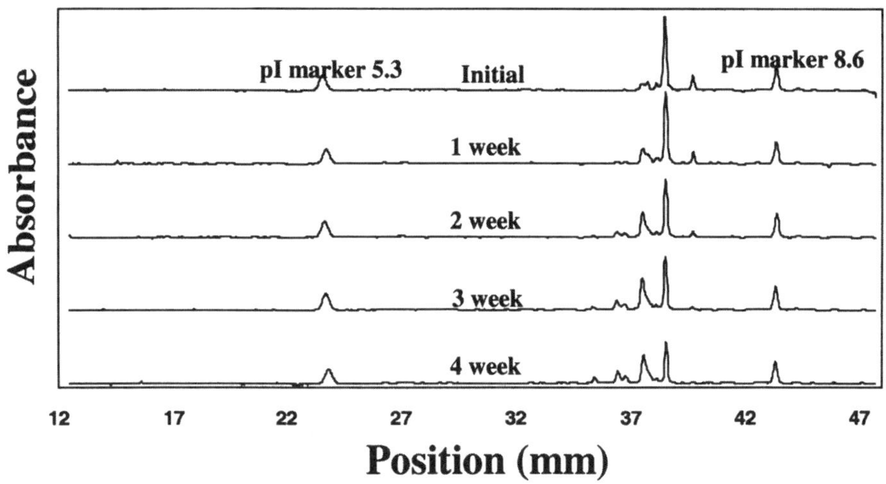

Fig. 6. Effect of incubation on CIEF results of α-hCG sample with two BioMarkers. Carrier ampholyte, pH 3.0–10.0.

biotin-labeled-BSA, and steptavidin *(21)*. First, whole-column-imaging CIEF of mixture of BSA and streptavidin was carried out. Experimental results show that neither new peaks nor any change in peak positions of BSA and streptavidin was observed in the mixture of the two proteins mixtures. This indicates that CIEF behavior of BSA and streptavidin are independent each other in the mixture, i.e., they have no strong interaction. However, when streptavidin was mixed with biotin-labeled-BSA, a new peak whose p*I* was between streptavidin and biotin-labeled-BSA was observed. This new peak corresponded to biotin-labeled-BSA streptavidin complex because biotin strongly interacted with streptavidin *(21)*.

In addition to the model protein samples, the method has been further applied to study immunoreaction in peanut allergenic proteins–rabbit IgG antibody system *(22)*. Two main proteins, *Ara h I* and *Ara h II*, were found in a peanut allergen solution. Their p*I* values were determined to be about 4.5–4.7 and 5.1–5.6, respectively. Value of p*I* for antibody rabbit IgG of the peanut allergen was 6.1–6.5. **Figure 7** shows absorption images of mixtures of peanut antigen (Ag) and antibody (Ab) after 200 s of isoelectric focusing. The lowest line in **Fig. 7** is the absorbance image of peanut Ag, where both *Ara h I* and *Ara h II* were present. When 15 μL of peanut Ag was mixed with 30 μL of Ab, the peak of *Ara h II* disappeared completely. On the other hand, the peak of *Ara h I* was still observed. This suggests that all *Ara h II* had bound with the Ab and precipitated as Ag–Ab complex; however, only a part of *Ara h I* had bound with the Ab and precipitated. When the amount of Ab was increased to 60, 90, and 120 μL, the peak area of *Ara h I* was further decreased and eventually disappeared. On the other hand, peak of Ab was increased. In the mixture, *Ara h I* and *Ara h II* competitively bound with the Ab. If the selectivity or affinity of the two peanut Ag binding to Ab were the same, the peaks of the two peanut Ag should be decreased proportionally when the amount of Ab was increased. Earlier disappear of *Ara h II* than *Ara h I* in **Fig. 7** suggest that *Ara h II* might bind with the Ab more easily than *Ara h I*. In other words, the peanut allergen *Ara h II*'s affinity in the immunoreaction might be larger than that of *Ara h I*. Although the quantitative analysis of the immunoreaction is difficult at present because of problems with protein solubility and detection sensitivity, it is clear that the method is a fast and unique tool to investigate protein–protein interaction. With the improvement of the detection sensitivity, the quantitative analysis of the protein–protein interaction is anticipated in the future.

3.3.2. Whole-Column-Imaging Detection of Protein in Microfluidic Devices

Whole-column-imaging detection is also ideal for CE or CIEF in a chip format. This has been demonstrated with a microchannel fabricated in a quartz chip *(23,24)*. The microchannel fabricated by photolithography and a chemical

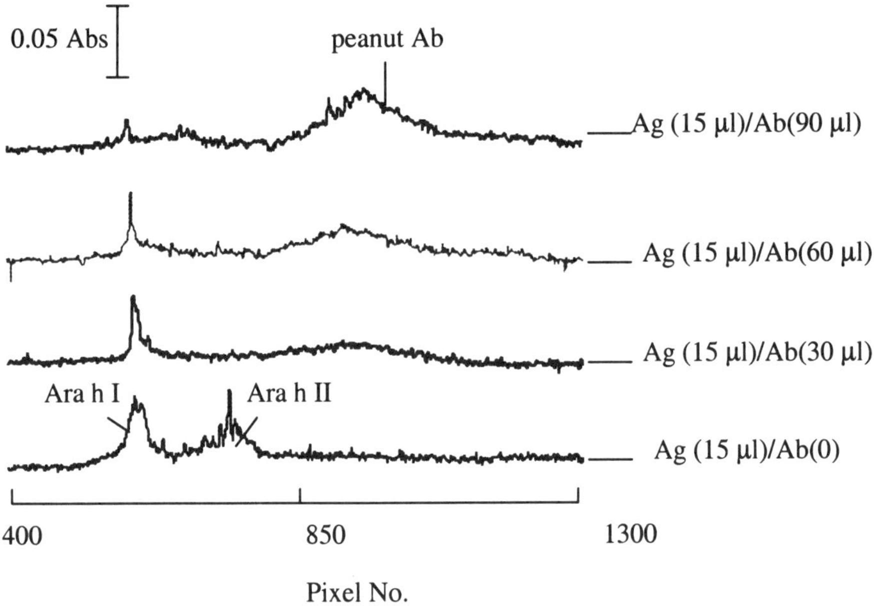

Fig. 7. Absorbance images of the mixture of peanut Ag and Ab obtained after isoelectric focusing time of 200 s. Before the CIEF, the mixture of 15 μL of peanut Ag with different amount (0, 30, 60, 90, 120 μL) of Ab in 4% carrier ampholyte, pH 3.0–10.0) containing 0.35% methylcellulose and 20% glycerol was placed into a microcentrifuge with a rotating speed of 10,000 rpm for 2 min.

etching process was 40-mm long, 100-μm wide, and 10-μm deep. Protein myoglobin and a p*I* marker were mixed with 4% carrier ampholyte solution, and the mixture was introduced into the microchannel. After an electric voltage of 3 kV was applied across the microchannel, the protein and p*I* marker were focused well in about 10 min. The detection limit was about 0.3 mg/mL or 24 pg for p*I* marker, and 30 mg/mL or 2.4 ng for myoglobin.

Recently, a novel microfabrication technique using screen printing for the preparation of microfluidic cartridge for CIEF has been reported *(25)*. Screen printing is the process of applying prepolymeric inks through a patterned screen or stencil onto a suitable substrate. This method was investigated for the production of microfluidic devices for CIEF at competitive costs. In this method, 80 double parallel strips of polymer coating with a 50-μm thickness, 39-mm length and 1-mm gap were printed onto 508 × 508 mm thin plastic sheet. Closed channels were made by bonding two units of the printed double strips face to face with epoxy glue, forming a microchannel with dimensions of 39 × 1 × 0.10 mm. Cartridges suitable for whole-column-detection CIEF were fabricated using the printed microchannels. Electroosmotic flow and analyte adsorption were controlled by dynamic coating of the channel with methylcellulose solution. Validity of the constructed microfluidic device for CIEF was proved with four pI markers (p*I*s: 5.3, 6.4, 7.4, and 8.4).

3.3.3. CIEF Without Carrier Ampholytes

Carrier ampholytes are commonly used in gel IEF or CIEF with a relatively high concentration up to 8%. In protein purification with an IEF process, the purified proteins have to be further separated from the carrier ampholytes. Also, carrier ampholytes may interact with some protein samples, reducing the sensitivity of UV detection, and complicate the matrix or backgrounds when using mass spectroscopy for characterization. Accordingly, it is ideal to carry out IEF or CIEF without carrier ampholytes. Recently, CIEF without carrier ampholytes has been demonstrated with the whole-column-imaging detection *(26)*. **Figure 8** presents one example of CIEF without carrier ampholyte. CIEF without carrier ampholytes is explained to be related with electrolysis of water in the anode and cathode *(26)*.

CIEF utilizing different Joule heating along the axis of a tattered channel has also been investigated *(27)*. A simple microfabrication technique was developed for the preparation of a tapered microchannel, in which a tapered channel was cut into a plastic sheet (thickness was 120 μm), and the channel was closed by sandwiching the plastic sheet between two glass microscope slides. The length of the microchannel was 5 cm. The width of the separation channel was 0.4 mm at the narrow end and 4 mm at the wide end. The channel was coated with polyacrylamide to eliminate EOF during focusing. Two electrolyte vials were mounted on top of each ends of the channel with the wide end of the channel connected to the cathodic vial and the narrow to the anodic vial. Important parameters that determined the feasibility of thermally generated pH gradient in a tapered channel were analyzed, such as control of EOF and hydrodynamic flow, selection of power supply mode, and prevention of local overheating and air bubble formation. Tris-HCl buffer that has a high pKa dependence on temperature was used both to dissolve proteins and as the electrolytes. Concentrating and focusing of dog, cat, and human hemoglobin with whole-column-detection CIEF system was demonstrated.

3.3.4. Application in 2D Separation Techniques

Whole-column-imaging CIEF combines the high-resolution isoelectric focusing with real-time detection, which makes it an ideal method as the second dimensional separation in 2D separation techniques. Recently, the first step toward the 2D HPLC-CIEF separation, i.e., online coupling of HPLC to CIEF with whole-column-imaging detection, has been shown *(28)*. Further-

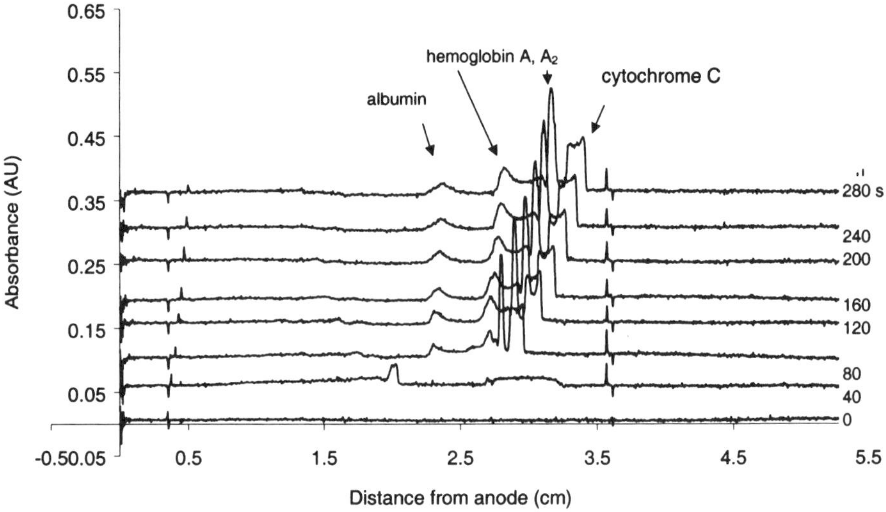

Fig. 8. Carrier ampholyte free CIEF separation of four protein (albumin, 550 µg/mL; hemoglobin A, 20 µg/mL; hemoglobin A2, 20 µg/mL, cytochrome-*c*, 160 µg/mL) at 100 V/cm. The anolyte is 100 m*M* phosphoric acid, and the catholyte is 100 m*M* sodium hydroxide.

more, a comprehensive 2D separation of CE/CIEF and MEKC/CIEF with whole-column-imaging detection is also reported ***(29)***. A novel interface was developed for coupling CIEF to CE or MEKC by using a 10- and 8-port switch valve. **Figure 9** shows the schematics of the interface and CE (or MEKC)/CIEF 2D system. Samples were electrokinetically injected into the CE capillary, and separated by CE or MEKC (first separation dimension). The CE effluent was transferred into the CIEF cartridge for CIEF (second separation dimension) after desalted and mixed with carrier ampholyte in one of the two conditioning loops. Satisfactory results of CE/CIEF 2D separation were obtained with standard protein mixture of myoglobin, carbonic anhydrase, and oval albumin. The 2D separation system was further applied to separation of trypsin digest of trypsinogen; its results were shown in **Fig. 10**.

3.3.5. Axially Illuminated Fluorescence Whole-Column-Imaging Detection

Recently, an ultrasensitive whole-column-fluorescence-imaging CIEF with axial illumination was developed ***(30)***. The excitation light is introduced from one end of the capillary, and propagated in the capillary by total internal reflection. **Figure 11** shows the dynamic focusing process of the two naturally fluorescent proteins R-phycoerythrin and green fluorescence protein (GFP)

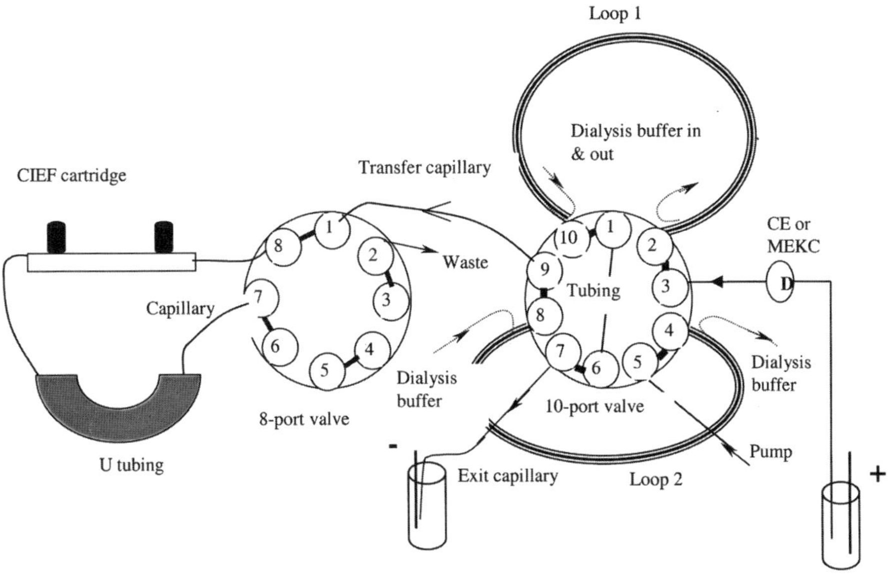

Fig. 9. Schematics of CE (or MEKC)/CIEF 2D separation system with an interface consisting of a 8- and 10-port valve.

obtained by axially illuminated whole-column-imaging detection. Detection of 10^{-13} mol/L or subattomol proteins is easily accomplished even without optimizing the optical detection system. Because of the extremely high sensitivity, this whole-column-fluorescence-imaging detector is expected to be the most powerful tool for CIEF analysis of trace biomolecules in a single cell.

Recently, noncovalently labeling for fluorescence detection of proteins has been investigated to facilitate the application of laser-induced fluorescence (LIF) detection ***(31)***. Noncovalently labeling fluorescent dyes, NanoOrange, Sypro red, Sypro orange, and Sypro tangerine, were explored for the coupling of BSA and hemoglobin. Labeled proteins were studied by two complementary detection methods, whole-column-UV and LIF detection instruments. The studies using a commercial CIEF instrument with UV detection confirmed that the noncovalently labeled BSA can be focused into a well defined peak. The p*I* value of this labeled BSA does not shift significantly compared to the calculated p*I* value of nonlabeled BSA. The axial LIF detection system confirmed the formation of fluorescent-labeled BSA. An improvement of detection sensitivity of at least 10 times was achieved using LIF with noncovalently labeling dye in comparison to using a UV absorption instrument.

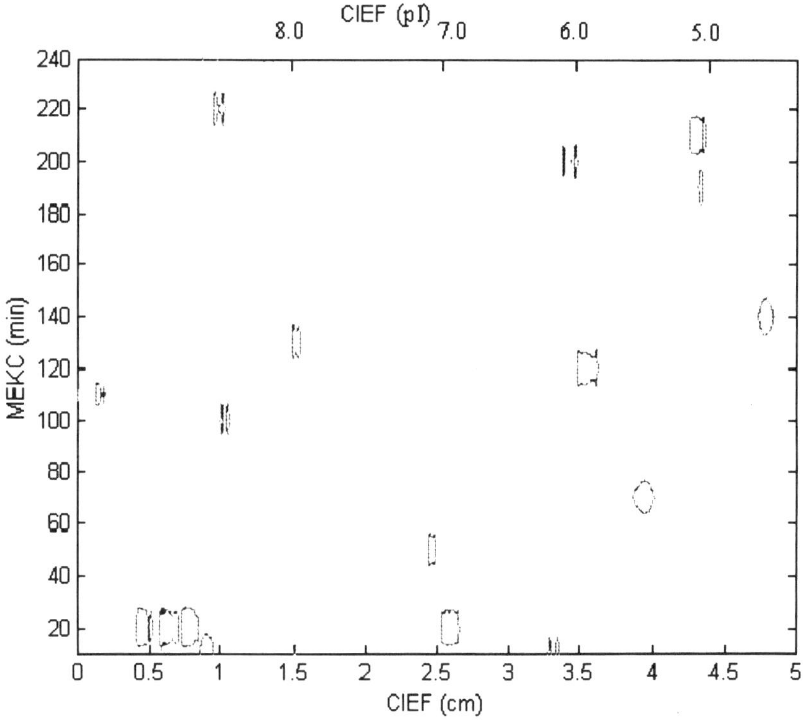

Fig. 10. MEKC/CIEF 2D separation of trypsinogen digest. MEKC conditions: 50 μm capillary, pH 8.0, 50 m*M* Tris-HCl with 10 m*M* CHAPS, 100 V/cm. CIEF conditions: 305 μm × 5 cm TFE capillary; 3 kV. Dialysis conditions: 0.5% CAs in 8 *M* urea; flow rate, 2 mL/min.

3.3.6. Miniaturization of Whole-Column-Imaging CIEF

To date, either a laser or a lamp (for example, a Xe lamp) has been used as a light source in the whole-column-imaging CIEF. In the commercial instrument shown in **Fig. 1**, 280-nm light filtered from the Xe lamp was first focused to a fiber bundle. The light from the fiber bundle was then focused to a whole-separation capillary by cylindrical lenses. The optical arrangement, light source, and related power system are much larger than the CIEF system. In view of the miniaturization, a light-emitting diode (LED) is expected to be an ideal light source. Recently, the miniaturization of the whole-column-imaging CIEF with a 1.2-cm capillary and a LED light source has been reported ***(32)***. Good CIEF results were obtained for both p*I* makers and protein myoglobin with the LED light source. The whole-column-imaging CIEF instrument was greatly simplified and miniaturized by the use of the LED. Its further development by integrating the LED, capillary or microchannel, and detector into one chip was also discussed.

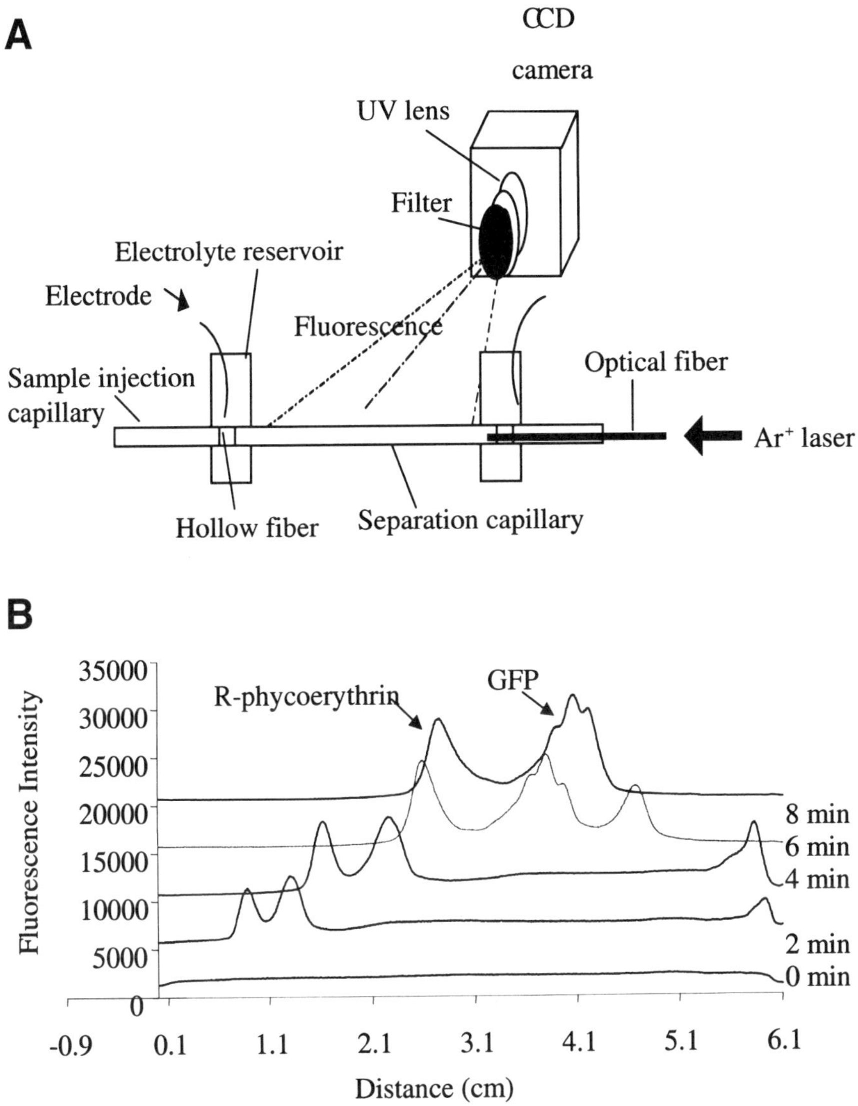

Fig. 11. Illustration of axially illuminated whole-column fluorescence-imaged CIEF (**A**) and separation examples of two fluorescent proteins 1.7×10^{-10} *M* R-phycoerythrin and 1.8×10^{-8} *M* green fluorescent protein (**B**).

3.3.7. Whole-Column-Imaging CE and Its Application in Protein-Binding Reaction

Whole-column-imaging detection also could be used for other CE modes such as CZE. Recently, whole-column-imaging CE with a short capillary has been examined in detail *(18)*. For the short capillary, hydrodynamic flow caused by a subtle height difference between the anodic and cathodic reservoirs affected the sample migration in the capillary greatly. Three sample injection methods, including use of a cross connection, sealing of the capillary ends with a gel, and use of a gel-filled capillary, have been discussed. **Figure 12** shows one CE example of two protein samples in a short capillary. It is clear that peak height decreased and peak width increased along the electromigration distance. Therefore, higher sensitivity can be obtained in a short capillary rather than a long capillary. **Figure 12** also shows the effect of the injection time on the CE separation behavior. The two proteins have been completely separated before 80 s when the injection time was 2 s (*see* **Fig. 12B**). However, when the injection time was 10 s, they were completely separated at about 120 s (*see* **Fig. 12A**).

The whole-column-imaging CE has also been applied for the study of conjugation reaction of protein cytochrome-*c* with sodium dodecylsulfate (SDS) *(18)*. **Figure 13** shows the direct visualization of the conjugation reaction in the capillary. Cytochrome-*c* electromigrated from anodic (left) to cathodic (right) end, whereas SDS electromigrated reversely. **Figure 13A** is the normal CE result of cytochrome-*c*. **Figures 13B**, **C**, and **D** show the conjugation reactions in the capillary with different concentration ratios of cytochrome-*c* to SDS. When the concentration ratio was 4 mg/mL:1 mg/mL (*see* **Fig. 13B**), the conjugation reaction neutralized a part of the cytochrome-*c* positive charge, thus electromigration of the SDS-cytochrome-*c* conjugate was slower than that of the protein alone. As a result, the front of the cytochrome-*c* sample zone was stacked as shown in the absorbance image at 160 s in **Fig. 13B** (SDS has low absorbance at 280 nm and thus its peak did not appear). However, at this concentration ratio, SDS did not neutralize all positive charge in cytochrome-*c*. Accordingly, the cytochrome-*c* sample zone still electromigrated toward the cathode after the conjugation or binding with the SDS. When the concentration ratio was 4 mg/mL:2 mg/mL, SDS neutralized all positive charges of cytochrome-*c* and formed a neutral conjugation product. Therefore, peak position after the conjugation reaction did not change (*see* **Fig. 13C**). On the other hand, when the concentration ratio was 4 mg/mL:5.6 mg/mL, the conjugation product electromigrated toward the anode (*see* **Fig. 13D**). This suggests that the reaction product was negatively charged. **Figure 13** shows that the method is a fast and convenient method for study of protein conjugation reaction. It has

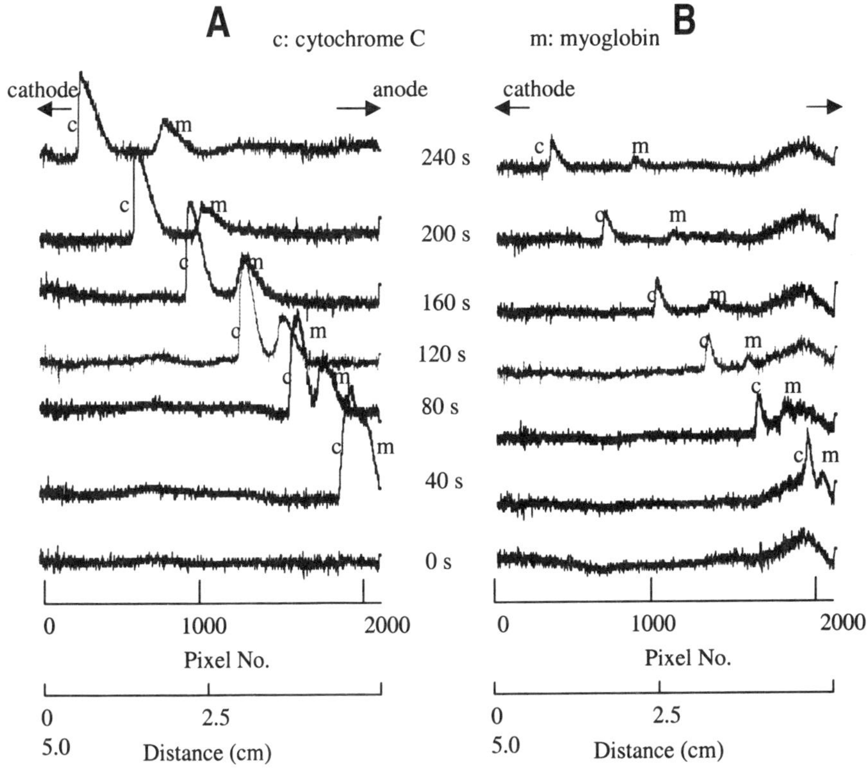

Fig. 12. Whole column absorbance images taken in the CE process of proteins with the short capillary (od 200 μm; id 100 μm, length 5 cm). Protein concentrations were 4 mg/mL. The sample was injected from the right (anodic) end, and the injection time was 10 s (**A**) and 2 s (**B**), respectively. The pH of the buffer and CE electric field were 2.5 and 100 V/cm, respectively.

also been used for studying conjugation reaction of protein with dye *(18)*. In addition to the conjugation reaction, the method has further been applied to *in situ* monitoring of electrophoretic protein desorption from capillary wall *(18)*. Experimental results showed that the technique is a unique tool for the study of interaction between analyte and inner wall of capillary.

4. Notes

1. Conditioning of the cartridge: the CIEF cartridge should be conditioned using 0.5% methylcellulose for more than 2 h to obtain the optimal resolution. However, for the column cartridges preconditioned by the manufacturer, further conditioning is not necessary. A properly conditioned cartridge should be able to

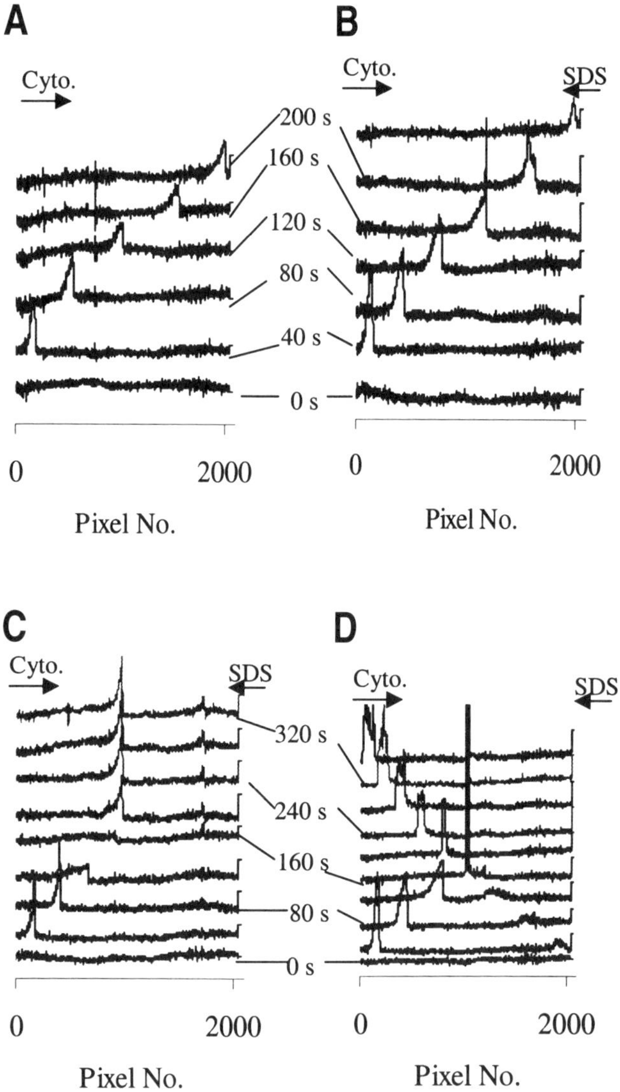

Fig. 13. Whole-column-imaging CE study of cytochrome-*c* binding reaction-in-capillary with SDS. A capillary (od 200 μm; id 100 μm; length 5 cm) was used. Concentration of cytochrome-*c* was 4 mg/mL; concentration of SDS mg/mL was 0 (**A**), 1 (**B**), 2 (**C**), 5.6 (**D**), respectively. Cytochrome-*c* was injected from the left (anodic) end, and SDS was from right (cathodic) end. The injection time was 10 s, and the pH of the buffer was 2.5. Electric field in CE was 100 V/cm.

separate hemoglobin A and hemoglobin A1c components in the standard sample of the *i*CE280's Chemical Test Kit. The peaks of the two components are shown in the kit's manual.

2. Hold the cartridge only by the two electrolyte tanks at all times. In order to avoid contamination of optical pass of the cartridge, do not touch any other surfaces of the cartridge.
3. Concentration of the sample: first, try 0.1 mg/mL sample concentration. If the highest peak height is less than 0.05 or greater than 0.3, the sample concentration should be adjusted. If the major peak height is lower than 0.05, a minor peak with the height that is 5% of the major peak height may not be able to be observed owing to the noise level of the detector. On the other hand, if a peak is higher than 0.3, it may be out of linear range of the UV absorption detector.
4. Selection of carrier ampholyte: because the p*I* value of the sample is unknown, it is first run alone in a wide-pH-range carrier ampholyte (in these examples, Pharmalyte pH 3.0–10.0).
5. Rinsing of the cartridge between runs: between runs the cartridge should be rinsed for 3 min by 0.5% methylcellulose to eliminate any carry over effect and regenerate the column surface.
6. Selection of two p*I* markers for p*I* calibration: ideally, the two marker peaks should bracket the sample peaks. Nonprotein, small-molecule p*I* markers (such as BioMarkers series [Bio-Rad +1-510-724-7000, P/N: 148-2100]) are strongly recommended because of their simplicity of peak pattern and high sensitivity at 280 nm.
7. Protein sample precipitation: in all IEF methods, including CIEF methods, the biggest difficulty is protein precipitation during focusing, especially for hydrophobic proteins. Smeared sample bands in slab gel IEF and nonreproducible peak patterns in CIEF are the sign of sample precipitation during focusing. This is mainly caused by the low ionic strength of IEF media at the end of the focusing process and by confining proteins at zero net charge status for a long time. The whole-column-imaging detection CIEF reduces the difficulty because its high analysis speed and elimination of the mobilization process that is necessary for the conventional CIEF methods (during mobilization process of the conventional CIEF methods, focusing voltage is always applied to prevent peak broadening). As long as the focusing process is complete, the detection finishes at the same time. From our experience, antibodies usually have no precipitation problem in IEF analysis when an *i*CE280 Analyzer is used. However, some other proteins may still have the precipitation problem during analysis using the Analyzer. To stabilize protein sample during focusing, some additives can be used. Sucrose, glycerol, and sorbitol enhance the solubility of proteins. They are usually used at 10% concentration in sample solution. Many nonionic or zwitterionic detergents, such as TritonX100, Tween-80, and CHAPS, stabilize hydrophobic proteins in aqueous solutions. The optimal concentration of these detergents in sample solution is in the 0.5 to 3% range. The hydrophobic proteins can also be run in denatured conditions, such as in 8 *M* urea.

8. Samples in high-ionic strength: the tolerance of the *i*CE280 Analyzer for salt is about 15 m*M* NaCl. Usually, salts in a sample matrix is not a problem for IEF analysis using an *i*CE280 Analyzer as long as the sample-in-salt matrix is diluted 20× by running buffer containing carrier ampholytes and methylcellulose. However, if the sample concentration is lower than 1 mg/mL, the sample peak may be too low to detect by the Analyzer after the 20× dilution. In this case, concentrating or desalting step is necessary for the sample before IEF analysis using the *i*CE280 Analyzer.
9. Capillary coating stability: the FC coating of the *i*CE280's separation column is stable in up to 0.5 *M* NaOH. The lifetime of the column is well above 200 runs. The lifetime of the column can be judged by resolution for the standard sample in *i*CE280's Chemical Test Kit or by user's standard samples.

References

1. Li, S. F. Y. (1992) *Capillary Electrophoresis, Principles, Practice, and Applications*. Elsevier, Tokyo, Japan.
2. Harrison, D. J., Manz, A., Fan, Z., Ludi H., and Widmer H. M. (1992) Capillary electrophoresis and sample injection systems integrated on a planar glass chip. *Anal. Chem.* **64,** 1926–1932.
3. Jacobson, S. C., Hergenroder, R., Koutny, L. B., and Ramsey, J. M. (1994) *Anal. Chem.* **66,** 2369–2373.
4. Effenhuser, S. C., Bruim, G. J. M., Paulus, A., and Ehrat, M. (1997) Integrated capillary electrophoresis on flexible silicone microdevices: analysis of DNA restriction fragments and detection of single dna molecules on microchips. *Anal. Chem.* **69,** 3451–3457.
5. Schmalzing, D., Adourian, A., Koutny, L., Ziaugra, L., Matsudaira, P., and Ehrlich, D. (1998) DNA sequencing on microfabricated electrophoretic devices. *Anal. Chem.* **70,** 2303–2310.
6. Palm, A., Lindh, C., Hjerten, S., and Pawliszyn, J. (1996) Capillary-zone electrophoresis in agarose gels using absorption imaging detection. *Electrophoresis* **17,** 766–770.
7. Kleparnik, K., Mala, Z., Pribyla, L., Blazkova, M., Vasku, A., and Bocek, P. (2000) Ultrafast detection of microsatellite repeat polymorphism in endothelin 1 gene by electrophoresis in short capillaries. *Electrohpresis* **21,** 238–246.
8. Sanders, G. H. W. and Manz, A. (2000) Chip-based microsystems for genomic and proteomic analysis. *Trends Anal. Chem.* **19,** 364–378.
9. Hjerten, S. and Zhu, M. (1985) Adaptation of equipment for high-performance electrophoresis to isoelectric focusing. *J. Chromatogr.* **346,** 265–270.
10. Wu, J. and Pawliszyn, J. (1992) Universal detection for capillary isoelectric focusing without mobilization using concentration gradient imaging system. *Anal. Chem.* **64,** 224–227.
11. Wu, J. and Pawliszyn, J. (1992) Capillary isoelectric focusing with a universal concentration gradient imaging system using a charge-coupled photodiode array. *Anal. Chem.* **64,** 2934–2941.

12. Wu, J. and Pawliszyn, J. (1994) Dual Detection for Capillary Isoelectric Focusing with Refractive Index Gradient and Absorption Imaging Detectors. *Anal. Chem.* **66,** 867–873.
13. Wu, J. and Pawliszyn, J. (1994) Imaging detection methods for capillary isoelectric focusing. *Am. Lab.* **26,** 48–52.
14. Wu, J. and Pawliszyn, J. (1995) Diode laser-based concentration gradient imaging detector for capillary isoelectric focusing. *Anal. Chim. Acta* **299,** 337–342.
15. Wu, X-Z., Wu, J., and Pawliszyn, J. (1995) Fluorescence imaging detection for capillary isoelectric focusing. *Electrophoresis* **16,** 1474–1978.
16. Wu, J. and Pawliszyn, J. (1995) Absorption spectra and multicapillary imaging detection for capillary isoelectric focusing using a charge coupled device camera. *Analyst* **120,** 1567–1571.
17. Wu, J., Tragas, C., Watson, A., and Pawliszyn, J. (1998) Capillary isoelectric focusing with whole column detection and a membrane sample preparation system. *Anal. Chim. Acta* **383,** 67–78.
18. Wu, X.-Z. and Pawliszyn, J. (2002) Whole-column imaging capillary electrophoresis of proteins with a short capillary. *Electrophoresis* **23,** 542–549.
19. Wu, J., Watson, A. H., and Torres, A. R. (1999) Protein analysis using imaged capillary isoelectric focusing. *Am. Biotech. Lab.* **17,** 24–26.
20. Wu, X.-Z., Wu, J., and Pawliszyn, J. (2001) Whole-column-imaging detection for capillary isoelectric focusing and capillary electrophoresis. *LCGC* **19,** 526–545.
21. Wu, X.-Z., and Pawliszyn, J. (2001) Application of whole column imaging capillary isoelectric focusing in study of protein-protein interaction. *Anal. Sci.* **17(Suppl.),** ***i189–i192.***
22. Wu, X.-Z., Huang, T., Mullet, W. M., Yeung, J. M., and Pawliszyn, J. (2001) Determination of isoelectric point and investigation of immunoreaction in peanut allergenic proteins-rabbit IgG antibody system by whole-column imaged capillary isoelectric focusing. *J. Microcolumn Seperation* **13,** 322–326.
23. Mao, Q. and Pawliszyn, J. (1999) Capillary isoelectric focusing with whole column imaging detection for analysis of proteins and peptides. *J. Biochem. Biohys. Methods* **39,** 93–110.
24. Mao, Q. and Pawliszyn, J. (1999) Demonstration of isoelectric focusing on an etched quartz chip with UV absorption imaging detection. *Analyst* **124,** 637–641.
25. Huang, T., Ertl, P., Wu, X.-Z., Mikkelsen, S., and Pawliszyn J. (2002) Microfabrication of microfluidic device for isoelectric electric focusing. *Sensors and Materials* **14,** 141–149.
26. Huang, T., Wu, X.-Z., and Pawliszyn, J. (2000) Capillary isoelectric focusing without carrier ampholytes. *Anal. Chem.* **72,** 4758–4761.
27. Huang, T. and Pawliszyn, J. (2002) Microfabrication of a tapered channel for isoelectric focusing with thermally generated pH gradient. *Electrophoresis* **23,** 3504–3510.
28. Tragas, C. and Pawliszyn, J. (2000) On-line coupling of high performance gel filtration chromatography with imaged capillary isoelectric focusing using a membrane interface. *Electrophoresis* **21,** 227–237.

29. Li, S. and Pawliszyn, J. (2002) Comprehensive two dimensional separation based on coupling micellar electrokinetic chromatography with capillary isoelectric focusing. *Analyst* **127,** 1159–1163.
30. Huang, T. and Pawliszyn, J. (2000) Axially illuminated fluorescence imaging detection for capillary isoelectric focusing on Teflon capillary. *Analyst* **125,** 1231–1233.
31. Sze, N. S. K., Huang, T., and Pawliszyn, J. (2002) Laser-induced fluorescence detection of non-covalently labelled protein in capillary isoelectric focusing, *J. Separation Sci.* **25,** 1119–1122.
32. Wu X-Z., Sze N.S.K., & Pawliszyn J. (2001) Miniaturization of capillary isoelectric focusing. *Electrophoresis* **22,** 3968–3971.

13

Capillary Electrophoresis–Electrospray Ionization Mass Spectrometry of Amino Acids, Peptides, and Proteins

Mehdi Moini

Summary

Separation in capillary electrophoresis (CE) is based on the movement of charged compounds inside a background electrolyte under an applied potential. Because the mechanism of separation of CE differs from that of conventional high-performance liquid chromatography (HPLC), where separation is based on the analyte's hydrophobic properties, CE is often used as a complementary technique to HPLC. In addition, because CE is performed in narrow capillaries at atmospheric pressure, it is used as an alternative to HPLC, capable of handling small sample volumes while providing shorter analysis times with higher efficiency. For the analysis of amino acid, protein, and peptide mixtures in small volume samples such as in single cells, CE has rapidly evolved as a preferred separation technique. The combination of a high-efficiency separation technique, such as CE, with mass spectrometry (MS) detection provides a powerful system for the analysis of complex biological mixtures. In this chapter, a theoretical and practical approach to achieving high-performance CE–MS is discussed and the utility of CE–MS for the analysis of amino acids, peptides, and proteins is demonstrated.

Key Words

Amino acids; amino acid enantiomers; capillary electrophoresis; electrospray ionization; mass spectrometry; peptides; proteins.

1. Introduction

Separation in capillary electrophoresis (CE) is based on the movement of charged compounds inside a conductive solution under an applied potential. Because the mechanism of separation of CE is based on the electrophoretic mobility of the analytes (which is dependent on the analytes' charge and shape) and differs from that of conventional high-performance liquid chromatogra-

From: *Methods in Molecular Biology, vol. 276: Capillary Electrophoresis of Proteins and Peptides*
Edited by: M. A. Strege and A. L. Lagu © Humana Press Inc., Totowa, NJ

phy (HPLC), where separation is based on the hydrophobic properties of the analytes, CE is often used as a complementary technique to HPLC. In addition, because CE is performed in a narrow capillary at atmospheric pressure, it is also used as an alternative to HPLC, capable of handling small sample volumes while providing shorter analysis times with a higher efficiency. For the analysis of amino acid, protein, and peptide mixtures in small volume samples such as in single cells ***(1–6)***, CE has rapidly evolved as a preferred separation technique.

A variety of detection systems have been employed as CE detectors. These techniques can be divided into two general categories: non-mass spectrometric techniques and mass spectrometric techniques. Among the non-mass spectrometric techniques, electrochemical detection and laser-induced fluorescence (LIF; ***7–16)*** offer the highest sensitivity. Voltammetry and wavelength resolved fluorescence ***(17,18)*** can also provide some structural information, but their chemical identification capability is limited when compared to mass spectrometric techniques. Mass spectrometric techniques provide accurate molecular weight (mol wt) information as a means of chemical identification, a feature that is especially useful when dealing with complex mixtures. The combination of a high-efficiency separation technique, such as CE, with mass spectrometry (MS) detection provides a powerful system for the analysis of complex biological mixtures. Both electrospray ionization (ESI) and matrix-assisted laser desorption ionization (MALDI) ***(19)*** have been used for interfacing CE to MS. ESI, however, is the most suitable and the most commonly used ionization technique for on-line CE–MS analysis (and is the only ionization technique discussed here).

Recently, CE–MS and its application to the analysis of complex mixtures have been reviewed ***(20–23)***. This chapter emphasizes the practical aspects of on-line CE–MS using ESI (CE–ESI/MS).

1.1. Electrochemical Nature of CE, ESI, and CE–ESI/MS

Electrochemistry plays an important role in the operation of CE and ESI. An understanding of the electrochemical nature of CE and ESI can aid in achieving robust CE–ESI/MS operation with high separation efficiency, as well as provide remedies for the negative consequences of the electrochemical nature of CE–ESI/MS.

1.1.1. CE

In CE, the electrophoretic current (i_{CE}) inside the capillary is generated by the movement of charged background electrolyte (BGE) species under the action of an electric field. The current is controlled by several factors including the cross section of the column (*S*), the magnitude of the electrical field (*E*), and the conductivity (*k*) of the BGE and is expressed by ***(24)***

$$i_{CE} = SEk = SEF \sum_j z_j \hat{u}_j c_j \quad (1)$$

where z_j is the charge of component j, $\hat{u}_j$ is the effective mobility of component j, and c_j is the concentration of component j. For example, the total CE current for a 0.1% acetic acid solution (pH of 3.5) using a 75-cm-long 75-µm-id column with a separation voltage of 30 kV was calculated to be 3.4 µA. The total CE current (i_{CE}) is the vector sum of all ion currents within the capillary. Under a specific set of experimental conditions (constant temperature, BGE concentration, capillary diameter, and separation voltage), the CE current is fixed. Because only electrons can move through the external wire that supplies potential to the electrodes, oxidation and reduction reactions proceed, respectively, at the anode and cathode to maintain the CE current and, therefore, the electroneutrality of the cell. In the absence of a species (including the electrodes) with a redox potential lower than that of the aqueous BGE, reactions 2 and 3 (below) will proceed at the anode and cathode, respectively, to maintain the CE current ***(25)***. At pH 7.0:

$$O_2\ (g) + 4H^+ + 4e^- \rightarrow 2H_2O \qquad E^o_{red} = +2.42\ V\ (vs\ SHE) \quad (2)$$

$$2H_2O + 2e^- \rightarrow H_2\ (g) + 2OH^- \qquad E^o_{red} = -0.828\ V\ (vs\ SHE) \quad (3)$$

The consequences of these reactions include a pH increase at the cathode, a pH decrease at the anode, and the formation of bubbles at both electrodes owing to the production of gas. The low flow rates associated with nanotechniques make them particularly vulnerable to the negative effects of these electrochemical reactions. For example, in sheathless nano-CE–ESI/MS, the pH change of the BGE and/or the formation of bubbles have been shown to have a significant effect on selectivity and resolution ***(23,26–29)***. The extent of these reactions depends on the CE current, which is governed by **Eq. 1**. Reducing the conductivity of the BGE and the capillary id minimizes the negative effects of the CE electrochemical reactions by decreasing the CE current.

1.1.2. ESI

In ESI, the application of a high voltage (1–5 kV in positive ionization mode) to a conductive solution exiting a capillary that is pointed toward a counter electrode (such as the MS inlet) at low potential (0–200 V) initiates the formation of a Taylor cone at the ESI tip (the capillary outlet), which is enriched with positive electrolyte ions. Excess positive charge in the Taylor cone is caused by the electrophoretic separation of positive and negative ions at the electrospray electrode and the electrochemical oxidation of water at this electrode (anode), which pumps an excessive quantity of protons into the solution. The emission of positively charged droplets from the tip of the Taylor cone along with solvent evaporation from the charged droplets lead to the formation of positively charged ions. The ES current (i_{ES}) depends on several factors including the solution conductivity, the BGE flow rate, and the magnitude of the electric field at the ESI tip ***(30,31)*** and is given by

$$i_{ES} = A_H V^{v} E^{\varepsilon} \sigma^{n} \quad (4)$$

where A_H is a constant and depends on the dielectric constant and surface tension of the solvent, V^v is the BGE flow rate, E^{ε} is the electric field at the ESI tip, and σ^n is the conductivity of the BGE. Similarly to CE, the extent of the electrochemical reactions and their possible consequences (pH change, bubble formation, and ES electrode degradation) depend on the ES current. The ES current is usually approx 1 μA or less.

1.1.3. CE–ESI/MS

CE and ESI/MS represent two electrical circuits with two sets of electrodes, the CE inlet and outlet electrodes, and the ESI emitter and MS inlet electrodes. CE–ESI/MS overlays these two separate circuits forming a three electrode system in which the CE outlet electrode and the ES emitter electrode are shared between the two circuits (hereafter called the shared electrode) ***(32)***. Therefore, under CE–ESI/MS two electrochemical reactions occur simultaneously at the shared electrode. Depending on the polarity and magnitude of the voltage at the shared electrode compared with that at the CE and MS inlet electrodes, the electrochemical reactions at the shared electrode can be either both reductive (the electrode is giving off electrons), both oxidative (the electrode is accepting electrons), or one reductive and the other oxidative. The total current flowing into the shared electrode is, therefore, a vector sum of the currents flowing through both the CE and ESI circuits. When electrochemical reactions at the shared electrode are either both reductive or both oxidative, the power supply that provides voltage to the shared electrode must be able to supply or sink ***(33)***, respectively, enough current to satisfy both the CE and ESI circuits. For current demanding applications (e.g., when a highly conductive BGE is used or when running under multi-ESI conditions) ***(34)***, a high-current power supply is needed. To protect the MS electronics from arcing at the ESI needle, the ESI power supply of most mass spectrometers provides voltage to the ESI needle through a current-limiting resistor (several mega ohms), which is designed to provide just enough current for the ESI process (~1 μA). When CE is added to this system, the current in the CE circuit will be added to the ESI current. Therefore, under CE–ESI/MS the actual ESI voltage at the shared electrode may be higher or lower (under forward or reverse polarity mode, respectively) than what is measured at the power supply. ESI voltages above the optimum value decrease sensitivity, whereas voltages below the optimum value destabilize or seize the ESI process (*see* **Note 1**). For maximum sensitivity, the ESI voltage must be optimized under CE–ESI/MS conditions.

Another important consequence of the electrochemical nature of ESI is analyte oxidation at low flow rates under a high ESI voltage. As is shown in **Eq. 4**, the i_{ES} is proportional to the ESI voltage. Under very high current densities (high ESI voltages) and low BGE flow rates, where the redox reaction (reaction 2) at the anode is unable to supply the current required at the ESI electrode, electrolysis reactions of water with higher redox potentials (reactions 5 and 6) will occur to supply the necessary current ***(31)***.

$$H_2O_2 + 2H^+ + 2e^- \leftrightarrow 2H_2O \qquad E^o_{red} = -1.776\ \text{V (vs SHE)} \tag{5}$$

$$O\ (g) + 2H^+ + 2e^- \leftrightarrow H_2O \qquad E^o_{red} = -2.42\ \text{V (vs SHE)} \tag{6}$$

Interactions of reactive species generated in these reactions with peptides are proposed to be the primary factor responsible for the oxidation of peptides at low flow rates. Analyte oxidation significantly reduces the sensitivity of detection by diluting the analyte signal over several oxidized species ***(35)***. The extent of these reactions depends on i_{ES}, which itself depends on the electric field at the ESI tip. Because it is the geometry of the tip (and, therefore, the electric field at the ESI tip) that dictates the voltage necessary for ESI operation ***(30)***, sharpening the capillary outlet (by hydrofluoric acid [HF] etching, for example) can significantly enhance sensitivity by reducing the voltage required for stable ESI operation. This will decrease the ESI current and result in reduced analyte oxidation. In order to minimize analyte oxidation, it is important to set the ESI voltage very close to the ESI onset voltage (V_{on}) but not low enough to cause ESI instability.

1.2. High-Performance CE–ESI/MS

High separation efficiency and high-sensitivity CE–MS analysis depend on several factors including the CE capillary, the BGE, the CE to MS interface, and the mass spectrometer.

1.2.1. Parameters Related to the Capillary

Parameters related to the CE capillary include the capillary length, the capillary inner diameter (id), the capillary wall thickness, the sharpness of the capillary tip, and the chemical composition of the inner wall of the capillary.

1.2.1.1. Capillary Length

A practical measure of resolving power in CE is

$$R = t_m/W_{1/2} \quad (7)$$

where t_m is the migration time of the peak and $W_{1/2}$ is its full width at half maximum (FWHM). According to this equation, as long as the rate of increase of $W_{1/2}$ is proportionally less than that of t_m, increasing the migration time increases the resolving power of the CE. In the absence of analyte–wall interactions and diffusion (factors that can increase $W_{1/2}$ as a result of increasing the migration time), the injection plug width is the major factor that affects $W_{1/2}$ (*see* **Note 2**). This is especially true for derivatized capillaries, where analyte–wall interactions are eliminated, and for the analysis of proteins, where diffusion is minimal and it has been shown that even an analysis time of 1 h does not significantly deteriorate peak widths ***(36)***. In the absence of electro-osmotic flow (EOF), t_m is given by **Eq. 8**:

$$t_m = L/\nu_{ep} = L^2/\mu_{ep}V \quad (8)$$

where ν_{ep} is the analyte's electrophoretic velocity, L is the capillary length, μ_{ep} is the analyte's electrophoretic mobility, and V is the magnitude of the separation voltage across the capillary. According to **Eq. 8**, in the absence of EOF, the most efficient way to increase the migration time of the analytes is to increase the capillary length. However, in CE–ESI/MS the presence of EOF toward the capillary outlet is necessary for maintaining stable ESI and for achieving high separation efficiency. Therefore, in addition to the length of the capillary, the EOF rate also affects t_m. Because the presence of EOF toward the capillary outlet shortens the migration times of the analytes, the optimum resolution in high-performance CE–ESI/MS is achieved at the lowest possible EOF, while maintaining stable ESI ***(23)***. The disadvantage of using long capillaries in conjunction with a low EOF, however, is long analysis times.

1.2.1.2. Capillary Inner Diameter

Experimentally, the highest sensitivity and resolution has been achieved using narrow capillaries ***(37)***. This is because of the lower BGE flow rates of narrow capillaries, which, for the same amount of sample injected, causes the analytes to be less diluted upon exiting the capillary. Because ESI is a concentration-sensitive ionization technique, a higher analyte concentration translates into a greater sensitivity of detection. In addition, narrower capillaries have narrower outlets and after sharpening their tips, they generate finer droplets, which enhances analyte ionization efficiency. Moreover, because narrower capillaries dissipate heat more efficiently, they enhance separation efficiency by reducing analyte diffusion and by maintaining the plug profile flow of the BGE (by preventing viscosity variation across the diameter of the CE capillary) *(23)*. Also, because of the high sensitivity of narrow capillaries, samples can be injected in a narrower plug, which eliminates the peak broadening associated with wide injection plugs. As a result, peaks generated with narrow capillaries (<20-µm id) are usually very narrow (~2–3 s wide), and to obtain quantitative peak information (10 data points across the electrophoretic peak), a fast mass spectrometer such as a TOF-MS is required. For example, **Fig. 1** (top and bottom) shows the separation of a cytochrome-*c* digest using a 10-µm-id and a 50-µm id capillary with 61 and 362 fmol of sample injected, respectively. In these experiments, an ion-trap MS with an acquisition time of 1 s was used. As is shown, the narrower capillary provided narrower peaks and a higher sensitivity. However, the mass spectrometer was not fast enough to quantitatively handle the narrow peaks (1–2 s FWHM) generated using the 10-µm id capillary. This resulted in electrophoretic peaks that were not quantitatively represented. For comparison, the FWHM for the peaks of the 50-µm id capillary were approx 4 s (notice peak intensity variations in the two capillaries).

The process of enhancing sensitivity by using narrower capillaries apparently does not continue indefinitely. At very low nL/min flow rates (using a sharpened capillary outlet tip) the ionization efficiency of the analyte from charged droplets and their transfer efficiency to the MS approaches 100%. Under this condition, ESI becomes a mass sensitive ionization technique and, therefore, a further reduction of capillary diameter (<5-µm id) does not necessarily enhance the sensitivity of detection. On the other hand, frequent capillary blockage and difficulties in interfacing CE to MS using these narrow capillaries limit the application of these capillaries and make them impractical for routine analysis.

1.2.1.3. Capillary Wall Thickness

Increasing the capillary wall thickness (and, therefore, the capillary outer diameter [od]) increases the ruggedness of the capillary but reduces the heat dissipation efficiency of the capillary. It also increases capillary tip sharpening and interface preparation time. For these reasons, 150-µm od capillaries are usually preferred over 360-µm od capillaries.

1.2.1.4. Capillary Tip

The dimensions of the capillary outlet tip have a strong effect on the performance of CE–ESI/MS. The sharpness, the diameter of the opening, and the front end surface area of the capillary outlet tip all affect the size of the emitted charged droplets, the voltage required for the initiation of the ESI process (onset

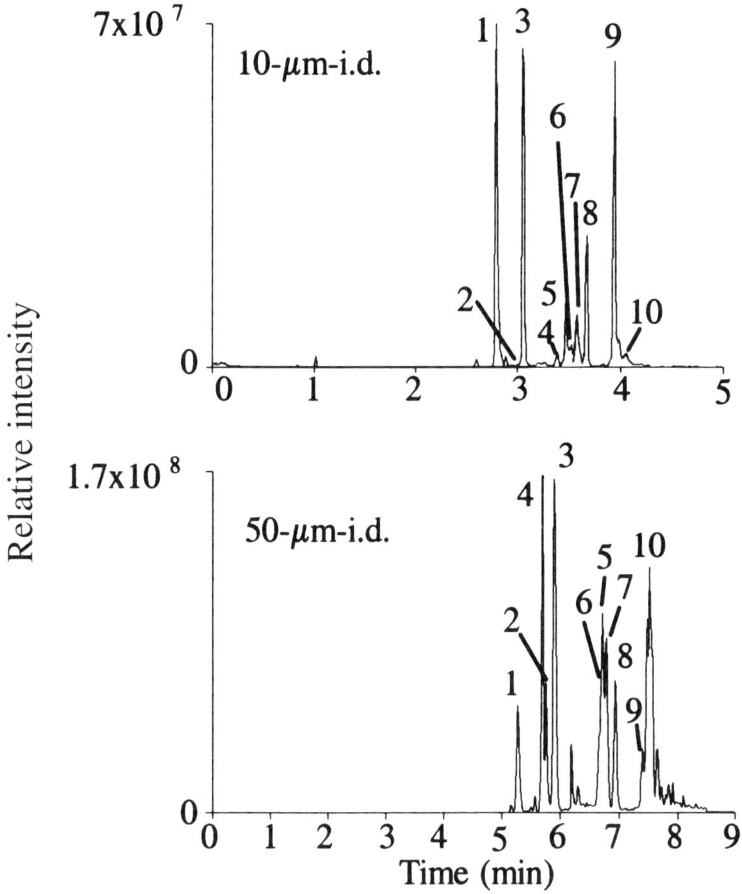

Fig. 1. The top and bottom panels show base peak electropherograms of the tryptic digest of cytochrome-*c* using a 10- and a 50-µm id capillary with 61 and 362 fmol of sample injected, respectively. In these experiments, an ion-trap MS with an acquisition time of 1 s was used. The labeled peaks are: 1, Acetyl-GDVEK; 2, EDLIAYLK; 3, TCQAPGFTYTDANK; 4, EETLMEYLENPK; 5, YIPGTK; 6, MIFAGIK; 7, MIFAGIK; 8, GITWK; 9, TGPNLHGLFGR; 10, CAQCHTVEK (heme). Adapted with permission from **ref.** ***37***.

voltage), and the sensitivity of detection. Having a narrow opening (≤20 µm id) and a thin wall (≤20 µm) at the capillary outlet tip are essential for achieving high efficiency (separation and sensitivity) CE–ESI/MS operation. Theoretically, the voltage required to initiate the electrospray process from the capillary tip is given by

$$V_{ON} \approx 2 \times 10^5 (\gamma r_c)^{1/2} \ln (4d/r_c) \quad (11)$$

where γ is the surface tension of the BGE (for water $\gamma = 0.073$ N/m^2), r_c is the capillary radius at the tip (one-half of the capillary od at the tip), and *d* is the distance between the capillary tip and the counter electrode (MS inlet). For example, for a capillary with a tip radius of 20 µm that is positioned 2 mm away from the MS inlet, $V_{ON} \approx 2 \times 10^5 (0.073 \times 20 \times 10^{-6})^{1/2} \ln (4 \times 2 \times 10^{-3}/ 20 \times 10^{-6}) \approx 1450$ V. Experimentally, for CE–ESI/MS analysis of peptides, the highest S/N is achieved when using ESI voltages very close to the onset voltage. At higher voltages, other processes such as corona discharge and analyte oxidation degrade the sensitivity of detection of the analytes of interest ***(35)***. In our laboratory, a solution of 49% HF is used for tip sharpening whereas N_2 gas is passed through the capillary to minimize inner wall etching.

In addition to sharp tips, adequate liquid flow at the capillary tip is essential for stable ESI operation. The minimum flow required for stable ESI operation depends on the id of the capillary opening and the sharpness of the tip, which together determine the electrical field at the tip. Capillaries with narrower openings, sharper tips, and lower liquid flow rates require lower onset voltages than capillaries with wider openings, duller tips, and higher liquid flow rates. Unstable ESI results in an uneven accumulation of liquid at the tip and a pulsation of liquid exiting the tip, resulting in a noisy background and low resolution ***(32,38)***. The tip of the capillary can be pulled to a narrower and sharper opening, which results in enhanced sensitivity and more stable ESI. Pulling the tip, however, increases the chance that the capillary will clog at the tip because particles that are small enough to enter the capillary can be too big to exit the narrow opening of the pulled tip. Therefore, narrow capillaries (~15–20 µm id) with etched tips provide high sensitivity with minimal clogging.

1.2.1.5. Chemical Composition of the Inner Wall Surface

In CE–ESI/MS analysis, because the composition of the BGE is limited to those that are compatible with ESI, analyte–wall interactions cannot always be eliminated by choosing from the variety of BGEs that are available for conventional CE analysis. Analyte–wall interactions, therefore, must often be eliminated by capillary wall surface derivatization. Of course, the use of a BGE that can eliminate this interaction without column derivatization is desirable because column derivatization is usually time-consuming, expensive, and short-lived. Also, the derivatization reagent or its degradation products can contribute to the background chemical noise. Underivatized capillaries have been used for the CE–ESI/MS analysis of polynucleotides, which are negatively charged at basic pH values, in conjunction with a volatile BGE such as ammonium acetate at pH approx 10.0 under negative ESI ***(39)***. In addition, 1 *M* formic acid and 30 m*M* 18-crown-6-tetra carboxylic acid have been used, respectively, for the analysis of amino acids and their D/L enantiomers using bare fused-silica capillaries *(40,41)*. However, for peptides and proteins, optimum sensitivity and resolution is achieved in acidic conditions (such as a 0.1% acetic acid solution, pH ~3.5) under positive ESI mode. At these pHs, however, the silanol groups (pH ~2.0) on the capillary inner wall are negatively charged whereas the proteins/peptides are positively charged. The interaction between the positively charged proteins/peptides and the negatively charged silanol groups results in a severe degradation of separation efficiency and can even result in a complete loss of protein peaks owing to their irreversible adsorption onto the capillary wall. Optimum separation and sensitivity for proteins and peptides is achieved using an acidic BGE (pH 2.0–4.0) in conjunction with a derivatized capillary. A variety of surface wall derivatization techniques have been used in CE–ESI/MS *(23)*. The most common derivatization technique for CE–ESI/MS is treatment with trimethoxyaminopropylsilane (APS; ***42***). The reactions between the methoxy groups of APS and the –OH of the silanol groups attached to the inner wall reverse the wall polarity by leaving amino groups (which are positively charged at low pH values) dangling from the capillary wall. Under this condition, the EOF flow is reversed and the CE must be operated in reverse polarity mode (negative high voltage at the inlet) to obtain EOF towards the outlet of the capillary.

1.2.2. Background Electrolyte

There are a variety of BGEs that are commonly used with CE in conjunction with UV detection such as sodium phosphate, borate, acetate, and so on. These BGEs, however, have a high salt content and are not compatible with ESI. This is because there is a competition under ESI between salts and analyte for available charge, causing the analyte signal to be suppressed when excessive salts or other easily ionizable compounds are present. Moreover, salts often form adducts with the analytes of interest, which can decrease the sensitivity of detection and complicate compound identification by spreading the analyte signal over multiple adduct compounds. Commonly used BGEs for CE–ESI/MS analysis include aqueous solutions of high vapor pressure acids such as acetic acid and formic acid for positive ionization mode, or aqueous solutions of ammonium acetate or ammonia for negative ionization mode. In addition to their high vapor pressure, these BGEs offer several advantages including: (1) they have a low conductivity, which enhances resolution by reducing CE current and, therefore, heat generation in the capillary; (2) they tend to charge analytes in solution rather than suppress them, thereby enhancing the sensitivity of detection under ESI; (3) under ESI, they generate a low mol wt background chemical noise that does not interfere with the high mol wt analyte signals; (4) adduct formation with analytes is minimal and they, therefore, do not reduce the sensitivity of detection by spreading the analyte signal over a variety of species. The concentration of acidic and basic BGEs are usually 10–30 m*M* with a pH = 3.0–4.0 and 8.0–10.5, respectively. For the analysis of amino acids using underivatized capillaries under positive ionization mode, a 1 *M* formic acid solution (pH ~1.8) is commonly used as the BGE ***(41)***.

1.2.3. CE–MS Interfacing

An important feature of any CE–MS interface is the method by which electrical current is provided to the outlet electrode. Over the past 15 yr, a variety of CE–ESI/MS interfaces have been introduced. These interfaces are divided into three general categories: sheath-flow interfaces, sheathless interfaces, and split-flow interfaces. A comprehensive review of CE–MS interfaces was recently published ***(23)*** and only a brief description is given here.

1.2.3.1. Sheath-Flow Configuration

In the sheath-flow configuration, the electrical connection to the CE outlet is achieved with a liquid sheath that either mixes with the CE BGE at the CE outlet through a conductive coaxial tubing, or enters a nanospray tip via a liquid junction. Sheath-flow configurations with coaxial tubing, in which the outlet of the CE capillary is simply inserted into an ESI emitter (a piece of stainless steel tubing, commonly referred to as an ESI needle), have several advantages, including simple fabrication, simple implementation, and good reliability. These advantages make them the most widely used interfaces for routine CE–ESI/MS analysis. The sheath-flow configurations, however, have several disadvantages as well: (1) they have a low sensitivity of detection attributable to dilution of the analyte by the sheath liquid; (2) there is competition between the species present in the sheath liquid and the analyte for available charge in the ESI process; (3) there are adverse effects on separation, solubility, and molecular conformation, which vary according to the sheath liquid composition.

1.2.3.2. Sheathless Configuration

In sheathless interfaces, the electrical connection to the capillary outlet is achieved by direct metal to liquid contact at or near the tip of the CE capillary outlet. Because there is no sheath liquid to dilute the CE effluent, the major advantage of the sheathless interface is a high sensitivity of detection. A variety of sheathless interfaces have been introduced. (1) One is the attachment of a metal-coated ESI tip to the CE capillary outlet. The major disadvantage of the coated tip technique is the degradation of the tip caused by the electrochemical reactions occurring at the tip and electrical discharge between the tip and the MS inlet. (2) Another is the attachment of a nanospray tip to the CE capillary outlet with polysulphone microdialysis tubing or a low dead volume Tee. Because the capillary id is usually smaller than the wall thickness, the major disadvantage of this technique is the existence of significant dead volume where the two capillaries are attached. In addition, there are problems with the degradation of the stainless steel Tee, the formation of gas bubbles caused by the electrochemical reactions occurring at the Tee junction, and the detachment of the microdialysis tube from the capillary and/or the ESI emitter. (3) A third sheathless interface involves the insertion of a wire into the CE capillary through the CE capillary outlet, or through a small opening near the CE outlet where it is sealed in place using epoxy. Wire designs are robust and there is no dead volume associated with them; however, their main disadvantage is the formation of gas bubbles in the capillary caused by the electrochemical reactions that occur at the electrode.

1.2.3.3. Split-Flow Interface

To remedy some of the problems associated with sheath–flow and sheathless configurations, a split-flow CE–ESI/MS interface was introduced, in which the electrical connection to the CE capillary outlet was achieved by diverting part of the CE BGE out of the capillary through an opening near the capillary outlet (*see* **Fig. 2**) *(37)*. The BGE exiting the opening (the crack) contacts a metal sheath, which acts as the shared electrode. In cases where the ESI source uses a metal needle, the voltage contact to the CE BGE was achieved by simply inserting the outlet of the CE capillary, which contains an opening, into the existing ESI needle (thereby greatly simplifying CE to MS interfacing). Owing to the concentration sensitive nature of ESI, splitting a small percentage of the CE flow has a minimal effect on the sensitivity of detection. In addition, because the liquid is flowing through an opening and out of the capillary, there is no dead volume associated with this interface. Moreover, bubble formation caused by electrochemical reactions at the electrode does not affect CE–ESI/MS performance because the actual metal-to-liquid contact occurs outside of the CE capillary. The sensitivity associated with the split-flow CE–MS interface, the ease of fabrication, the universality, and the lack of dead volume make this design a superior CE–ESI/MS interface.

2. Materials

1. Mass spectrometer.
2. CE instrument.
3. HPLC pump.
4. Dentist's drill.
5. Gas chromatography (GC) oven.
6. Column wash bomb.
7. Microscope.

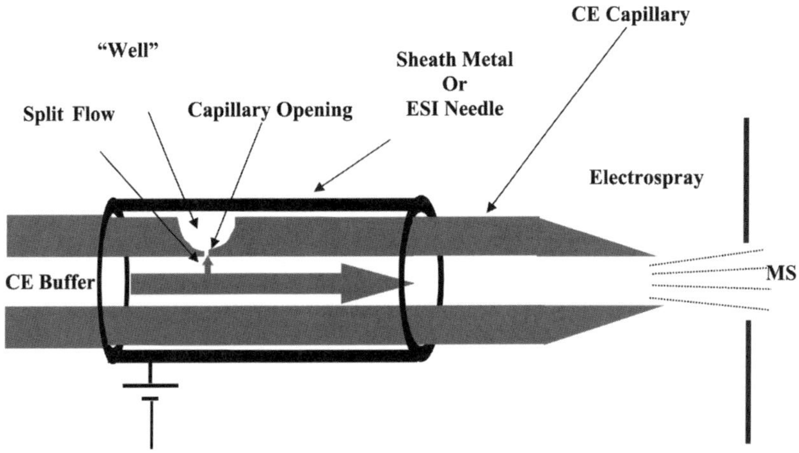

Fig. 2. Schematic of the split-flow interface. Adapted with permission from **ref. *23***.

8. Fused-silica capillaries.
9. Ceramic glass cutter.
10. Stainless steel tubing.
11. Platinum and stainless steel wires.
12. 49% Hydrofluoric acid (HF).
13. Sodium bicarbonate.
14. Mixture of proteins.
15. Mixture of peptides.
16. Mixture of amino acids.
17. 0.1% Acetic acid.
18. 10% 3-Aminopropyltrimethoxysilane in toluene.
19. 1 *N* Sodium hydroxide (NaOH).
20. 1 *N* Hydrochloric acid (HCl).
21. Trypsin.
22. 1 *M* Ammonium hydroxide.
23. 1 *M* Formic acid.
24. (+)-(18-Crown-6)-2,3,11,12-tetracarboxylic acid (18-C-6-TCA).

3. Methods

3.1. Procedure for APS Derivatization of Fused-Silica Capillaries

Note: pressures up to 500 psi are required inside the column wash bomb depending on the capillary id. High-purity nitrogen gas, approx 50 and 500 psi, are used for 50 and 20-μm-id capillaries, respectively.

1. Capillaries are cut to the desired length (50–130 cm), but multiple capillaries can be derivatized at the same time.

2. Using a column wash bomb, the CE capillaries are rinsed with 1 *N* HCl for 4 h (to clean the surface walls from any residual cations).
3. Rinse with water for 0.5 h.
4. Rinse with 1 *N* NaOH for 4 h (to hydrolyze siloxane to silanol).
5. Rinse with water for 0.5 h.
6. Rinse with methanol for 0.5 h.
7. Dry capillaries in a GC oven overnight (12–16 h) at 300°C (while passing He gas through them at ~30 psi).
8. Derivatize the capillaries by passing 10% APS in toluene through the column for 4 h (*see* **Note 3**).
9. Rinse the capillaries with dry toluene for 0.5 h.
10. Rinse the capillaries with methanol for 0.5 h.
11. Air out the methanol.
12. Let the capillaries sit for at least overnight.
13. Before use, rinse the capillaries with the BGE for 0.5 h.

3.2 Procedure for Making a Split-Flow Interface

The split-flow interface design is achieved by making a small hole (a few micrometers in diameter) or a crack, using a dentist's drill under a microscope, into the wall of the capillary 2–3 cm from the capillary outlet. To facilitate observation of the opening in the capillary wall, methanol is forced through the capillary from the inlet side using an HPLC pump. The drilling is stopped immediately after observing liquid exiting the opening. A variation of this technique can be applied to make a crack in the wall of narrow capillaries (<50 µm). In this case, the drilling process is stopped just before the id of the capillary is exposed. Depending on the capillary id and od, a 25- or 50-µm diameter stainless steel wire is then placed underneath the capillary on a hard surface, opposite the side of the capillary containing the "well" (*see* **Fig. 2**). The "well" is then aligned directly on top of the wire and pressure is applied on both sides of the "well" (by sliding two fingers on the capillary toward the "well") until a flow of liquid is observed, indicating the formation of a small crack. After the crack is made, the outlet tip of the CE capillary is sharpened using 49% HF, and the overall performance of the capillary is tested using a peptide mixture (*see* **Note 4**).

3.3 Application of CE–ESI/MS to the Analysis of Amino Acid, Peptide, and Protein Mixtures

3.3.1 CE–ESI/MS Analysis of Underivatized Amino Acids

The detection, identification, and quantitation of amino acids is important in many areas of science including biological and biochemical analysis ***(43,44)***, medical diagnostics ***(45,47)***, and food analysis ***(45,48,49)***. Gas chromatography (GC) ***(50)***, liquid chromatography (LC) ***(51)***, GC–MS ***(52)***, and CE ***(45)*** have all been used for the detection, identification, and quantitation of amino acids. However, for efficient separation and sensitive detection using these methods, amino acids generally require derivatization prior to their analysis. This is a labor-intensive and time-consuming process. HPLC–MS *(53)* and sheath–flow CE–ESI/MS *(48,54)* techniques have been introduced for the analysis of underivatized amino acids. However, because of the high liquid flow (>5 µL/min) associated with these techniques, their main disadvantage is a low absolute sensitivity. Recently, we have introduced a sheathless CE–ESI/MS technique for the analysis of underivatized amino acids utilizing a bare-fused silica capillary with 1 *M* formic acid as the BGE *(41)*. **Figure 3** shows the CE–ESI/MS electropherogram of the separation of the 20 standard protein amino acids using a 130-cm-long underivatized CE capillary. For this experiment, approx 400 fmol of each amino acid was injected. As is shown in **Fig. 3**, all amino acids were separated and detected. The use of a nonderivatized column is certainly a big advantage because capillary derivatization techniques are usually time-consuming, labor intensive, expensive, and short lived. The limits of detection for the amino acids were at the low femtomole level.

3.3.2. CE–ESI/MS Analysis of Underivatized Amino Acid Enantiomers

One aspect of amino acid analysis that has been particularly challenging in the past is the separation and detection of amino acid enantiomers *(55–57)*. Interest in the role of D-amino acids in mammalian systems has surged in the past decade due to findings that suggest neuronal and neuroendocrine roles of some D-amino acids *(58–60)*. The analysis of amino acid enantiomers is also important for food quality analysis where the enantiomeric ratio of amino acids can be used as a reliable parameter to assess food quality *(49)*. An increase in the ratio of D-amino acids to L-amino acids within foods may be indicative of extensive processing, contamination, adulteration, or aging *(49)*. The enantiomeric ratio of amino acids within a biological sample can also be used in biological dating applications *(61)*. An advantage of amino acid enantiomer separation using CE is that many enantiomers can be separated on-line by simply using a BGE that contains a chiral selector. One chiral selector reagent that has been found to be compatible with ESI/MS is (+)-(18-Crown-6)-2,3,11,12-tetracarboxylic acid (18-C-6-TCA) *(41)*. In fact, the acidity of 18-C-6TCA renders it suitable as a CE BGE without the need for additional acid. 18-C-6-TCA is a macrocyclic polyether ring system consisting of several oxygens joined by ethylene bridges. The structure of this molecule is shown in the inset of **Fig. 4**. The polyethylene ring forms a cavity with the oxygens roughly forming a plane on the inner side of the cavity *(56)*. The cavity formed by the ring of 18-C-6-TCA can form complexes with cations of suitable sizes. The ammonium cation of an amino acid can form a complex with the polyethylene ring of 18-C-6-TCA

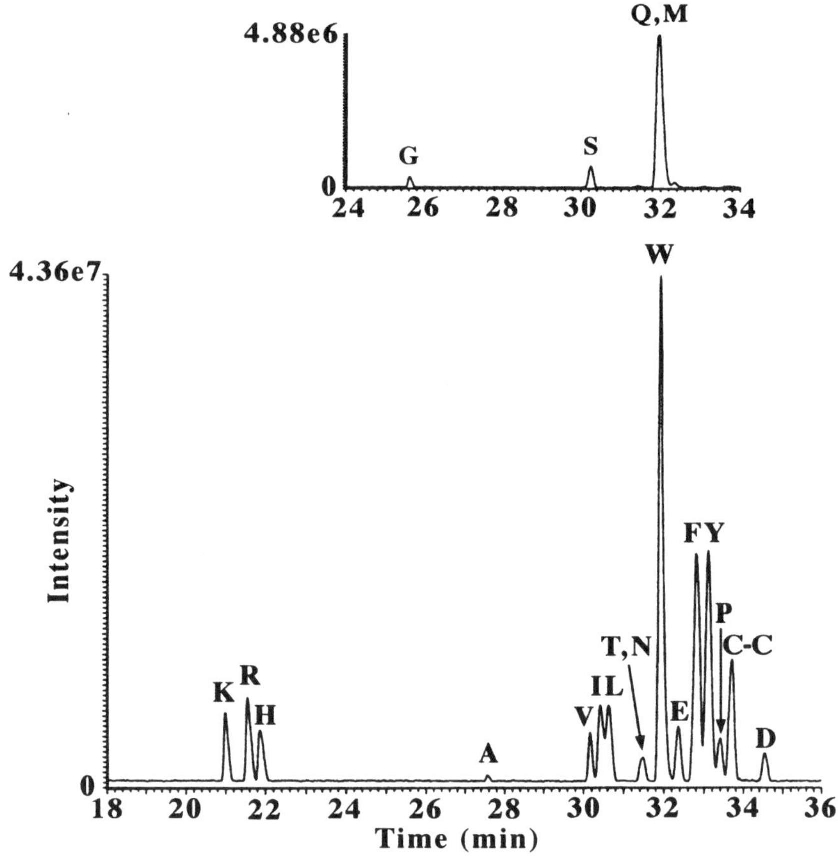

Fig. 3. CE–ESI/MS base peak electropherogram of the separation of the 20 standard protein amino acids using a 20-μm id, 130-cm-long capillary. No pressure was applied at the CE inlet during this separation. Forward polarity with 30-kV separation voltage was used. Approximately 400 fmol of each amino acid was injected. Adapted with permission from **ref. *41***.

through three $^{+}NH \cdots O$ hydrogen bonds in a tripod arrangement ***(62–65)***. The N–C* bond lies perpendicular to the plane defined by the oxygens of the polyethylene ring system (C* represents the chiral center of the amino acid).

The carboxyl group substituents of 18-C-6-TCA allow for enantiorecognition. They lie perpendicular to the plane defined by the polyethylene ring system and differentially impede complexation through stereospecific steric hindrance ***(56)***. Thus, D- and L-amino acid enantiomers have different affinities for complexation with 18-C-6-TCA. It has been experimentally determined that a 30-m*M* solution of 18-C-6-TCA provides the optimum resolution for amino acid enantiomer separation under CE–ESI/MS ***(41)***. **Figure 4** demonstrates the separation of eleven amino acid enantiomers using a 30-m*M* 18-C-6-TCA solution as the BGE/chiral selector reagent. Approximately 500 fmol of each amino acid was injected. As is shown (*see* **Fig. 4**), almost all D/L enantiomers are baseline separated. However, compared to the absolute intensities of the amino acids analyzed under 1 *M* formic acid (*see* **Fig. 3**), the absolute intensities of the amino acids analyzed under 30 m*M* 18-C-6-TCA (*see* **Fig. 4**) are on the average approx four times lower (after normalization with respect to the amount injected in each experiment) ***(41)***.

3.3.3. High Mass Accuracy Peptide Mapping

Tandem mass spectrometry (MS/MS) in conjunction with capillary or micro HPLC or CE has recently become a desired technique for peptide mapping ***(66–69)***. For the separation of complex biological mixtures, CE has several advantages over HPLC including higher resolution capabilities, the ability to analyze smaller sample quantities, and a higher absolute sensitivity when combined with ESI/MS ***(12,70)***. As a result, the combination of CE and ESI/MS has proven to be applicable to a wide range of biologically important mixtures ***(23,71)***. For example, CE time-of-flight MS (CE–TOF-MS) has been successfully applied to the analysis of standard mixtures of peptides and proteins using both amine-coated and bare fused-silica capillaries in conjunction with both sheath flow and sheathless CE–MS techniques ***(23,72)***.

Most CE–MS analyses to date have used mass spectrometers that are capable of operating with only unit mass resolution and low mass accuracy. This is because until recently, only magnetic sector ***(73,74)*** and Fourier transform ion cyclotron resonance (FTICR) ***(75)*** MS were capable of generating high resolution, high mass accuracy mass spectra under ESI. Currently, in addition to the magnetic sector instrument and FTICR, modern quadrupole and TOF-MS are capable of providing high resolution (>5000) and high mass accuracy (<10 ppm). Modern FTICR MS can provide ultrahigh resolution ($>10^5$) and superior mol wt information with a mass accuracy of sub-ppm, but they do so on a 1-s/acquisition time-scale. As shown above, this time-scale is too long for a quantitative analysis of the narrow peaks generated using narrow capillaries. Under these conditions, it is the fast acquisition time-scale (<0.1 s) of the TOF-MS that makes it spectrometer of choice for the CE–MS analysis of complex biological mixtures ***(76–80)***.

One of the most commonly used techniques for protein identification by MS is peptide mapping in conjunction with a protein database search. In order to identify a protein using a protein database, several pieces of information are needed, including the average mol wt of the protein and the masses of several of the peptide fragments obtained through the enzymatic digestion of the pro-

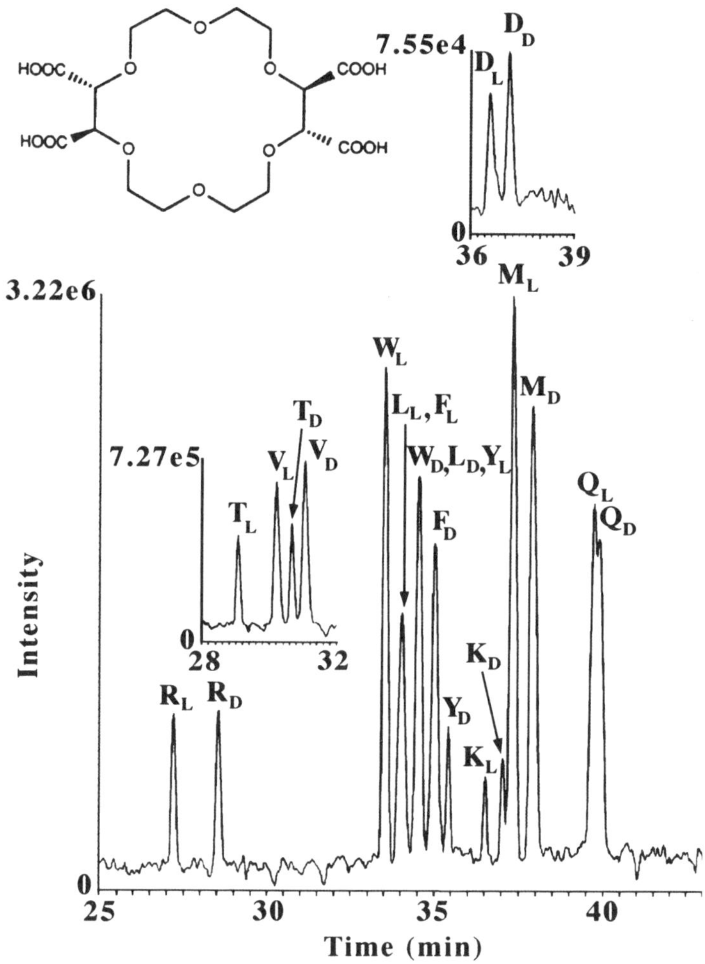

Fig. 4. The selected ion electropherogram of the separation of eleven amino acid enantiomers using a 30-m*M* 18-C-6-TCA solution as the BGE/chiral selector reagent and a 20-μm id, 130-cm-long capillary. This analysis was carried out with 5 psi of inlet pressure. This pressure was necessary to maintain stable ESI due to a lowered electroosmotic flow under 30 m*M* 18-C-6-TCA conditions. Approximately 500 fmol of each amino acid was injected. Inset shows the chemical structure of 18-C-6-TCA. Adapted with permission from **ref. *41***.

tein of interest. The enzymatic digestion of proteins followed by CE–MS or HPLC–MS analysis of the resulting fragments have been routinely used for protein identification using ESI. However, under ESI conditions these peptide fragments are often multiply charged and to obtain the masses of these multiply protonated peptides, one must know both the peptides' respective *m/z* values and charge states. Because the peptide fragments formed under enzymatic digestion, such as tryptic digestion, are usually small (<3000 Da) and their charge states are generally between 1 and 4, a mass spectrometer with a resolving power of approx 5000 can easily identify the charge states and, therefore, the mol wt of these peptides.

Although knowing the mol wt of a protein and several of its peptide fragments can greatly increase the odds of identifying a protein using a protein database, this information may not be sufficient to uniquely identify the protein of interest. This is the case, for example, when dealing with modified proteins or when only low sample quantities are available, such that the number of peptide fragments that are detectable are too few to unambiguously identify the protein of interest *(81)*. In these situations, accurate mass measurements of the peptide fragments can be used as yet another important factor in reducing the list of most-likely proteins provided by a protein database and in increasing the confidence level in the database search results *(82)*. In fact, most protein databases allow the user to select error limits associated with the masses of the protonated peptides used in the search, such that a higher peptide mass accuracy corresponds to a shorter list of most-likely proteins.

In some cases, when the amino acid composition of peptides are the same but their sequences are different, even accurate mass measurements of a few peptide fragments may not be enough to identify an unknown protein since different peptides may have the same mass. In such cases, collision-induced dissociation (CID) of these peptides can be used to obtain total or partial determination of the amino acid sequences of these peptides *(83)*. However, because several amino acids differ in mass by only 1–2 Da, it is ultimately desirable to use a mass spectrometer that, in CID mode, can easily provide the mass accuracy needed to distinguish these amino acids and, therefore, their corresponding peptides from each other. To demonstrate this point, the CE–MS analysis of horse heart cytochrome-*c* using a high resolution, high mass accuracy TOF-MS is shown later *(82)*. First, the accurate average mol wt of the intact proteins were measured. The measured average mol wt of horse heart cytochrome-*c* was 12,359.2±0.5 Da (calculated mass 12,359.80 Da). Second, the accurate monoisotopic masses of the peptide fragments derived from the tryptic digestion of horse heart cytochrome-*c* were measured (*see* **Table 1**). In order to calculate the accurate masses of the peptide fragments, two pieces of information are needed: (1) the accurate, monoisotopic mass of each peak and (2) the number of charges associated with the peak. The resolution provided by the TOF-MS was sufficient to distinguish isotopic peaks of all peptides studied and, therefore, to determine their individual charge states. The procedure for measuring the accurate mass of each peak is shown in **Fig. 5**, in which the ion

Table 1
Comparison of Mass Measurement Errors Using the Single-Sprayer, Single-Nozzle and the Dual-Sprayer, Dual-Nozzle Design

m/z (calculated)	Mass error Single nozzle	Dual Nozzle	Sequence
589.283	<1.0	–6.1	Acetyl-GDVEK
964.536	7.3	–1.4	EDLIAYLK
735.847	12.2	+1.0	TGQAPGFTYTDANK
748.354	10.7	+3.6	EETLMEYLENPK
678.383	<1.0	–1.5	YIPGTK
779.449	11.5	–3.5	MIFAGIK
634.385	3.2	–3.2	IFVQK
604.346	5.0	–1.3	GITWK
584.815	17.1	+3.4	TGPNLHGLFGR
545.217	11.0	–0.4	CAQCHTVEK(heme)

In the single-nozzle experiment, the calibration compound was added to the mixture and its electropherograms were averaged with that of the unknown to generate a single spectrum. In the dual sprayer, dual-nozzle experiment, one sprayer and nozzle were used for sample introduction whereas the other sprayer and nozzle were used for reference compound introduction. Acetyl DVEKGKKIFVQKCAQC HTVEK(heme)GGKHKTGPNLHGLFGRKTGQAPGFTYTDANKNKGITWKEETL MEYLENPKKYIPGTKMIFAGIKKKTEREDLIAYLKKATNE (104 amino acids, calculated average mol wt 12359.80 Da)

electropherogram of the peak of the unknown was added to the ion electropherograms of the calibration compounds to generate a single spectrum. The masses of the two reference compounds were then used as internal reference masses to measure the accurate mass of the unknown peptide. This procedure was repeated for all protein fragments listed in **Table 1**. The measured masses of the protonated peptide fragments were then used to search the database. It was found that a higher peptide mass accuracy significantly reduces the number of possible matches and, therefore, simplifies protein identification ***(82)***.

3.3.4. Enhancing Mass Accuracy Using Multispray, Multiinlet MS

In the experiment mentioned earlier, a solution containing the reference compounds was mixed with the cytochrome-*c* tryptic digest solution and the mixture was hydrodynamically injected into the CE inlet. For the accurate mass measurement of each peptide fragment, the peaks of the ion electropherogram for each peptide fragment and those for the two reference peaks (*m/z* 524.266 and *m/z* 1221.991 in **Fig. 5**) were averaged to obtain a single spectrum. The accurate masses of the reference peaks were then used as internal reference

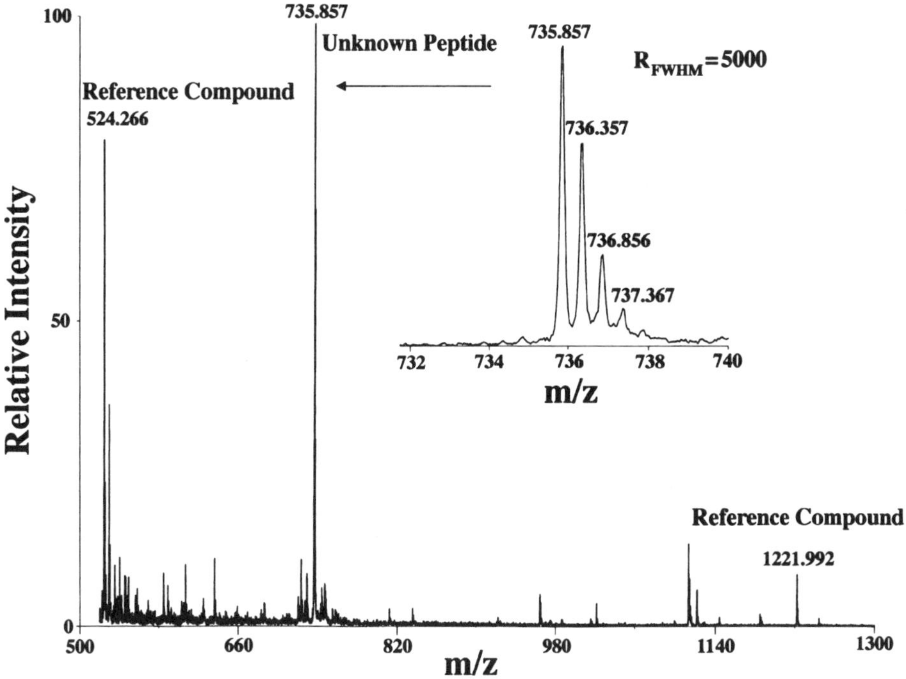

Fig. 5. The procedure for measuring the accurate mass of each peak in **Fig. 4**, in which the ion electropherogram of the peak of the unknown was added to the ion electropherograms of the calibration compounds to generate a single spectrum. The masses of the two reference compounds were then used as internal reference masses to measure the accurate mass of the unknown peptide. This procedure was repeated for all protein fragments listed in **Table 1**. Adapted with permssion from **ref.** ***82***.

masses for the accurate mass determination of the peptides. In most cases, the error between the calculated and measured masses was approx 10 ppm using this technique of calibration.

In order to reduce the mass error to <5 ppm, we recently introduced a multi-spray, multiinlet TOF-MS in which the atmospheric pressure sampling inlet (nozzle) of the TOF-MS was modified by replacing its single nozzle with multiple atmospheric pressure nozzles ***(34)***. This allowed multiple streams of liquid to be introduced into the MS in parallel (one electrosprayer for each nozzle) with minimum analyte interaction between streams. To obtain a higher mass accuracy by providing an internal reference on each scan (acquisition), and to evaluate the suitability of the TOF-MS for molecular formula confirmation, a dual–ESI sprayer, dual-nozzle version of this design was used (*see* **Fig. 6**) ***(34)***. A significant improvement in mass accuracy was observed when this tech-

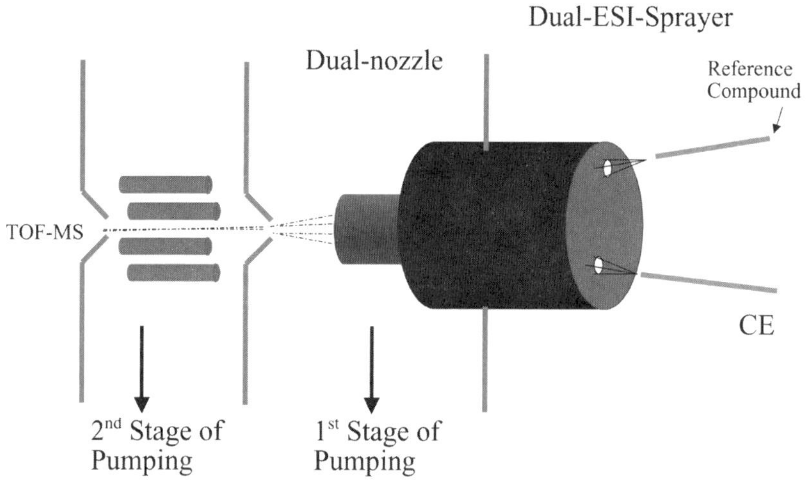

Fig. 6. Schematic of the dual-sprayer, duel-nozzle TOF-MS. Adapted with permission from **ref. *34***.

nique was used. Using an HPLC as a device for the introduction of the first liquid stream (the sample), and a syringe pump as a device for the introduction of the second liquid stream (the reference standard), the accurate mass of the tryptic digest of cytochrome-*c* was measured. The range of mass errors was from –6.1 to +3.6 ppm, a significant improvement over the previously reported mass accuracy for this digest using single-nozzle TOF-MS. Similar mass accuracy is expected for CE–MS analysis. **Table 1** is the comparison between mass measurement errors using the two different calibration procedures. In the single-nozzle experiment, the calibration compound was added to the mixture and the resulting electropherograms were averaged with that of the unknown to generate a single spectrum. In the dual-sprayer, dual-nozzle experiment, one sprayer and nozzle were used for sample introduction while the other sprayer and nozzle were used for reference compound introduction. Therefore, every spectrum contained a pair of reference compounds. Under this condition, a significant mass accuracy improvement was observed (*see* **Table 1**).

For the mass accuracy experiments a P/ACE System 2100 (Beckman Instruments, Fullerton, CA) CE instrument was employed. A 0.01 *M* acetic acid running BGE solution (pH 3.4) was used. Samples were injected hydrodynamically (1.0 psi, 5 s) into the CE capillary (injecting approx 5 nL of sample solution which equates to 120 fmol of each protein digest). A field strength of –398 V/cm was applied to the capillary for separation while a voltage of +1.8 kV was applied to the CE outlet/ESI electrode to close the CE electric circuit and to generate electrospray. A Mariner ESI–TOF-MS (Perseptive Biosystems, Inc.,

Framingham, MA) with a reflectron TOF mass analyzer (1 m total flight length) and with a resolving power of approx 5000 was used. The mass spectrometer was operated in the *m/z* range of 500–1500 (350–1500 for the CID experiments) at a rate of 8000 acquisitions/s, which led to the generation of a single spectrum every second. The nozzle temperature was set at 150°C.

3.3.5. Protein Analysis

Recently, there have been several mass spectrometric approaches to analyzing whole proteins in complex protein mixtures such as cell lysates using MALDI–MS ***(84,85)***, HPLC faction collection followed by gas phase concentration ***(86)***, and capillary isoelectric focusing (CIEF) FTICR–MS ***(87)***. By far, the CIEF/FTICR–MS has been able to analyze the largest number of proteins in one run (up to 900) with a mass accuracy high enough to predict protein modification.

CE–ESI/MS is also used for the analysis of whole proteins. In most cases, however, the analysis has been limited to simple mixtures ***(88–91)*** at relatively high concentrations. Because the injection volume in conventional CE is in the order of a few nanoliters, and the mass spectrometer's detection limit for proteins is generally in the low femtomole range, protein concentrations have to be in the μ*M* (μg/mL) range to be detected. This concentration is seldom available at the endogenous level. For this reason, a variety of on-column sample concentration techniques have been developed to compensate for the high concentration detection limit of CE–ESI/MS. Some on-line concentration techniques that are used with CE–ESI/MS, such as partial packing or membrane preconcentration, can cause band broadening, which deteriorates separation efficiency. For this reason, these on-line concentration techniques are generally not suitable for complex protein analysis. Whereas the resolving power of conventional CE is lower than that of CIEF (in which proteins with a *pI* difference of only 0.003 units can be separated), the high percentage of ampholytes present in CIEF (~1%) can cause signal suppression under ESI, resulting in a lower sensitivity of CIEF–ESI/MS compared to CE–ESI/MS. In addition, some catholyte solutions that are used with cIEF (such as 20 m*M* NaOH) can suppress the signals of proteins with high *pI* values.

Because the number of CE–ESI/MS studies of protein mixtures is very limited, the capability of CE–ESI/MS in analyzing complex protein mixtures at endogenous levels is still open for investigation. Recent results in our laboratory show that long capillaries (>70 cm) can provide high resolution and high sensitivity for the analysis of protein mixtures using a BGE of 0.1% acetic acid in a mixture of water and acetonitrile (50/50, v/v). For example, **Fig. 7** shows the electropherogram of a mixture of five proteins utilizing the split-flow interface with an APS derivatized, 10-μm id, 73-cm-long capillary. As is shown, the

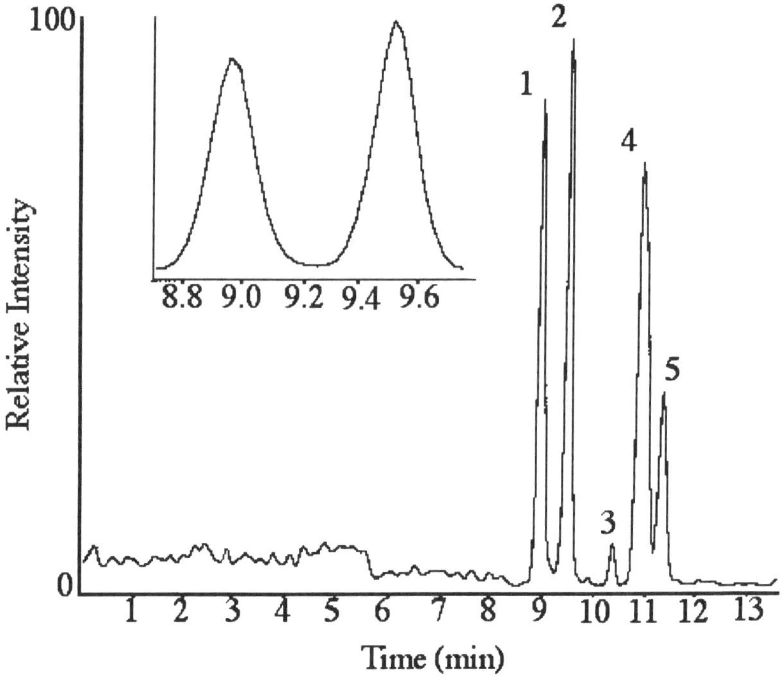

Fig. 7. The CE–ESI/MS electropherogram of the mixture of five proteins utilizing the split-flow interface. An APS derivatized, 10-μm id, 73-cm-long capillary was used. The BGE was a solution of water + acetonitrile (50/50, v/v) containing 0.1 % acetic acid. As is shown, the protein peaks are approx 9-s wide (FWHM) with almost no tailing (inset). The peak marked 1 is bovine insulin, 2 is bovine β-lactoglobulin, 3 is bovine CA, 4 is ribonuclease, and 5 is cytochrome-*c*. A 1-nL injection of the protein solution contained 5 fmol of cytochrome-*c*, 20 fmol of bovine β-lactoglobulin, 4 fmol of bovine CA, 6 fmol of bovine insulin, and 31 fmol of ribonuclease.

protein peaks are approx 9 s wide (FWHM) with almost no tailing (*see* **Fig. 7**, inset). More recently, we have applied this technique to the analysis of a more complicated protein mixture (~50 proteins) using a longer capillary (>1 m) and the results are very promising.

The importance of using long (>1 m) and narrow (<20-μm id) capillaries for complex protein analysis was recently demonstrated by the analysis of carbonic anhydrase isoforms (CA) in red blood cells (RBCs) ***(92)***. The most abundant proteins in RBCs are: the α (average mol wt 15,126 Da) and β (average mol wt 15,865 Da) chains of hemoglobin (each at ~900 amol/cell), carbonic anhydrase I (CAI, average mol wt 28,780 Da; ~7 amol/cell) ***(93,94)*** and carbonic anhydrase II (CAII, average mol wt 29,156 Da; ~0.8 amol/cell) ***(95,96)***.

Detection of CA in blood is challenging because in RBCs, it coexists with a more than 100-fold molar excess of hemoglobin. For example, using a 55-cm-long, 15-μm id capillary, CA I was separated from and migrated before the α and β chains of hemoglobin (Hb) and was detected at a detection limit of three cells (~20 amol) ***(92)***. With this short capillary, however, the peak of CA II comigrated with the β-chain of Hb, and its signal was masked by the high chemical background noise associated with the approx 1000× molar excess of the β-chain. By using a 120-cm-long capillary, however, CA II was separated from the α and β chains of hemoglobin and CAI and was detected at low amol levels (*see* **Fig. 8**).

3.4. Multielectrode Sheathless CE–ESI/MS

It is important to note that because the maximum voltage available on most commercial CE instruments is 30 kV, long capillaries have a low electric field strength. Therefore, even higher separation efficiencies and shorter analysis times are expected by using ultrahigh separation voltages (>50 kV). To achieve ultrahigh separation voltages (>50 kV), several approaches have been used. In one approach, an ultrahigh voltage (>100kV) power supply was used ***(97)***. Recently, we proposed the use of a multielectrode CE capillary for ultrahigh voltage CE–ESI/MS analysis ***(98)***.

In conventional CE and CE–ESI/MS two electrodes are used: one as the inlet electrode and the other one as the outlet/ESI electrode. In multielectrode CE–MS, additional electrode(s) are inserted into the capillary somewhere in between the inlet and the outlet electrodes. The inserted electrode nearest to the outlet is used as the CE outlet/ESI electrode. The multielectrode capillaries (APS-derivatized, 360-μm od, 75-μm id, 75-cm long) were fabricated by cutting two holes into the capillary, one 38 cm from the outlet and the other 3 cm from the outlet, and inserting a platinum wire (25-μm diameter) into each hole (*see* **Fig. 9**).

For these three-electrode experiments, a 25-m*M* hydroquinone in 0.1% acetic acid separation BGE was used to suppress bubble formation inside the capillary owing to redox reactions of water at the added electrodes ***(32,38)***. Three experiments were performed using this three-electrode CE–MS design. In all cases, the inlet and outlet voltages were kept constant at –30 kV and +2.5, respectively, with the midpoint electrode voltage at: (1) ground potential; (2) initially at ground potential and then decreased to –20 kV after the compounds of interest had passed the second electrode; or (3) at constant +20 kV. In the absence of any voltage applied to the middle electrode, the electric field strength in the capillary was 451 V/cm. In experiment one (*see* **Fig. 10A**), where the middle electrode was grounded, The electric field strength in the capillary was 881 V/cm at the beginning of the separation and approx 70 V/cm

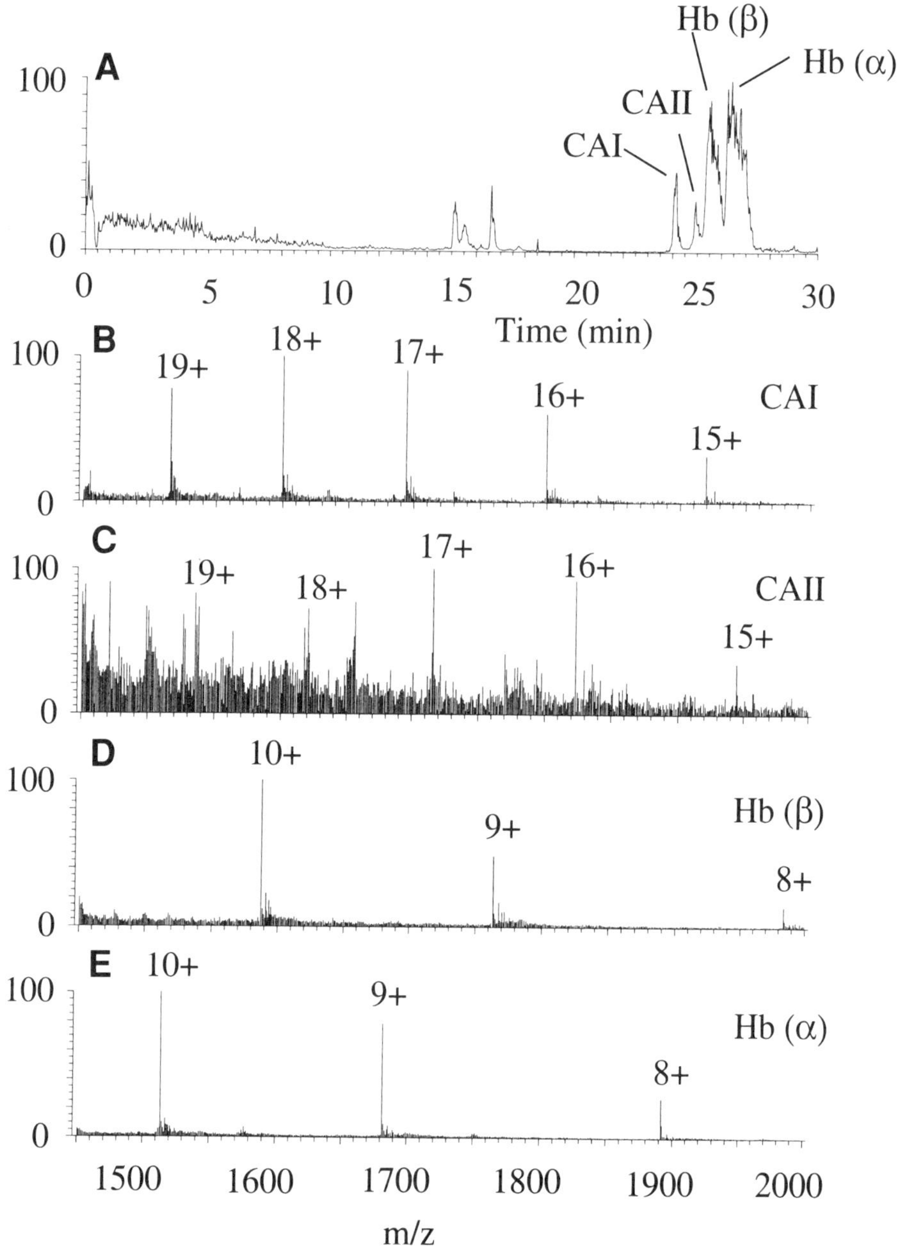

Fig. 8. CE–ESI/MS analysis of lysed RBCs using a 120-cm-long capillary. **(A)** The base peak electropherogram of the major proteins in RBCs. **(B–E)** Respectively, the mass spectra of CAI, CAII, and the β and α chains of Hb. Adapted with permission from **ref.** ***92***.

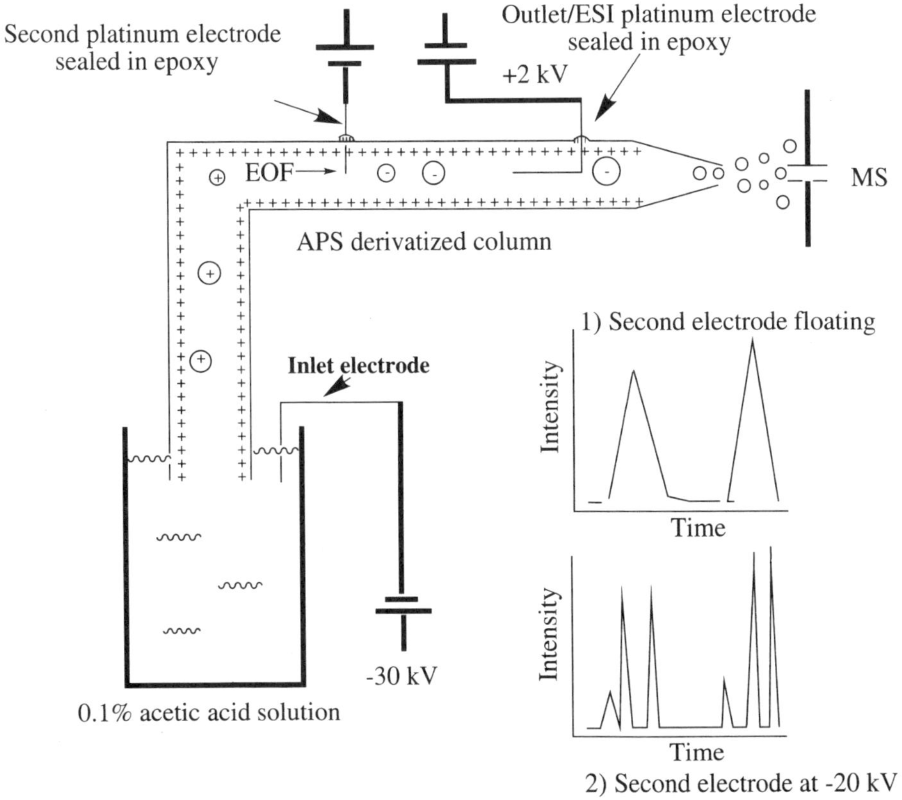

Fig. 9. A schematic of the multi-electrode capillary (APS-derivatized, 360-μm od, 75-μm id, 75-cm-long), which was fabricated by cutting two holes into the capillary, one 38 cm from the outlet and the other 3 cm from the outlet, and inserting a platinum wire (25-μm dia.) into each hole.

after the analyte ions passed the second electrode. In experiment 2 (*see* **Fig. 10B**), the electric field strength was 881 V/cm at the beginning of the separation and dropped to 571 V/cm after all analytes passed the second electrode. In experiment 3, (*see* **Fig. 10C**), the electric field strength was 1351 V/cm at the beginning of the separation and dropped to 571 V/cm (with reversed polarity) after all analytes passed the second electrode. The Sigma HPLC peptide standard mixture was used for analysis. Results show that the third experiment provided the highest separation efficiency, owing to the higher field strength across the capillary. The use of hydroquinone as a buffer additive for controlling the electrochemical reactions at these electrodes was deemed necessary ***(32,38)***. It is important to note that these experiments were performed using a 75-μm id, 75-cm-long capillary. Under conventional sheathless CE–MS, resolution of

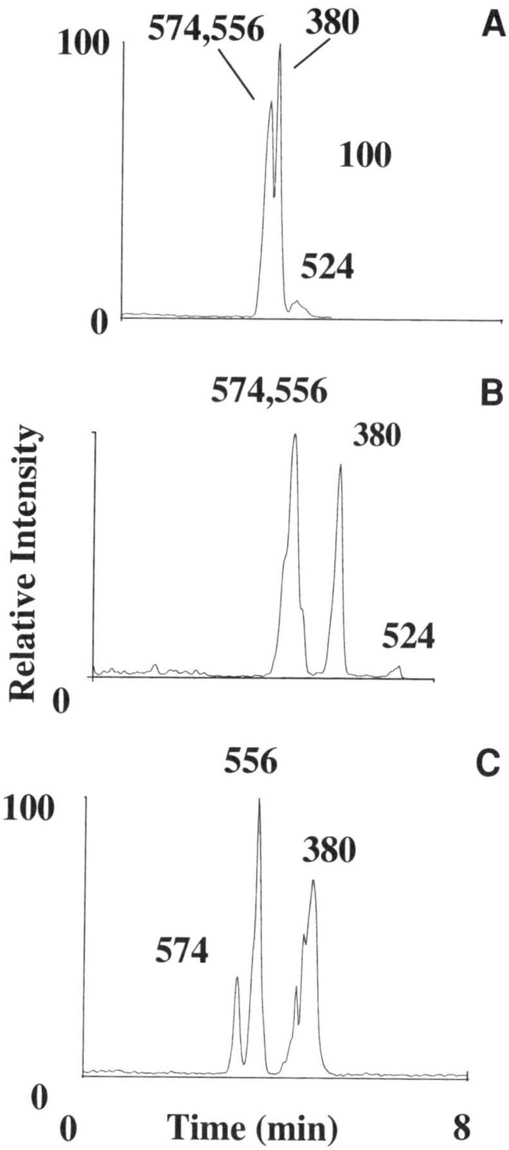

Fig. 10. Multiple in-capillary electrode CE–ESI/MS of the standard peptide mixtures. **(A)** Electropherogram of peptides taken with the middle electrode held at ground potential. **(B)** The same experiment with the middle electrode initially at ground potential, and then at -20 kV after 2.2 min. **(C)** The same experiment with the middle electrode held at +20 kV.

the order demonstrated in **Fig. 10C** is difficult to achieve with capillaries having an id greater than 50 μm. In addition, the results imply that sample concentration (concentration polarization) may be achieved on the middle electrode (*m/z* 524, **Fig. 10B**), which could potentially be exploited as a preconcentration method. Better separation efficiency using long capillaries has also been observed for peptide mixtures and even amino acid mixtures. The application of these capillaries to the analysis of complex protein mixtures is under investigation and the early results are very promising.

4. Notes

1. Prior to analysis, the ESI voltage should be optimized for maximum sensitivity using a dilute solution of a compound (similar to the analyte of interest) in BGE under CE–ESI/MS conditions (both CE and ESI voltages on).
2. Experimentally, resolution is maximized by optimizing injection conditions (injection pressure and injection time). For dilute samples, as much sample as possible is injected while avoiding band broadening caused by injection plug width.
3. APS is extremely reactive toward –OH. Therefore, the toluene must be anhydrous (in a septum-sealed container). Use dried (in a 110°C oven) glass syringes for preparing the 10% APS in toluene solution. Dry all secondary containers (at 110°C for ~ 1 h) before use. Immediately before taking the capillaries out of the GC oven, prepare the 10% solution of APS in dry toluene using a dry container (screw cap containers are preferred) and seal the container. Once the CE capillaries are removed from the oven, immediately open the container containing the 10% APS in toluene solution and place it inside the bomb. Quickly insert one end of the capillaries into the bomb, making sure that all of the inserted ends are submerged in the APS solution, and place the other end of the capillaries in a dry container containing dry toluene. Seal the bomb and apply pressure. Crystal formation at the ends of the capillaries or in the capillaries is indicative of the existence of moisture in the capillaries, containers, or chemicals. If any crystal is observed, discard the capillaries.
4. We have found that the HPLC standard peptide mixture sold by Sigma-Aldrich is a good peptide mixture for testing the overall performance of CE–ESI/MS. The test mixture contains five peptides: GY (MW 238), VYV (MW 379), YGGFL (leucine-enkaphalin, mol wt 555), YGGFM (methionine-enkaphalin, mol wt 573), and DRVYIHPF (angiotensin II, mol wt 1046). The contents of the vial are dissolved in 1 mL of pure water and a 100× dilution of this solution is used for performance evaluation. The mass spectrometer is usually scanned in the *m/z* range of 375–600 with scan speed of 1–2 scans/s. Once the baseline separation is achieved for the peptide mixture (*see* **Fig. 11**), the CE–ESI/MS is usually tested with a more complex mixture such as a tryptic digest of cytochrome-*c*. **Figure 12** shows ion electropherograms of the major peptide fragments of a tryptic digest of cytochrome-*c* in the m/z range of 500–1000.

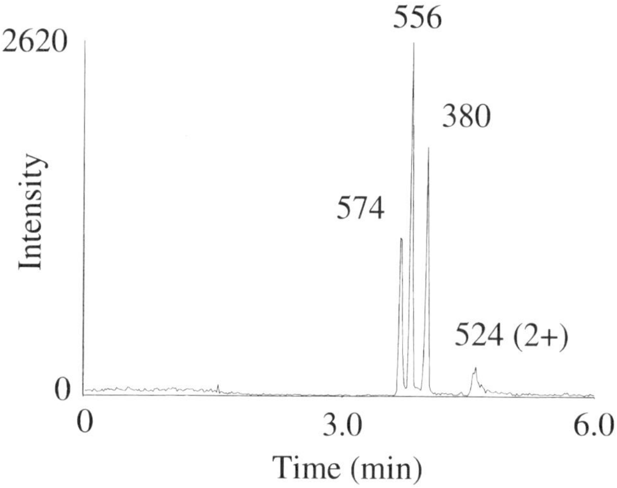

Fig. 11. CE–ESI/MS analysis of the standard peptide mixture. The *m/z* values 380, 556, 574, and 524 (2+) correspond to VYV, YGGFL (leucine-enkaphalin), YGGFM (methionine-enkaphalin,), and DRVYIHPF (angiotensin II). Adapted with permission from **ref. 23**.

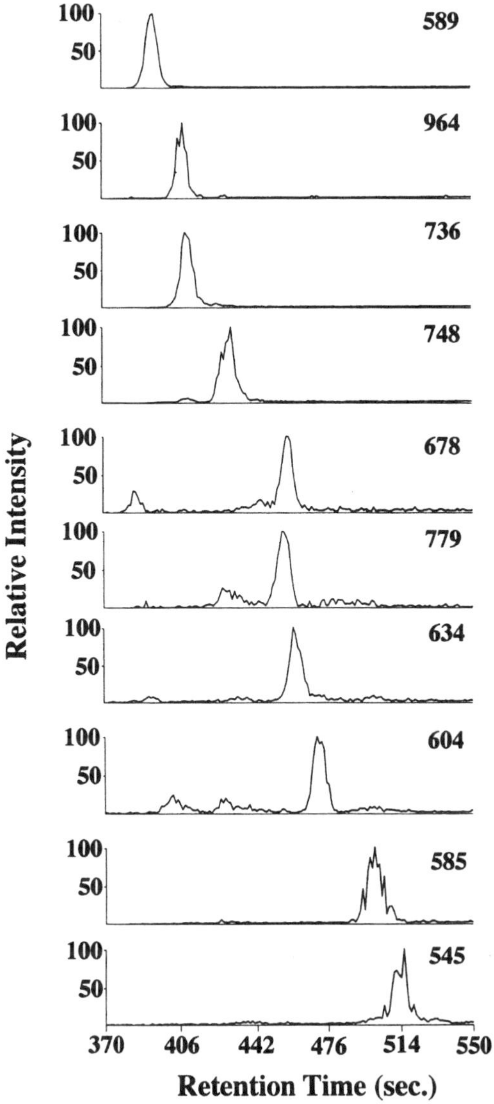

Fig. 12. CE–ESI/MS analysis of the tryptic digest of cytochrome-*c* using TOF-MS. Adapted with permission from **ref. *82***.

References

1. Swanek, F. D., Ferris, S. S., and Ewing, A. G. (1996) Capillary electrophoresis for the analysis of single cells: electrochemical, mass spectrometric, and radiochemical detection, in *Handbook of Capillary Electrophoresis* (Landers, J. P., ed.), CRC, Boca Raton, FL, pp. 495–521.
2. Kennedy, R. T., Oates, M. D., Cooper, B. R., Nickerson, B., and Jorgenson, J. W. (1989) Microcolumn separations and the analysis of single cells. *Science* **246,** 57–63.
3. Kennedy, R. T., St. Claire, R. L., White, J. G., and Jorgenson, J. W. (1987) Chemical analysis of single neurons by open tubular liquid chromatography. *Mikrochim. Acta.* **2,** 37–45.
4. Kennedy, R. T. and Jorgenson, J. W. (1989) Quantitative analysis of individual neurons by open tubular liquid chromatography with voltammetric detection. *Anal. Chem.* **61,** 436–441.
5. Oates, M. D., Cooper, B. R., and Jorgenson, J. W. (1990) Quantitative amino acid analysis of individual snail neurons by open tubular liquid chromatography. *Anal. Chem.* **62,** 1573–1577.
6. Cooper, B. R., Jankowski, J. A., Leszyczyszyn, D. J., Wrightman, R. M., and Jorgenson, J. W. (1992) Quantitative determination of catecholamines in individual bovine adrenomedullary cells by reversed-phase microcolumn liquid chromatography with electrochemical detection. *Anal. Chem.* **64,** 691–694.
7. Ewing, A. G., Wallingford, R. A., and Olefirowicz, T. M. (1989) Capillary electrophoresis. *Anal. Chem.* **61,** 292A–303A.
8. Olefirowicz, T. M. and Ewing, A. G. (1990) Capillary electrophoresis in 2 and 5 microns diameter capillaries: application to cytoplasmic analysis. *Anal. Chem.* **62,** 1872–1876.
9. Olefirowicz, T. M. and Ewing, A. G. (1991) Capillary electrophoresis for sampling single nerve cells. *Chemia.* **45,** 106–108.
10. Hogan, B. L. and Yeung, E. S. (1992) Determination of intracellular species at the level of a single erythrocyte via capillary electrophoresis with direct and indirect fluorescence detection. *Anal. Chem.* **64,** 2841–2845.
11. Olefirowicz, T. M. and Ewing, A. G. (1990) Dopamine concentration in the cytoplasmic compartment of single neurons determined by capillary electrophoresis. *J. Neurosci. Meth.* **34,** 11–15.
12. Jorgenson, J. W. and Lukacs, K. D. (1983) Capillary zone electrophoresis. *Science* **222,** 266–272.
13. Gordon, M. J., Huang, X., Pentoney, Jr., S. L., and Zare, R. N. (1988) Capillary electrophoresis. *Science* **242,** 224–248.
14. Gilman, S. D. and Ewing, A. G. (1995) Recent advances in the application of capillary electrophoresis to neuroscience. *J. Cap. Elec.* **2,** 1–13.
15. Yeung, E. S. (1994) Chemical analysis of single human erythrocytes. *Acc. Chem. Res.* **27,** 409–414.
16. Jankowski, J. A., Tracht, S., and Sweedler, J. V. (1995) Assaying single cells with capillary electrophoresis. *Trends Anal. Chem.* **14,** 170–176.

17. Cannon, Jr., D. M., Winograd, N., and Ewing, A. G. (2000) Quantitative chemical analysis of single cells. *Ann. Rev. Biophys. Biomol. Struct.* **29,** 239–263.
18. Zhang, X., Stuart, J. N., and Sweedler, J. V. (2002) Capillary electrophoresis with wavelength-resolved laser-induced fluorescence detection. *Anal. Bioanal. Chem.* **373,** 332–343.
19. Preisler J., Hu, P., Rejtar, T., and Karger, B. L. (2000) Capillary electrophoresis—matrix assisted laser desorption/ionization time-of-flight mass spectrometry using a vacuum deposition interface. *Anal. Chem.* **72,** 4785–4795.
20. Kennedy, R. T., German, I., Thompson, J. E., and Witowski, S. R. (1999) Fast analytical-scale separations by capillary electrophoresis and liquid chromatography. *Chem. Rev.* **99,** 3081–3131.
21. Pertsen, J. R. and Mohammad, A. D. eds., (2001) *Clinical and Forensic Applications of Capillary Electrophoresis*, Humana, Totowa, NJ.
22. Von Brocke, A., Nicholson, G., and Bayer, E. (2001) Recent advances in capillary electrophoresis/electrospray-mass spectrometry. *Electrophoresis* **22,** 1251–1266.
23. Moini, M. (2002) Capillary electrophoresis mass spectrometry and its application to the analysis of biological mixtures. *Anal. Bioanal. Chem.* **373,** 466–480.
24. Bocek, P., Deml, M., Gebauer, P., and Dolnik, V. (1988) *Analytical Isotachophoresis* (Radola, B. J., ed.), VCH, New York.
25. Dobos, D. (1975) *Electrochemical Data: A Handbook for Electrochemists in Industry and Universities.* Elsevier Scientific, New York.
26. Macka, M., Andersson, P., and Haddad, P. R. (1998) Changes in electrolyte pH due to electrolysis during capillary zone electrophoresis. *Anal. Chem.* **70,** 743–749.
27. Timperman, A., Tracht, S. E., and Sweedler, J. V. (1996) Dynamic on-column pH monitoring in capillary electrophoresis: application to volume-limited outlet vials. *Anal. Chem.* **68,** 2693–2698.
28. Bello, M. S. (1996) Electrolytic modification of a buffer during a capillary electrophoresis run. *J. Chromatogr.* **A744,** 81–91.
29. Corstjens, H., Billiet, H. A., Frank, J., and Luyben, K. C. (1996) Variation of the pH of the background electrolyte due to electrode reactions in capillary electrophoresis: theoretical approach and in situ measurement. *Electrophoresis* **17,** 137–143.
30. Kebarle, P. and Tang, L. (1993) From ions in solution to ions in the gas phase - the mechanism of electrospray mass spectrometry. *Anal. Chem.* **65,** 972A–86A.
31. Van Berkel, G. J. (1997) The electrolytic nature of electrospray, in *Electrospray Ionization Mass Spectrometry: Fundamentals, Instrumentation, and Applications* (Cole, R. B., ed.), Wiley, New York, pp. 65–105.
32. Smith, A. D. and Moini, M. (2001) Control of electrochemical reactions at the capillary electrophoresis outlet/electrospray emitter electrode under CE-ESI MS through the application of redox buffers. *Anal. Chem.* **73,** 240–246.
33. Bruinis, A. P. (1997) ESI source design and dynamic range considerations, in *Electrospray Ionization Mass Spectrometry: Fundamentals, Instrumentation, and Applications* (Cole, R.B., ed.), Wiley, New York, pp. 107–136.

34. Longfei, J. and Moini, M. (2000) Development of multi-ESI-sprayer, multi-atmospheric-pressure-inlet mass spectrometry and its application to accurate mass measurement using time-of-flight mass spectrometry. *Anal. Chem.* **72,** 20–24, and 885.
35. Bateman, K. P. (1999) Electrochemical properties of capillary electrophoresis-nanoelectrospray mass spectrometry. *J. Am. Soc. Mass Spectrom.* **10,** 309–317.
36. Chang, H. T. and Yeung, E. S. (1993) Oncolumn digestion of protein for peptide mapping by capillary zone electrophoresis with laser-induced native fluorescence detection. *Anal. Chem.* **65,** 2947–2951.
37. Moini, M. (2001) Design and performance of a universal sheathless capillary electrophoresis to mass spectrometry interface using a split-flow technique. *Anal. Chem.* **73,** 3497–3501.
38. Moini, M., Cao, P., and Bard, A. J. (1999) Hydroquinone as a buffer additive for suppression of bubbles formed by electrochemical oxidation of the CE buffer at the outlet electrode in capillary electrophoresis/electrospray ionization-mass spectrometry. *Anal. Chem.* **71,** 1658–1661.
39. Rindgen, D, Turesky, R. J., and Vouros, P. (1995) The application of CE/ESI/MS/MS for detection of IQ adducts of DNA, in *Proce. 43th ASMS Conf. Mass Spectrometry and Allied Topics*, p. 590.
40. Soga, T. and Heiger, D. N. (2000) Amino acid analysis by capillary electrophoresis electrospray ionization mass spectrometry. *Anal. Chem.* **72,** 1236–1241.
41. Schultz, C. L. and Moini, M. (2003) The analysis of underivatized amino acids and their D/L enantiomers using sheathless CE-MS. *Anal. Chem.* **75,** 1508-1513.
42. Moseley, M. A., Jorgenson, W. J., Shabanowitz, J., Hunt, D. F., and Tomer, K. B. (1992) Optimization of capillary zone electrophoresis/electrospray ionization parameters for the mass spectrometry and tandem mass spectrometry analysis of peptides. *J. Am. Soc. Mass Spectrom.* **3,** 289–300.
43. Teerlink, T. (1994) Derivatization of posttranslationally modified amino acids. *J. Chromatogr.* **B659,** 185–207.
44. Schegg, K. M., Denslow, N. D., Anderson, T. T., Bao, Y., Cohen, S. A., Mahrenholz, A. M., and Mann, K. (1997) Quantitation and identification of proteins by amino acid analysis: ABRF-96AAA collaborative trial, in *Techniques in Protein Chemistry VIII* (Marshak, D. R., ed.), Academic, San Diego, CA, pp. 207–216.
45. Prata, C., Bonnafous, P., Fraysse, N, Treilhou, M., Poinsot, V., and Couderc, F. (2001) Recent advances in amino acid analysis by capillary electrophoresis. *Electrophoresis* **22,** 4129–4138.
46. Vecchione, G., Margaglione, G., Grandone, E., et al. (1999) Determining sulfur-containing amino acids by capillary electrophoresis. A fast novel method for total homocyste(e)ine human plasma. *Electrophoresis* **20,** 569–574.
47. Nyhan, W. L. (1984) *Abnormalities in Amino Acid Metabolism in Clinical Medicine.* Appleton Century Crofts, East Norwalk, CT.
48. Soga, T. and Heiger, D. N. (2000) Amino acid analysis by capillary electrophoresis electrospray ionization mass spectrometry. *Anal. Chem.* **72,** 1236–1241.
49. Marchelli, R., Dossena, A., and Palla, G. (1996) The potential of enantioselective analysis as a quality control tool. *Trends in Food Sci. Tech.* **7,** 113–119.

50. Zumwalt, R. W., Kuo, K. C., and Gehrke, C. W., eds. (1987) Amino Acid Analysis by Gas Chromatography (Vol I-III). CRC, Boca Raton, FL.
51. Deyl, Z., Hyanek. J., and Horikova, M. (1986) Profiling of amino acids in body fluids and tissues by means of liquid chromatography. *J. Chromatogr.* **379,** 177–250.
52. Duncan, M. W. and Poljak, A. (1998) Amino Acid Analysis of Peptides and Proteins on the Femtomole Scale by Gas Chromatography/Mass Spectrometry. *Anal. Chem.* **70,** 890–896.
53. Kwon, J. -Y and Moini, M. (2001) Analysis of underivatized amino acid mixtures using high performance liquid chromatography/dual oscillating nebulizer atmospheric pressure microwave induced plasma ionization-mass spectrometry. *J. Am. Soc. Mass Spectrom.* **12,** 117–122.
54. He, T., Quinn, D., Fu, E., and Wang, Y. K. (1999) Analysis of diagnostic metabolites by capillary electrophoresis-mass spectrometry. *J. Chromatogr.* **B727,** 43–52.
55. Verleysen, K. and Sandra, P. (1998) Separation of chiral compounds by capillary electrophoresis. *Electrophoresis* **19,** 2798–2833.
56. Kuhn, R., Stoecklin, F., and Erni, F. (1992) Chiral separations by host-guest complexation with cyclodexterin and crown ether in capillary zone electrophoresis. *Chromatographia* **33,** 32–36.
57. Verleysen, K., Bosch, T. V. and Sandra, P. (1999) Comparison of highly sulfated α-, β-, and γ- cyclodextrins and 18-crown-6-tetracarboxylic acid for the enantiomeric separation of some amino acids and derivatives by capillary electrophoresis. *Electrophoresis* **20,** 2650–2655.
58. D'Aniello, A., Lee, J. M., Petrucelli, L., and Maddalena Di Fiore, M. (1998) Regional decreases of free D-aspartate levels in Alzheimer's disease. *Neurosci. Lett.* **250,** 131–134.
59. Schell, M. J., Cooper, O. B., and Snyder, S. H. (1997) D-Aspartate localizations imply neuronal and neuroendocrine roles. *Proc. Natl. Acad. Sci.* **94,** 2013–2018.
60. Schell, M. J., Brady Jr., R. O., Molliver, M. E., and Snyder, S. H. (1997) D-serine as a neuromodulator: regional and developmental localizations in rat brain glia resemble NMDA receptors. *J. Neurosci.* **17,** 1604–1615.
61. Lubec, G. and Lubec, B (1993) Racemization rates of amino acids for dating ancient samples. *Amino Acids* **4,** 1–3.
62. Kyba, E. P., Timko, J. M., Kaplan, L. J., de Jong, F., Gokel, G. W., and Cram, D. J. (1978) Host-guest complexation. 11. Survey of chiral recognition of amine and amino ester salts by dilocular bisdinaphthyl hosts. *J. Am. Chem. Soc.* **100,** 4555–4568.
63. Sousa, L. S., Sogah, G. D. Y., Hoffman, D. H., and Cram, D. J. (1978) Host-guest complexation. 12. Total optical resolution of amine and amino ester salts by chromatography. *J. Am. Chem. Soc.* **100,** 4569–4576.
64. Behr, J. P., Girodeau, J. M., Heyward, R. C., Lehn, J. M., and Sauvage, J. P. (1980) Molecular receptors. Functionalized and chiral macrocyclic polyethers derived from tartaric acid. *Helv. Chim. Acta.* **63,** 2096–2111.

65. Behr, J. P., Lehn, J. M., and Vierling, P. (1982) Molecular receptors. Structural effects and substrate recognition in binding of organic and biogenic ammonium ions by chiral polyfunctional macrocyclic polyethers bearing amino acid and other side-chains. *Helv. Chim. Acta.* **65,** 1853–1867.
66. Banks, J. F. and Gulcicek, E. E. (1997) Rapid peptide mapping by reversed-phase liquid chromatography on nonporous silica with on-line electrospray time-of-flight mass spectrometry. *Anal. Chem.* **69,** 3973–3978.
67. Wu, J., He, L., Li, M. X., Parus, S., and Lubman, D. M. (1997) Online capillary separations/tandem mass spectrometry for protein digest analysis by using an ion trap storage/reflectron time-of-flight mass detector. *J. Am. Soc. Mass Spectrom.* **8,** 1237–1246.
68. Roberts, G. D., Johnson, W. P., Burman, S., Anumula K. R., and Carr, S. A. (1995) An integrated strategy for structural characterization of the protein and carbohydrate components of monoclonal antibodies: application to anti-respiratory syncytial virus MAb. *Anal. Chem.* **67,** 3613–3625.
69. Yates, J. R. 3rd, Speicher, S., Griffin, P. R., and Hunkapiller, T. (1993) Peptide mass maps: a highly informative approach to protein identification. *Anal. Biochem.* **214,** 397–408.
70. Kitagishi, K. (1997) Comparison with other analytical methods, in *Handbook of Capillary Electrophoresis Applications* (Shintani, H. and Polonsky J., eds.), Chapman & Hall, New York, pp. 137–146.
71. Wahl, J. H., Udseth, H. R., and Smith, R. D. (1996) Capillary electrophoresis—mass spectrometry in peptide mapping, in *New Methods in Peptide Mapping for the Characterization of Proteins* (Hancock, W. S., ed.), CRC, Boca Raton, FL, pp. 143–179.
72. Banks, J. F. and Dresch, T. (1996) Detection of fast capillary electrophoresis peptide and protein separations using electrospray ionization with a time-of-flight mass spectrometer. *Anal. Chem.* **68,** 1480–1485.
73. Jiang, L. and Moini, M. (1995) Design and performance of a high resolution electrospray ion source for a magnetic sector mass spectrometer with a heated capillary inlet. *J. Am. Soc. Mass Spectrom.* **6,** 1256–1261.
74. Moini, M. (1994) Ultramark 1621 as a calibration/reference compound for mass spectrometry. II. Positive- and negative-ion electrospray ionization. *Rapid Commun. Mass Spectrom.* **8,** 711–714.
75. Laude, D. A., Stevenson, E., and Robinson, J. M. (1997) Electrospray ionization/fourier transform ion cyclotron resonance mass spectrometry, in *Electrospray Ionization Mass Spectrometry: Fundamentals, Instrumentation and Applications* (Cole, R.B., ed.), Wiley, New York, NY, pp. 291–319.
76. Chernushevich, I. V., Ens, W., and Standing, K. G. (1997) Electrospray ionization time-of-flight mass spectrometry, in *Electrospray Ionization Mass Spectrometry: Fundamentals, Instrumentation and Applications* (Cole, R.B., ed.), Wiley, New York, NY, pp. 203–234.
77. Lazar, I. M., Xin, B., et al. (1997) Design of a time-of-flight mass spectrometer as a detector for capillary electrophoresis. *Anal. Chem.* **69,** 3205–3211.

78. Fang, L., Zhang, R., Williams, E. R., and Zare, R. N. (1994) Online time-of-flight mass spectrometric analysis of peptides separated by capillary electrophoresis. *Anal. Chem.* **66,** 3696–3701.
79. Muddiman, D. C., Rockwood, A. L., Gao, Q., Severs, J. C., Udseth, H. R., and Smith, R. D. (1995) Application of sequential paired covariance to capillary electrophoresis electrospray ionization time-of-flight mass spectrometry: Unraveling the signal from the noise in the electropherogram. *Anal. Chem.* **67,** 4371–4375.
80. Wu, J. T., Qian, M. M. G., Li, M. M. X., Liu, L., and Lubman, D. M. (1996) Use of an ion trap storage/reflectron time-of-flight mass spectrometer as a rapid and sensitive detector for capillary electrophoresis in protein digest analysis. *Anal. Chem.* **68,** 3388–3396.
81. Yates, J. R., III, Zhou, J., Griffin, P. R., and Hood, L. E. (1990) Computer aided interpretation of low energy MS/MS mass spectra of peptides, in *Techniques in Protein Chemistry II* (Villafranca, J. J., ed.), Academic, New York, pp. 477–486.
82. Cao, P. and Moini, M. (1998) Capillary electrophoresis/electrospray ionization high mass accuracy time-of-flight mass spectrometry for protein identification using peptide mapping. *Rapid. Commun. Mass Spectrom.* **12,** 864–870.
83. Mann, M. and Wilm, M. (1994) Error-tolerant identification of peptides in sequence databases by peptide sequence tags. *Anal. Chem.* **66,** 4390–4399.
84. Chong, B.E., Wall, D.B., Lubman, D.M., and Flynn, S.J. (1997) Rapid profiling of E. coli proteins up to 500 kDa from whole cell lysates using matrix-assisted laser desorption/ionization time-of-flight mass spectrometry. *Rapid. Commun. Mass Spectrom.* **11,** 1900–1908.
85. Arnold, R. J. and Reily, J. P. (1999) Observation of Escherichia coli ribosomal proteins and their posttranslational modifications by mass spectrometry. *Anal. Biochem.* **269,** 105–112.
86. Reid, G. E., Shang, H., Hogan, J. M., Lee, G. U., and McLuckey, S. A. (2002) Gas-phase concentration, purification, and identification of whole proteins from complex mixtures. *J. Am. Chem. Soc.* **124,** 7353–7362.
87. Jensen, P. K., Pasa-Tolic, L., Anderson, G. A., et al. (1999) Probing proteomes using capillary isoelectric focusing-electrospray ionization fourier transform ion cyclotron resonance mass spectrometry. *Anal. Chem.* **71,** 2076–2084.
88. Johnson, D. H., Naylor, S., Rhode, E., and Tomlinson, A. J. (1998) Comparison of protein mixtures in aqueous humor by membrane preconcentration-capillary electrophoresis-mass spectrometry. *Electrophoresis* **19,** 2361–2370.
89. Cao, P. and Moini, M. (1998) Analysis of peptides, proteins, protein digests, and whole human blood by capillary electrophoresis/electrospray ionization-mass spectrometry using an in-capillary electrode sheathless interface. J. Am. Soc. Mass Spectrom. **9,** 1081–1088.
90. Naylor, S. and Tomlinson, A. J. (2001) Capillary electrophoresis-mass spectrometry of biologically active peptides and proteins, in *Clinical and Forensic Applications of Capillary Electrophoresis* (Mohammad, A. A. and Peterson, J. R., eds.), Humana, Totowa, NJ, pp. 317–353.

91. Cutalo, J. N., Deterding, L. J., Khaledi, M., and Tomer, K. B. (2002) Separation and characterization of human high-density apolipoproteins using a nonaqueous modifier in capillary electrophoresis-mass spectrometry. *Electrophoresis* **23,** 2296–2305.
92. Moini, M., Demars, S. M., and Huang, H. (2002) Analysis of carbonic anhydrase in human red Blood cells using capillary electrophoresis/ electrospray ionization-mass spectrometry. *Anal. Chem.* **74,** 3772–3776.
93. Valaskovic, G. A., Kelleher, N. L., and McLafferty, F. W. (1996) Attomole protein characterization by capillary electrophoresis-mass spectrometry. *Science* **273,** 1199–1202.
94. Andersson, B., Nyman, P. O., and Strid, L. (1972) Amino acid sequence of human erythrocyte carbonic anhydrase *B. Biochem. Biophys. Res. Commun.* **48,** 670–677.
95. Terzakis, J. A. and Santegada, E. (2000) Measurement of erythrocytes on diagnostic slides by scanning electron microscopy. *Anal. Quantitat. Cytol. Histol.* **22,** 244–246.
96. Henderson, L. E., Henriksson, D., and Nyman, P. O. (1973) Amino acid sequence of human erythrocyte carbonic anhydrase *C. Biochem. Biophys. Res. Commun.* **52,** 1388–1394. For a more recent CAII amino acid sequence, see the NCBI database at http://www.ncbi.nlm.nih.gov.
97. Hutterer, K. M. and Jorgenson, J. W. (1999) Ultrahigh-voltage capillary zone electrophoresis. *Anal. Chem.* **71,** 1293–1297.
98. Smith, A. D. and Moini, M. (2000) Advances in capillary electrophoresis/electrospray ionization mass spectrometry achieved through the in-capillary electrode technique, in *Proc. 48th ASMS Conf. Mass Spectrom. Alli. Top.*, pp. 1119–1120.

14

Capillary Isoelectric Focusing–Mass Spectrometry of Proteins and Protein Complexes

Suzana Martinović, Ljiljana Paša-Tolić and Richard D. Smith

Summary

Complex proteome samples require efficient separation and detection methods in order to characterize their protein components. On-line combination of capillary isoelectric focusing (CIEF) with electrospray ionization (ESI) mass spectrometry (MS) is shown as an effective method to analyze complex protein mixtures. Our experience with several microorganisms allowed us to establish successful experimental protocol. Here we use the example of *E. coli* whole-cell lysate for the CIEF separation and MS detection on the intact protein level. The protocol was further adapted for the analysis of the mixture of noncovalent complexes on the intact complex level.

Key Words

Capillary isoelectric focusing; protein; protein noncovalent complex; ampholyte; mass spectrometry; electrospray ionization; Fourier transform ion cyclotron resonance; stable isotope labeling.

1. Introduction

The growing demand for broad proteome measurements in the post-genomic era has greatly increased the need for more confident and sensitive protein identification. The on-line combination of capillary isoelectric focusing (CIEF) with electrospray ionization (ESI) mass spectrometry (MS) is emerging as an effective method for analyzing complex protein mixtures. Contemporary mainstream proteomics approaches are dominantly based upon analysis of enzimatically digested cellular lysates *(1,2)*. However, CIEF–MS analysis of intact proteins yields more direct information on posttranslational modifica-

From: *Methods in Molecular Biology, vol. 276: Capillary Electrophoresis of Proteins and Peptides*
Edited by: M. A. Strege and A. L. Lagu © Humana Press Inc., Totowa, NJ

tions, and is attracting more attention recently as a complementary proteomics technique.

In CIEF, the focusing of proteins at their respective isoelectric point (p*I*) is realized through the use of a polyampholyte mixture in free solution. When an electric field is applied to the extremities of the capillary, which are immersed in a low- and high-pH solution, the polyampholyte mixture creates a pH gradient through the length of the capillary (*see* **Fig. 1**). The proteins migrate through the gradient until they reach zero net charge (p*I* = pH). After focusing, protein bands are mobilized to the detector (ultraviolet [UV], MS).

CIEF allows effective use of small sample volumes, has high speed and resolution, and can be easily automated. The concentration effect associated with focusing provides attomole (amol) sensitivity when combined with MS ***(3)***. CIEF–MS has been successfully applied for the analysis of mixtures of various complexity: from a single protein (hemoglobin ***(4)***, alcohol dehydrogenase ***(5)*** variants) to whole cell lysates ***(3,6–10)***. It can be used for the analysis of proteins covering a wide range of masses, from smaller proteins of a few kDa ***(11)***, to much larger proteins and noncovalent complexes with $M_r > 100$ kDa ***(12)***.

Separation and detection of proteins from an *Escherichia coli* whole cell lysates (cytosolic protein fraction) will be used to illustrate the CIEF–ESI/MS methodology. A mixture of commercial noncovalent complexes will be used to further explain the adaptation of the CIEF–ESI/MS procedure to those more fragile compounds.

2. Materials

1. *E. coli* K12 strain MG1655.
2. M9 minimal medium.
3. Phosphate-buffered saline (PBS) buffer.
4. 0.1-mm Diameter zirconium/silica beads (Biospec Products, Inc., Bartlesville, OK).
5. Mini-Beadbeater (Biospec Products, Inc.).
6. Dual microdialysis system.
7. Bradford assay.
8. Protein complexes (standard complexes from Sigma).
9. 50-cm (50 μm id and 192 μm od) Fused silica capillary internally coated with linear polyacrylamide.
10. Pharmalyte 3-10 (Amersham Pharmacia Biotech, Piscataway, NJ).
11. CIEF equipment.
12. Phosphoric acid (20 m*M*).
13. Sodium hydroxide (20 m*M*).
14. A sheath liquid: 50% CH_3OH, 49% H_2O, 1% CH_3COOH, pH 2.6; or 10 m*M* ammonium acetate, pH 6.0.

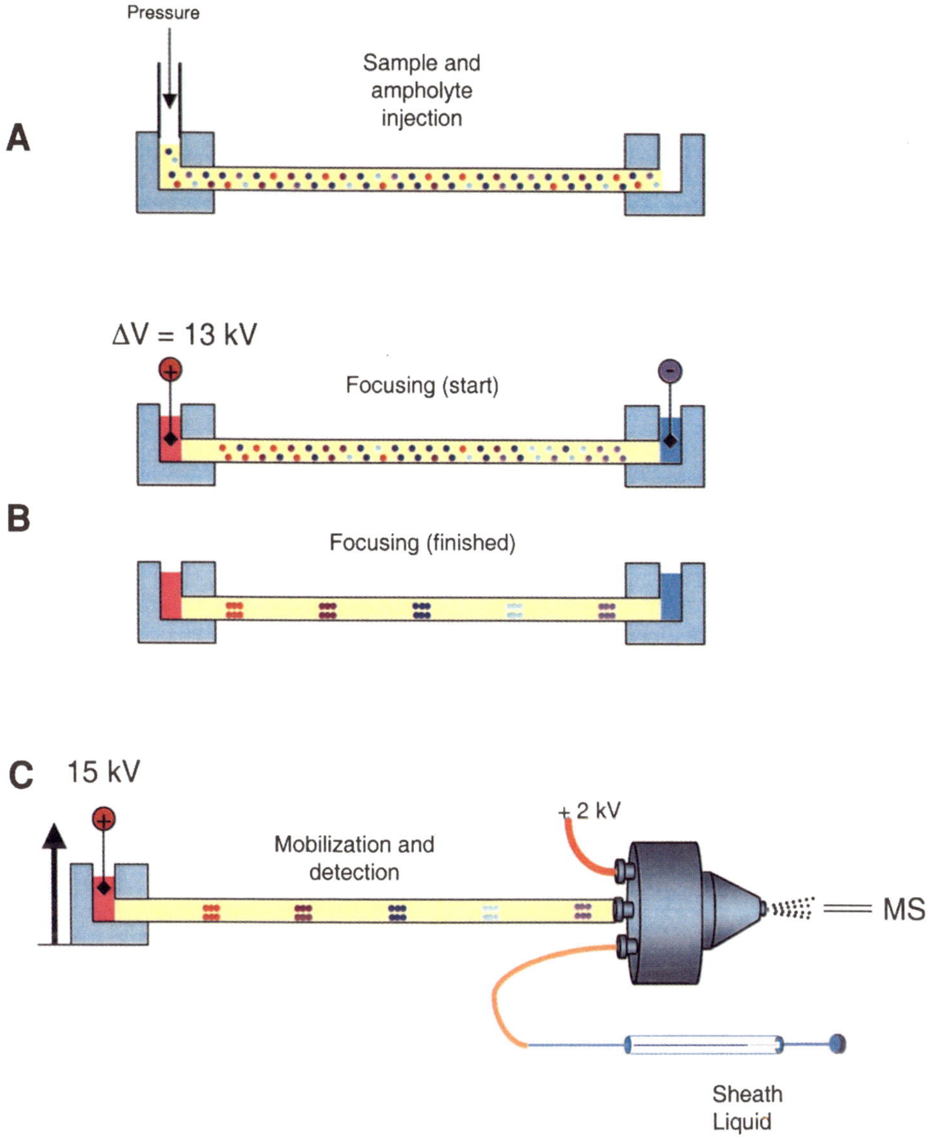

Fig. 1. Schematic CIEF arrangement and procedure. (**A**) sample injection by pressure. (**B**) Focusing. (**C**) Mobilization and MS detection.

15. High-voltage power supply (Glassman High-Voltage, Inc., Whitehouse Station, NJ).
16. Syringe pump (Harvard Apparatus 22, South Natick, MA).
17. Mass spectrometer (e.g., 7-Tesla Fourier Transform Ion Cyclotron Resonance (FTICR) mass spectrometer equipped with an Odyssey data station (Finnigan FTMS, Madison, WI).

3. Methods

The methods described below outline CIEF separation and MS detection of (1) the whole-cell lysate cytosolic proteins and (2) intact protein–protein noncovalent complexes.

3.1. CIEF–MS of Proteins From Cell Lysate

The procedure for CIEF–MS separation and detection of proteins from whole-cell lysate include: (a) sample preparation, (b) CIEF separation, and (c) MS detection and data analysis.

3.1.1. Sample Preparation

1. The *E. coli* K12 strain MG1655 was used (*see* **Note 1**). *E. coli* was grown in M9 minimal medium at 37°C with shaking at 225 rpm until the OD_{600nm} was approx 1. Cells were harvested by centrifugation at 10,000*g* for 30 s and kept frozen until the analysis.
2. The cells were resuspended in 200 µL PBS buffer and lysed by mechanical agitation at 4600 rpm for 60 s in the presence of 0.1-mm diameter zirconium/silica beads (Biospec Products, Inc.) using a Mini-Beadbeater (Biospec Products, Inc.).
3. The cell lysate was recovered and centrifuged at 10,000*g* for 5 min to remove any cellular debris.
4. The protein extract was processed using a dual microdialysis system ***(13)***. Low molecular mass (salts) and high-molecular mass components were removed using a modified dual microdialysis system composed of 300 kDA molecular mass cut-off regenerated cellulose membrane in the first microfabricated stage, followed by a second stage containing an 8 kDa molecular mass cutoff membrane (regenerated cellulose microdialysis fiber).
5. Dialysis was done against 10 m*M* NH_4OAc. Protein concentration was measured using the Bradford assay ***(14)***. Sample solution containing approx 0.25 mg/mL of protein isolated from *E. coli* was mixed with 0.5 % Pharmalyte 3–10 and vortexed. The sample was centrifuged for 2 min prior to CIEF.

3.1.2. CIEF of Complex Protein Mixture (Cell Lysate)

On-line CIEF–FTICR/MS analysis was performed using 50-cm-long (50 µm id and 192 µm od) fused silica capillaries coated with linear polyacrylamide ***(15)*** (*see* **Note 2**). The homemade CIEF setup schematically presented in **Fig. 1** was used.

3.1.2.1. Sample Injection Procedure (*see* **Fig. 1A**)

1. Mount the capillary between the cathode and anode reservoir using fingertight unions.
2. Pipet 10–15 μL of sample into the anode reservoir. Although the method uses only approx 1 μL of sample (volume of the capillary) for actual handling, 10–15 μL are presently used to avoid introduction of air.
3. Push the sample through the capillary with compressed air (at ~20 mBarr), until the sample comes out of the other end of the capillary (fixed in cathode reservoir).
4. Remove excess sample (can be reused) from the reservoir, and then clean the reservoir (flush with anolyte).

3.1.2.2. Focusing Procedure (*see* **Fig. 1B**)

1. Fill the electrolyte reservoirs: the inlet reservoir with 20 m*M* phosphoric acid as the anolyte and the outer reservoir with 20 m*M* sodium hydroxide as the catholyte, respectively.
2. Immerse the electrodes in the electrolytes.
3. Apply constant voltage (~13 kV) for 15 min with a high-voltage power supply. In parallel, start the sheath liquid flow during focusing in order to establish a stable flow at the start of mobilization (*see* **Note 3**).

3.1.2.3. Mobilization Procedure (*see* **Fig. 1C**)

1. Once the focusing is completed, turn off the voltages (work fast but securely).
2. Remove the capillary tip from the cathode reservoir and adjust it so that it protrudes from the coaxial liquid sheath flow electrospray interface (electrospray interface of a standard Finningan MAT, San Jose, CA) and extends 1 mm outside the electrospray sheath tube surrounding the capillary. A sheath liquid was delivered at a flow rate of 2 μL/min using the syringe pump.
3. Electrospray the sample and surrounding sheath liquid into the inlet of the mass spectrometer. As the electrospray interface (spraying tip) is at 2000 V, increase potential of the anolyte to 15 kV to keep the potential difference the same as during focusing. Once all voltages are applied and an electrospray is established, sample mobilization and MS detection may start.
4. Raise the inlet reservoir 5 cm above the level of the electrospray emitter to apply gravity (concurrently with cathodic mobilization) to mobilize protein bands toward the mass spectrometer.
5. Start recording mass spectra.

3.1.3. MS Dctection of Proteins From CIEF

3.1.3.1. Interface and Instruments Parameters

ESI and detection of ions may be performed using different mass spectrometers, all of which will have different levels of performance (e.g., sensitivity, speed, mass measurement accuracy, resolution, and so on). We used a 7-Tesla

FTICR mass spectrometer *(16)* operated with an Odyssey data station (Finnigan FTMS, Madison, WI). This mass spectrometer is equipped with three serial quadrupoles, which are separated by two conductance limits and are operated synchronously, to guide ions produced by the ESI source to a rectangular closed cell through four stages of differential pumping. The final pumping stage, which contains a 1-m-long third quadrupole, is efficiently pumped by a cryopump with two concentric cryopannels extending to the ICR analysis cell at a pressure below 10^{-9} torr.

The ESI source inlet to the MS consisted of a heated stainless steel "desolvation" inlet capillary, a 1-mm orifice diameter skimmer, and a short quadrupole segment (i.e., collisional quadrupole) added to the set of two original quadrupole ion guides. Source conditions were as follows: spraying capillary at 2 kV, heating capillary at 200 V and 160°C, skimmer at 25 V, and a source quadrupole at 15 V, with all quadrupoles operating in an radio frequency (rf)-only mode (~750 kHz, ~500 Vpp). Ion accumulation was accomplished by introducing N_2 into the FTICR trap at 10^{-5} torr via a piezoelectric valve (Lasertechniques Inc., Albuquerque, NM). Background pressure in the FTICR trap was maintained at 10^{-9} torr using a custom cryopumping assembly.

Mass spectra were acquired using a standard experimental sequence for ion injection and accumulation, pump-down, excitation, and detection. Typically, coherent ICR motion was excited by dipolar sweep excitation in $800 < m/z < 2000$ range, and transient was acquired for *m/z* >800 (*see* **Note 4**). The total spectrum acquisition time was about 4 s (*see* **Note 5**). ICR trap control, ion excitation, data acquisition, and storage were controlled through the Odyssey data station. During the MS experiment (about 30 min) several hundred mass spectra were typically recorded (*see* **Fig. 2**).

3.1.3.2. Data Analysis and Display of the Results

On-line CIEF–FTICR/MS analysis of cell lysates results in large and complex data sets. Interpretation of the results in the context of available genomic databases was done using software developed in our laboratory to assist protein identification.

Data processing was semiautomated using the ICR-2LS software *(17)*. Briefly, ICR-2LS converted the raw data into m/z spectra that were subsequently transformed to generate a table of neutral masses using an implementation of the THRASH algorithm originally developed by Horn et al. *(18)*. ICR-2LS data processing was performed on the complete CIEF–FTICR data set in a single batch run to generate an output data file (PEK file) file containing neutral masses for all sample components observed in the run.

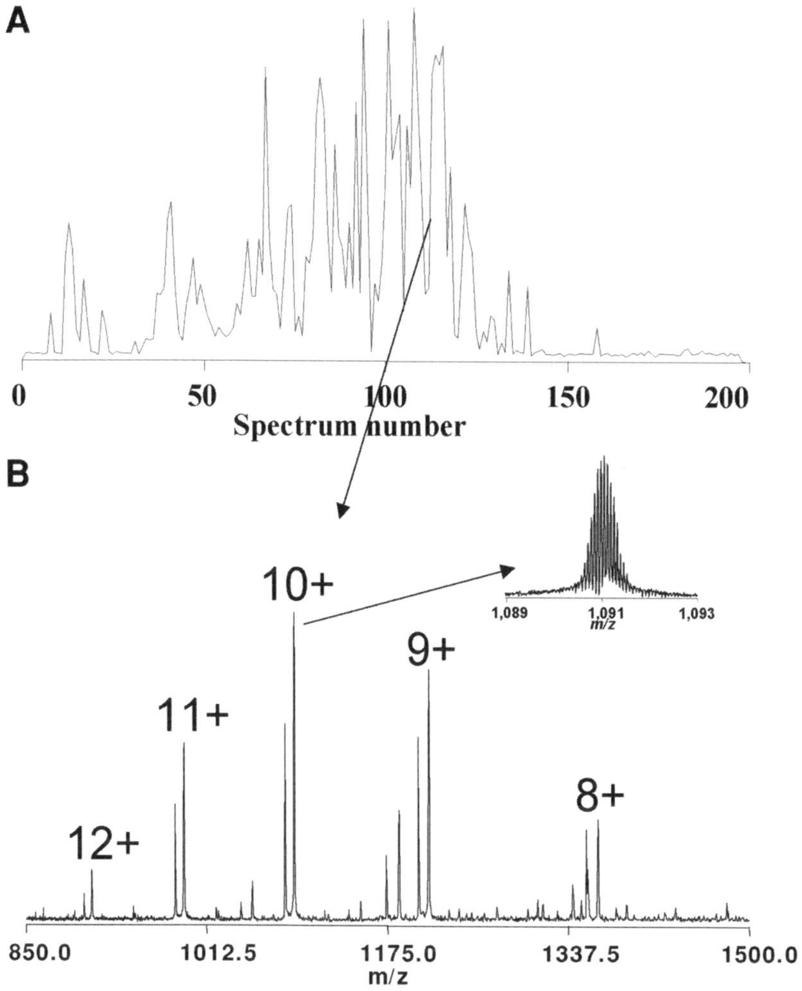

Fig. 2. (**A**) Total ion chromatogram reconstructed from the FTICR mass spectra obtained during approx 30-min-long CIEF mobilization of *E. coli* cellular lysate. (**B**) A representative mass spectrum. Inset shows an isotopic envelope detected for 11-kDa protein.

The PEK can be visualized in the form of a 2D display of neutral masses vs spectrum number, with "spot" size representing component intensity, using LaV2DG software developed at our laboratory (*see*, for example, **Fig. 3**). The neutral masses from the PEK file were also imported into the protein-search module of the ICR-2LS, where they can be searched for matches against the

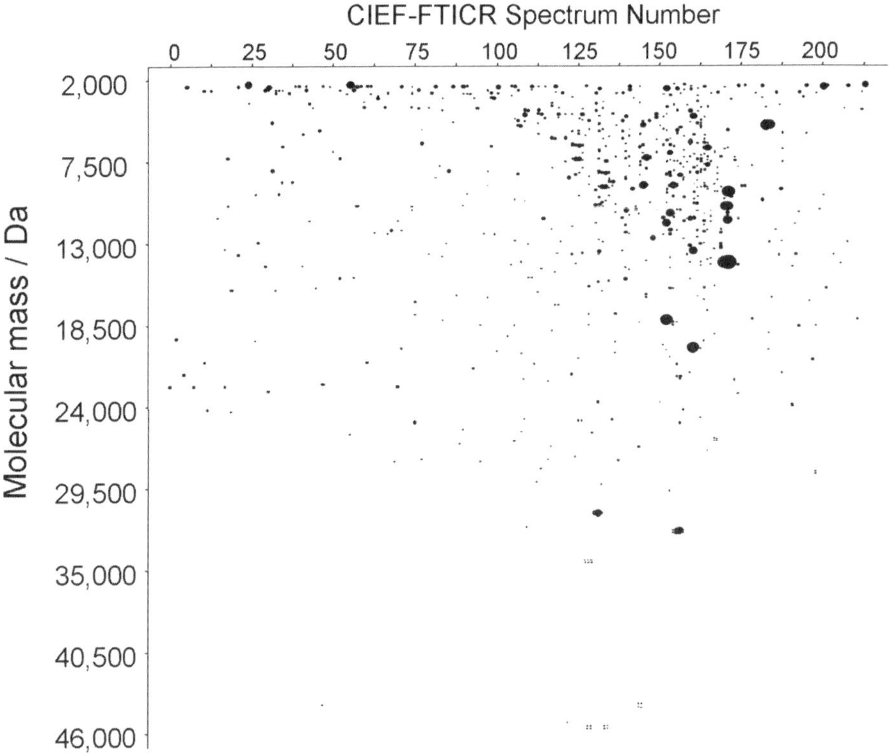

Fig. 3. Representative 2D display of the CIEF-FTICR analysis of *E. coli* lysate generated by plotting molecular mass vs spectrum number (which can be correlated to isoelectric point).

whole *E. coli* protein database, typically using the criteria of 10 ppm mass measurement accuracy for monoisotopic masses for proteins with $M_r < 25$ kDa, and 100 ppm for larger proteins having unresolved isotopic distributions. For database searching, common modifications were permitted, including methylations, acetylations and losses of leading methionine.

Our experience has shown that only a modest fraction of intact proteins can be initially identified based solely on mass and p*I* measurements (*see* **Note 6**). This led to the development of different stable isotope labeling schemes *(8,9)*, with an example given in **Fig. 4**. This figure shows two representative spectra obtained during the CIEF separation of a mixture of unlabeled and Leu-D_{10}-labeled proteins extracted from *E. coli* cultured on normal and $^{13}C^{15}N$-depleted media (*see* **Note 7**). Unlabeled and labeled forms of each protein (having the same isoelectric point) focus together during CIEF separation and are thus

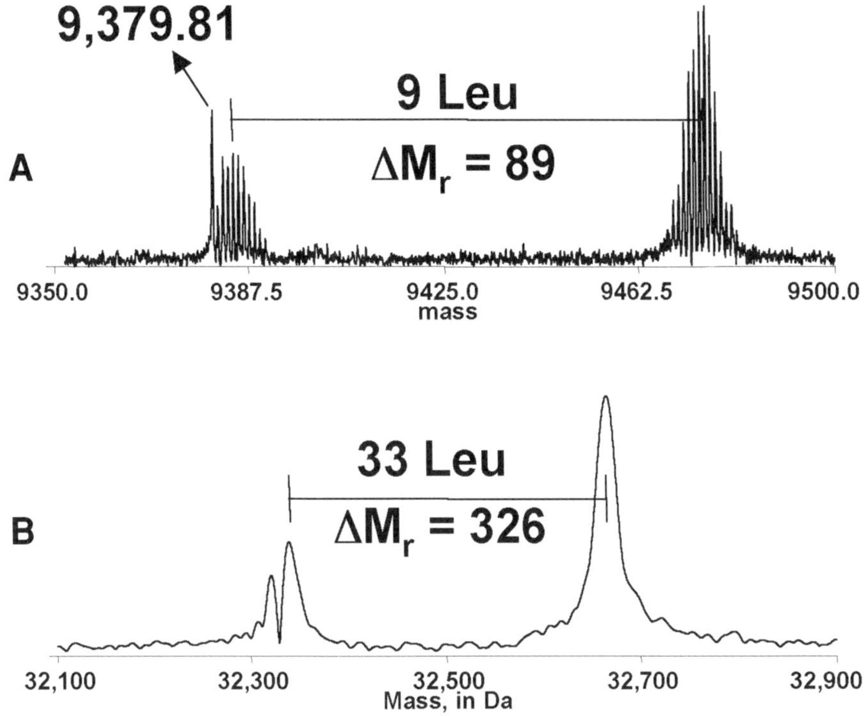

Fig. 4. (**A**) High-resolution CIEF-FTICR zero charge state mass spectra of unlabeled (normal and $^{13}C^{15}N$ depleted) and Leu-D_{10}-labeled forms of protein yciN isolated from *E. coli* grown in minimal medium and minimal medium supplemented with 0.1 mg/mL Leu-D_{10}. Low resolution and zero charge state spectrum (**B**) of unlabeled (normal and $^{13}C^{15}N$ depleted) and Leu-D_{10}-labeled forms of malate dehydrogenase. The mass difference between labeled and unlabeled species indicates the number of Leu residues in the protein.

observed within the same FTICR mass spectrum. Both protein versions display identical charge state distributions, allowing them to be easily assigned as two versions of the same protein.

The zero charge spectrum for the protein pair in **Fig. 4A** shows two average molecular masses of 9386 and 9475 Da. The mass difference of approx 89 Da corresponds to the incorporation of nine isotopically labeled Leu residues in this protein (*see* **Note 8**). A search of Genbank (National Center for Biotechnology Information [NCBI]) identifies this protein as protein yciN from the *E. coli* K12 strain MG1655 genome (Swiss Protein accession number P46132), which has a molecular mass of 9386 Da and contains nine Leu residues.

The mass spectrum for a larger protein for which isotopic resolution was not obtained is shown in **Fig. 4B**. Nevertheless, an accurate average molecular mass for both the unlabeled and isotopically labeled proteins could still be obtained from the charge state distributions. The average molecular mass of the unlabeled protein was 32,336 ± 1 Da, while the mass of the corresponding Leu-D_{10}-labeled protein was 32,662 ± 1 Da. The mass difference of approx 326 Da indicates the incorporation of 33 isotopically labeled Leu residues. This information allowed the protein to be unambiguously identified as malate dehydrogenase (P06994).

3.2. CIEF–MS of the Mixture of Noncovalent Protein Complexes

CIEF–MS analysis of the mixture of noncovalent protein complexes at the intact complex level, in combination with the analysis at the protein level, reveals molecular mass and composition of the complexes (i.e., identity and the number of building subunits). For an initial demonstration we will use a mixture of commercially available noncovalent protein complexes: creatine phosphokinase (CPK) from rabbit muscle and glyceraldehyde-3-phosphate dehydrogenase (GAPDH) from rabbit muscle, purchased from Sigma.

The identification procedure requires two experiments with the identical CIEF separation, but different ESI/MS conditions: (a) CIEF–MS analysis of the mixture of complexes with detection of protein units, and (b) CIEF–MS analysis of the same mixture of complexes with the detection of intact complexes. Combining the results from the two analysis (a) and (b) enables the characterization of separated complexes (i.e. size, subunits, homo- or heterocomplex).

3.2.1. CIEF–MS Detection of Protein Subunits

The procedures are identical to those described in **Subheadings 3.1.2.** and **3.1.3.** CIEF separation is in general performed under "native" conditions to preserve conformation of the proteins and their complexes. Use of the acidic sheath liquid together with the appropriate ESI source and MS conditions generally enables dissociation of complexes to their protein subunits (*see* **Fig. 5A**).

3.2.2. CIEF–MS With Detection of Intact Complexes

CIEF separation was performed as described in **Subheading 3.2.1.** except for the use of sheath liquid. For detection of intact complexes, one needs to preserve "native" conditions through both CIEF separation and MS detection. Conditions used to conserve noncovalently bounded complexes are as follows.

1. A sheath liquid composed of 10 m*M* ammonium acetate, pH 6.0, with 10–20% methanol provides stable on-line electrospray while also maintaining the integrity of complexes studied.

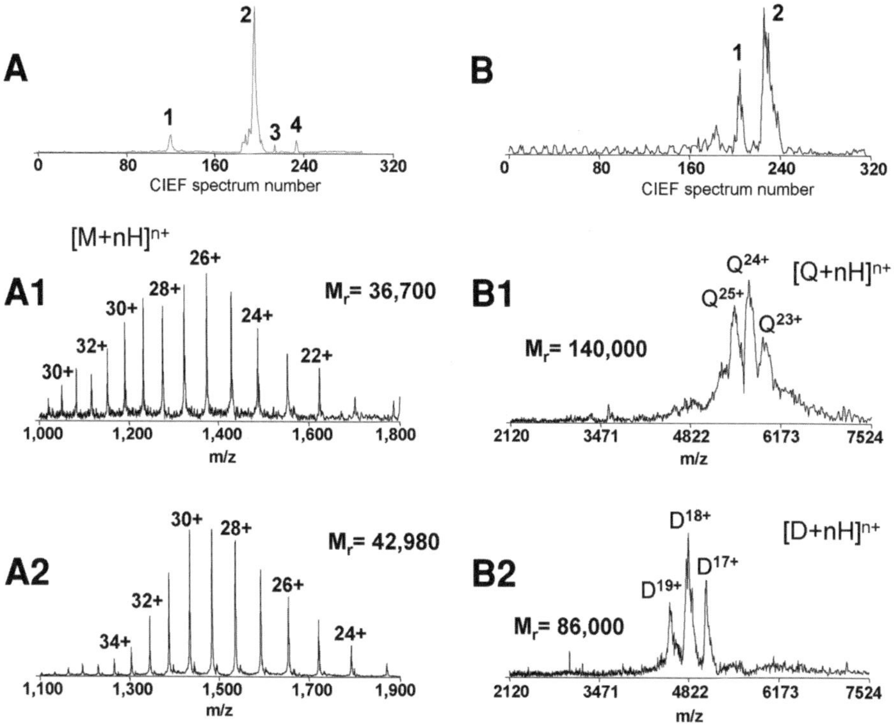

Fig. 5. (**A**) Separation of GAPDH (*1*) and CPK (*2*) protein complexes and detection of the dissociated monomeric units. Reconstructed total ion chromatogram and representative low resolution ESI-FTICR mass spectra of complex subunits (**A1** and **A2**). (Two lower mass proteins detected (*3*) and (*4*) were determined to be impurities in the CPK sample.) (**B**) Separation and detection of GAPDH and CPK intact noncovalent complexes from the mixture. Reconstructed total ion chromatogram and representative low resolution ESI-FTICR mass spectra (**B1** and **B2**).

2. The following "gentle" MS source conditions were also used: spraying capillary at 2 kV, heating capillary at 110 V and 90°C, lower inlet-skimmer voltage difference of 85 V to preserve intact protein complexes throughout the ESI processes.
3. FTICR "in trap" clean-up was employed to remove/reduce adduction (*see* **Note 9**).
4. Excitation and detection of ions were adapted for higher m/z ions, e.g., 2000 < m/z < 7000. Compared to the mass spectra of complex subunits obtained using the acidic sheath liquid condition, in the CIEF–MS analysis of the complexes, fewer charge states and lower charge state species were typically observed, both of which are correlated with a compact structure expected for the complex (*see* **Fig. 5B**).

3.2.3. Data Analysis

Identification of the masses for the monomeric subunits (*see* **Fig. 5A**) combined with the masses for the intact complexes (*see* **Fig 5B**), allowed characterization of the stoichiometry of GAPDH as homotetrametic (Q), and CPK as homodimeric (D).

4. Notes

1. We have also performed CIEF–MS analysis of cellular lysates from *Saccharomyces cerevisiae* (ATCC #200867), *Deinococcus radiodurans* R1, and *Shewanella oneidensis* MR1.
2. The capillary was effectively used five to seven times before its performance degraded. Use of different type of coatings showed increased stability through many CIEF runs with UV detection ***(19)***.
3. In our experiment, the capillary passes through the electrospray interface and its terminus functions as the ESI emitter. There are other interfaces available; however, we have found this design with the use of a sheath liquid flow to be the most effective.
4. Ampholytes coelute with proteins from CIEF capillary. They generally form singly charged ions and are usually detected in the lower *m/z* region. Although the charge envelope of the proteins is easily distinguished from the ampholyte signal, we started excitation and detection at *m/z* approx 800 to minimize "ampholyte noise." The best results are obtained if the lower *m/z* ampholytes are removed by adjustment of the rf level at one of the quadropoles during transmission to the FTICR cell.
5. Since the sample introduction takes only approx 0.25 s and total spectrum acquisition time is 4 s, most of the continuously mobilized sample is lost. Alternatively, ion accumulation is accomplished using an external quadropole as an accumulation device so that the ions are accumulated throughout most of the acquisition time, often increasing the effective duty cycle to >90%.
6. Calculated p*I* values of the identified proteins are within the range of three to ten, and generally follow the predicted elution order. It should be noted, however, that pI values are calculated for denatured proteins. Thus, they do not necessarily correctly reflect p*I*s of the native proteins (e.g., folding of the protein chain will most certainly bury some residues and thus change the p*I*).
7. In this particular case, unlabeled proteins came from both cultures (e.g., normal and $^{13}C^{15}N$ depleted), whereas Leu-D_{10}-labeled proteins were extracted only from *E. coli* grown on normal isotopic distribution media. Thus, each protein detected during CIEF separation of this mixture came in three versions: $^{13}C^{15}N$-depleted isotopic distribution (which allowed precise determination of monoisotopic mass) and normal isotopic distribution with Leu-H_{10} and Leu-D_{10} (which allowed determination of the number of Leu residues in the protein).
8. Although the isotopically labeled Leu contains ten deuteriums, the mass difference is not exactly 90 Da owing to the isotopic enrichment of the stable isotope being 97.9% and to the small mass defect of deuterium compared to hydrogen.

9. Extensive adduction of uncharacterized low-weight species to noncovalent complexes results in greater uncertainty in molecular mass determinations, as well as decreased sensitivity. “Soft” low-energy sustained off-resonance irradiation collisionally induced dissociation (SORI-CID) dissociates weakly bound adducted species and yields higher quality spectra (i.e., better resolution, a significant enhancement of the signal-to-noise ratio and improved mass measurement accuracy) ***(20)***. Careful control of SORI amplitude allows effective removal of adducts without significant dissociation of covalent or specific noncovalent interactions. Fundamental to the conventional application of SORI is that it requires appropriate selection of irradiation frequency so as to target a specific (narrow) m/z range, in contrast to the broad band nature of the SORI-CID clean-up procedure. Also, the irradiation amplitude required for dissociation will strongly depend on the stability of the complex. Therefore, when dealing with a mixture of noncovalent complexes with different stability, optimization experiments with a set of “standard” complexes with known relative stability should be performed prior to CIEF–MS experiment ***(12)***.

Acknowledgments

The authors want to thank to Dr. C. Masselon for constructive discussions. This research was supported by the U.S. Department of Energy, Office of Biological and Environmental Research. Pacific Northwest National Laboratory is operated by Battelle Memorial Institute for the U.S. Department of Energy under Contract DEAC06-76RLO 1830.

References

1. Pasa-Tolic, L., Lipton, M. S., Masselon, C. D., et al. (2002) Gene expression profiling using advanced mass spectrometric approaches. *J. Mass. Spec.* **37,** 1185–1198.
2. Goodlet D. R. and Yi, E. C. (2002) Proteomics without polyacrylamide: qualitative and quantitative uses of tandem mass spectrometry in proteome analysis. *Funct. Integr. Genomics* **2,** 138–153.
3. Pasa-Tolic, L., Jensen, P. K., Anderson, G. A., et al. (1999) High throughput proteome-wide precision measurements of protein expression using mass spectrometry *J. Amer. Chem. Soc.* **121,** 7949–7950.
4. Tang, Q., Harrata, A. K., and Lee, C. S. (1996) High-resolution capillary isoelectric focusing-electrospray ionization mass spectrometry for hemoglobin variants analysis *Anal. Chem.* **68,** 2482–2487.
5. Martinovic, S., Pasa Tolic, Lj., Masselon, C., Jensen, P. K., Stone, C. L., and Smith, R. D. (2000) Characterization of the human alcohol dehydrogenase isoenzymes by CIEF-ESI-FTMS *Electrophoresis* **21,** 2368–2375.
6. Tang, Q., Harrata, A. K., and Lee, C. S. (1997) Two-dimensional analysis of recombinant E. coli proteins using capillary isoelectric focusing electrospray ionization mass spectrometry. *Anal. Chem.* **69,** 3177–3182.

7. Yang, L., Lee, C. S., Hofstadler, S. A., Pasa-Tolic, L., and Smith, R. D. (1998) Capillary isoelectric focusing-electrospray ionization fourier transform ion cyclotron resonance mass spectrometry for protein characterization *Anal. Chem.* **70,** 3235–3241.
8. Jensen, P. K., Pasa Tolic, L., Anderson, G. A., et al. (1999) Probing proteomes using capillary isoelectric focusing-electrospray ionization fourier transform ion cyclotron resonance mass spectrometry *Anal. Chem.* **71,** 2076–2084.
9. Martinovic, S., Veenstra, T. D., Anderson, G. A., Pasa-Tolic, Lj., and Smith, R. D. (2002) Selective incorporation of isotopically labeled amino acids for identification of intact proteins on proteome-wide level. *J. Mass Spectrom.* **37,** 99–107
10. Jensen, P. K., Pasa-Tolic, L., Peden, K. K., et al. (2000) Mass spectrometric detection for capillary isoelectric focusing separations of complex protein mixtures. *Electrophoresis* **21,** 1372–1380.
11. Lyubarskaya, Y. V., Carr, S. A., Dunnington, D., et al. (1998) Screening for high-affinity ligands to the *Src* SH2 domain using capillary isoelectric focusing-electrospray ionization ion trap mass spectrometry *Anal. Chem.* **1,** 4761–5770.
12. Martinovic, S, Berger, S., Pasa Tolic, Lj., and Smith, R.D. (2000) Separation and detection of intact non-covalent protein complexes from the mixtures by on-line capillary isoelectric focusing - mass spectrometry. *Anal. Chem.* **72,** 5356–5360.
13. Liu, C. L., Hofstadler, S. A., Bresson, J. A., et al. (1998) Rapid on line dual microdialysis with ESI-MS for direct analysis of complex biological samples and microorganism lysates. *Anal. Chem.* **70,** 1797–1801.
14. Bradford, M. M. (1976) A rapid and sensitive method for quantitation of microgram quantities of protein utilizing the principle of protein-dye binding. *Anal. Biochem.* **72,** 248-254.
15. Kilar, F. and Hjerten, S. (1989) Fast and high resolution analysis of human serum transferrin by high performance isoelectric focusing in capillaries. *Electrophoresis* **10,** 23–29.
16. Winger, B. E., Hofstadler, S. A., Bruce, J. E., Udseth, H. R., and Smith, R. D. (1993) High-resolution accurate mass measurements of biomolecules using a new electrospray ionization ion cyclotron resonance mass spectrometer. *J. Am. Soc. Mass Spectrom.* **4,** 566–577.
17. The ICR-2LS suite of data analysis software is publicly available for noncommercial use. Requests and inquiries should be directed to: Gordon.Anderson@pnl.gov.
18. Horn, D. M., Zubarev, R. A., and McLafferty, F. W. (2000) Automated reduction and interpretation of high resolution electrospray mass spectra of large molecules. *J. Am. Soc. Mass Spectrom.* **11,** 320–332.
19. Shen, Y. and Smith, R. D. (2000) High-resolution capillary isoelectric focusing of proteins using highly hydrophilic-substituted cellulose-coated capillaries. *J. Microcolumn Separations* **12,** 135–141.
20. Pasa-Tolic, L., Bruce, J. E., Lei, Q. P., Anderson, G. A., and Smith, R. D. (1998) In-trap cleanup of proteins from electrospray ionization using soft sustained off-resonance irradiation with fourier transform ion cyclotron resonance mass spectrometry. *Anal. Chem.* **70,** 405–408.

15

Integrated System for Rapid Proteomics Analyses Using Microfluidic Devices Coupled to Nanoelectrospray Mass Spectrometry

Jianjun Li, Tammy-Lynn Tremblay, Jed Harrison, and Pierre Thibault

Summary

This chapter presents an integrated and modular microsystem providing rapid analyses of low femtomole of in-gel digests for proteomics applications. Enhancement of sample throughput is facilitated using an autosampler, a microfabricated device comprising a large (2.4-μL total volume) separation channel together with a low-dead-volume interface to nES mass spectrometry. Sample preconcentration is achieved by packing C_{18} reverse phase or immobilized metal affinity chromatogaphy (IMAC) beads into the large channel of this microfluidic device to adsorb peptides or enrich the sample in phosphopeptides prior to capillary electrophoresis separation and MS detection.

This integrated microfluidic systems enables a sample throughput of up to 12 samples/h with a detection limit of approx 5 n*M* (25 fmol inj.). Replicate injections of peptide standards indicated that reproducibility of migration time was typically 1.2–1.8%, whereas relative standard deviation (RSD) values of 9.2–11.8% were obtained on peak heights. The application of this device is demonstrated for 2D gel spots obtained from protein extracts of human astrocyte cells and for excised bands of membrane proteins from *Neisseria meningitidis.* A stepped acetonitrile gradient can be incorporated with the present microfluidic system to enhance selectivity during sample analysis.

Key Words

Gel-isolated proteins; human astrocytes; microfluidics; nES; *Neisseria meningitidis;* phosphopeptides; quadrupole/time-of-flight; tandem mass spectrometry.

1. Introduction

Proteomics was originally defined as the protein complement of any given cell, tissues or organism *(1)*. However, this definition has now been expanded to include the identification of protein isoforms, splice variants, posttransla-

From: *Methods in Molecular Biology, vol. 276: Capillary Electrophoresis of Proteins and Peptides*
Edited by: M. A. Strege and A. L. Lagu © Humana Press Inc., Totowa, NJ

tional modifications, interacting partners, and higher-order complexes under different environmental conditions *(2)*. Such research endeavors also include the monitoring of protein expression, which by itself represents a significant analytical challenge given the changes in protein content and distribution from cell to cell, the dynamic range of expression (more than 10^6-fold) and the wide diversity of protein modifications (e.g., splice variants, co- and posttranslational modifications, and proteolytic processing).

Two-dimensional polyacrylamide gel electrophoresis (2D-gel) offers the capability of resolving thousands of protein spots based on their respective isoelectric point (p*I*) and molecular weights (MWs) and has been traditionally used to monitor differential protein expression across different cellular extracts *(3)*. Although this technique is time (2–3 d of gel electrophoresis) and sample consuming (>100 μg of protein extract) it experienced significant popularity for its unparalleled separation performance and its ability to provide relative protein quantitation via visible and fluorescent staining reagents ***(4,5)***. The identification of spots of interest is typically achieved from in-gel tryptic digestion followed by MS analyses of the corresponding tryptic peptides and database searching ***(6–8)***. However, not all protein classes are amenable to 2D-gel separation, and membrane components or proteins of extreme p*I* (p*I* > 10, p*I* < 3) and MW (>150 kDa, <10 kDa) are typically under-represented. Alternate approaches to 2D gel include protein separation via 1D SDS-PAGE or multidimensional chromatography (2D-LC) of the proteolytic fragments derived from the intact proteins ***(9,10)***.

At present, gel-based protein separation still represent a sizable proportion of all samples analyzed by numerous proteomics core facilities. Although peptide mass fingerprinting can be use for sample of low complexity (<3 proteins/band or spot) and for small and well characterized proteomes, unambiguous identification of gel-isolated proteins typically relies on sensitive tandem mass spectrometry (MS/MS) techniques to derive partial amino acid sequences. Nano-scale liquid chromatography coupled with nanoelectrospray MS-MS (nano-LC-MS/MS) currently represents the most rugged workhorse system for fmol detection of in-gel protein digests ***(11,12)***.

While most proteomics platform uses nano-LC-MS/MS systems, the coupling of electrophoretic-based microfluidic systems to MS also offers an efficient means of handling nanoliter sample-volumes while maintaining high separation and sensitivity on devices of small footprint. In contrast to high-performance liquid chromatography (HPLC) these microfabricated devices do not involve low-dead-volume switch valves, actuators, syringe pumps, or precise gradient-flow delivery systems. Rather, microfluidic devices rely on the precise control of voltages and electrical fields across small separation channels, which are typically achieved with voltage controllers and power supplies of modest cost.

The coupling of microfluidic devices to MS is relatively recent, and early reports of such input sample system for nanoelectrospray MS (nES-MS) were published independently in 1997 by the research groups of B. Karger ***(13)***, M. Ramsey ***(14)***, and R. Aebersold ***(15)***. Two different interface designs have been proposed to couple microfluidic devices to nES-MS ***(13–17)***. The first design is characterized by nES emission directly from the exposed channel on the side of the microfabricated surface, ***(13)*** while the second configuration involves the insertion of a capillary transfer tube into the microchip ***(15,16)***. In the second configuration, the transfer capillary can act as the nES emitter, provided that the electrical contact be made at the tip of sprayer using a conductive surface such as a gold-coated layer deposited on the tapered fused-silica capillary needle (<15 μm i. d) ***(17)***. Alternatively, a concentric sheath flow or a liquid junction can be used to maintain the electrical contact of the spraying solution ***(15,16)***.

It is noteworthy that most chip-MS systems used nES ionization. However, the coupling of microfluidic devices to matrix-assisted laser desorption ionization (MALDI) using off-line deposition techniques has also been reported ***(18–19)***. Low femtomole MS detection of peptides is commonly achievable using both MALDI and nES ionization techniques. However, the pL-nL injection volumes typically available with microfluidic devices have set the concentration detection limits of chip-MS systems to the low mM range. In order to improve these detection limits, enhancement of sample loading has been achieved using on-chip stacking ***(17)*** and sample preconcentration via reverse-phase adsorbant ***(20,21)*** thus enabling low-nanomolar detection of peptide standards. Recent microfluidic devices have incorporated plastic or glass nES emitters within the structure of the microchip ***(17,22)***. Such devices can also be used in conjunction with disposable adsorption preconcentration tips to desalt samples prior to nES-MS analyses.

In an effort to accelerate the sample throughput of the combined chip-MS systems, previous reports have combined pressure-driven autosampler with the microfluidic devices ***(21)***. Such advances were made possible through the fabrication of a large sample channel providing a convenient port of low flow resistance enabling the introduction of sample on and off the chip without perturbing the fluids in the analysis manifold. These systems not only integrate sequential injection and sample preconcentration but also enable electrophoretic separation of complex protein digests on an array of intersecting channels. Previous reports have described the separation of peptide standards and tryptic digests with a throughput of up to 12 samples/h and detection limits of approx 5 n*M* ***(21)***. Other forms of sample preconcentration include the integration of on-chip affinity chromatography using antibody beads to bind target antigens, or immobilized metal affinity chromatography to selectively enrich phosphorylated peptide fragments ***(21)***.

This chapter presents an integrated microsystem that provides rapid analyses of trace-level tryptic digests for proteomics application. This modular microsystem includes an autosampler, a microfabricated device comprising a sample introduction port, a separation channel, together with a low-dead-volume capillary transfer line providing a convenient nES interface to mass spectrometry. In the present investigation we have used the large channel of this microfluidic to integrate C_{18} adsorbant packing prior to electrophoretic separation and MS analyses. Application of this integrated system is demonstrated for the analysis of trace-level tryptic peptides obtained from gel-isolated proteins of human astrocyte cell extracts.

2. Materials

2.1. Growth of Bacterial Strains

1. Bacterial strain *N. meningitidis* strain L3 (*see* **Note 1**).
2. Bacterial plates: chocolate agar plates from Quelab of Montreal, Canada; 5% sheep blood agar plates.
3. Liquid growth media: 3.7% (w/v) brain heart infusion (BHI) 10 mg/L haemin, 2 mg/L nicotinamide adenine dinucleotide (NAD); Difco bacto Todd Hewitt (DBTH) broth from Difco.
4. Phosphate-buffered saline (PBS): 6.7 m*M* potassium phosphate, pH 7. 4, containing 150 m*M* NaCl and 0. 02% (w/v) NaN_3.
5. Bacterial killing solution: 0. 5% (w/v) phenol in PBS.
6. IF-75 fermenter from New Brunswick scientific.

2.2. Growth of Human Astrocytes

1. Primary fetal human astocyte (FHA) cultures were generously provided by Dr. J. Antel, Montreal Neurological Institute (Montreal, Quebec, Canada).
2. Dullbecco's modified Eagle's medium (DME).
3. Fetal bovine serum (Hyclone, Logan, UT).
4. Poly-L-lysine-coated dishes.
5. Phosphate-buffered saline (PBS).

2.3. Protein Isolation and Purification (see Notes 2 and 3)

2.3.1. Total Protein Extraction

1. Urea (Bio-Rad, Hercules, CA).
2. Thiourea (Sigma, St. Louis, MO).
3. 3-(3-Cholamidopropyl)dimethylammonio.-1-propanesulfonate (CHAPS; Sigma).
4. DL-Dithiothreitol (DTT; Sigma).
5. Biolytes (Bio-Rad).
6. Acetone (Fisher Scientific, Fair Lawn, NJ).

2.3.2. Hydrophobic/Hydrophilic Protein Extraction (Phase Separation)

1. Triton X-114 (Octylphenoxypoly-ethoxyethanol; Sigma).
2. Tris (Bio-Rad).
3. Sodium chloride (BDH Inc., Toronto, Canada).
4. Butylated hydroxytoluene (Sigma).
5. Acetone (Fisher Scientific).
6. Mammalian cocktail protease inhibitor (Sigma)

2.3.3. 2D Gel Electrophoresis

1. Immobilized pH-gradient (IPG) strips (Bio-Rad).
2. Sodium dodecyl sulfate (SDS; (Bio-Rad).
3. Iodoacetamide (Sigma).
4. Diothiothreitol (DTT; Sigma).
5. Urea (Bio-Rad).
6. Glycerol (Fisher Scientific).
7. Tris (Bio-Rad).
8. 30% Acrylamide (National Dignostics, Atlanta, GA).
9. *N,N,N',N'*-tetramethylethylenediamine (TEMED) (Bio-Rad).
10. Ammonium persulfate (APS; Bio-Rad).
11. Electrophoresis buffer: (Biorad's 10X Tris/glycine/SDS buffer).
12. Biorad's resolving gel buffer (Bio-Rad).

2.4. Gel Staining

2.4.1. Sypro Ruby Staining

1. Methanol (Fisher Scientific).
2. Acetic acid (BDH Inc.).
3. Sypro ruby (Bio-Rad).

2.4.2. Silver Staining

1. Ethanol (Fisher Scientific).
2. Sodium thiosulfate (Sigma).
3. Silver nitrate (Sigma).
4. Sodium carbonate monohydrate (BDH Inc.).
5. Formaldehyde (BDH Inc.).
6. Acetic acid (BDH Inc.).

2.5. Silver Destain and In-Gel Digestion

1. Trypsin (Promega, Madison, WI).
2. Potassium ferricyanide (Sigma).
3. Sodium thiosulfate (Sigma).
4. Ammonium bicarbonate (Fisher Scientific).
5. Methanol (Fisher Scientific).
6. Acetic acid (BDH Inc.).

7. Acetonitrile (Fisher Scientific).
8. Savant preconcentrator (Savant Instruments Inc., Farmingdale, NY).
9. Progest automated digestion unit (Genomics solutions, Ann Arbour MI).

2.6. Microfluidic Device

1. Corning 0211 glass (Corning Glass Works, NY).
2. Septa 77 (Chromatographic Specialties, Brockville, Ontario, Canada).

2.7. Low-Dead-Volume Connection

1. Crystal Bond 509 (CB) (Aremco Products, Valley Cottage, NY).
2. Staedtler Lumocolor permanent pen (Staedtler, Germany).
3. Hot plate (Corning Glass Works, Corning, NY).
4. Tungsten carbide drills (Tycom, Mississauga, ON, Canada).
5. Acetone (EM Science, Gibbstown, NJ).

2.8. Nanoelectrospray Emitter

1. 24K Bright English gold plating salts (Grobet File Co. of America, Inc., Carlstadt, NJ).
2. Fused-silica capillary (Polymicro Technologies, Phoenix, AZ).
3. Teflon tubing (LC Packing, San Francisco, CA).
4. Laser puller (Sutter Instruments, Novato, CA).

2.9. BCQ Coating

1. (Acryloylamino)propyl trimethylammonium chloride or BCQ (Chemische Fabrik Stockhausen, Krefeld, Germany).
2. 7-Oct-1-enyltrimethoxysilane (United Chemical Technologies Inc., Bristol, PA).
3. TEMED (Bio-Rad).
4. Sodium hydroxide (BDH Inc.).
5. Ammonium persulfate (BDH Inc.).
6. Acetic acid (BDH Inc.).

2.10. On-Line Adsorption Preconcentration and Affinity Selection Chip-CE–MS

1. 40 μm C_{18} reverse-phase packing material is excised from SepPak cartridges (Waters, Milford, MA).
2. 5 μm Poros C_{18} (Applied Biosystems, Framingham, MA).
3. IMAC beads are purchased from Pharmacia Biotech (Baie d'Urfe, Quebec, Canada).

3. Methods

3.1. Growth of Bacterial Strains (see Note 2)

1. Resuscitate bacterial strains from frozen stocks on 5% sheep blood agar plates and incubate overnight at 37°C.

2. Select colonies from plates and cultivate in 60 L fermenter batch of DBTH broth.
3. Harvest cells by low speed centrifugation (5000*g*).
4. Decant centrifuged material.
5. Resuspend cell pellet in bacterial killing solution and stir for 16 h.
6. Collect bacterial cell mass by centrifugation (5000*g*).

3.2. Growth of Human Astrocytes

1. FHA cells are grown in (DME containing 10% FBS in an atmosphere of 5% CO_2 / 95% air at 37°C.
2. Prior to proteome analyses, transfer cells on to poly-L-lysine coated dishes for 6 d.
3. Wash cells three times with physiological saline (PBS) for 10 min with gentle shaking.
4. Collect and pellet the cells by centrifugation (5000*g*).

3.3. Protein Isolation and Purification

3.3.1. Total Protein Extraction

1. Resuspend cell pellet in lysis buffer (7 *M* urea, 2 *M* thiourea, 4% CHAPS and 1% DTT (*see* **Note 3**).
2. Shake at room temperature for 1 h.
3. Centrifuge cell lysate for 10 min at 12,000*g* to pellet unbroken cells.
4. Precipitate proteins with 10 vol of acetone (to remove excess salts).
5. Centrifuge to pellet proteins, air-dry pellet, and resuspend proteins in IEF buffer (same as lysis buffer).
6. Protein content in supernatant was then evaluated by Bradford assay.

3.3.2. Hydrophobic/Hydrophilic Protein Extraction (Phase Separation)

3.3.2.1. Triton X-114 Condensation

1. In a small beaker, weigh 10 g of Triton X-114. Slowly transfer Triton in 350 mL condensation solution (10 m*M* Tris-HCl, pH 7.4, 150 m*M* NaCl, 72.6 μ*M* butylated hydroxytoluene) agitating at 4°C. Complete to 500 mL with cold condensation solution by rinsing beaker a few times.
2. Cover and agitate solution at 4°C for 5–6 h.
3. Transfer solution in a separatory funnel and incubate overnight at 30°C.
4. Slowly recover detergent phase (bottom phase) in a 100 mL graduated cylinder (perform this step at 30°C).
5. Transfer detergent phase in 250 mL condensation solution agitating at 4°C. Complete to 500 mL with cold condensation solution by rinsing graduated cylinder a few times.
6. Repeat last four steps twice.
7. Determine Triton X-114 final concentration by reading absorbance at 280 nm of a 1:1000 dilution in 1% cold SDS solution (2.28 A_{280} = 1 mg/mL = 0.1% Triton X-114).

3.3.2.2. Protein Extraction (Phase Separation)

1. Resuspend pelleted cells in solution A (10 m*M* Tris-HCl, pH 7.4: 0.15 *M* NaCl) and add mammalian cocktail proteases inhibitor (Sigma). Add one third of the volume of solution B (40 m*M* Tris-HCl, pH 7.4; 600 m*M* NaCl; 4% triton X-114). (*See* **Note 3**.)
2. Incubate on ice for 1 h with frequent vortexing.
3. Centrifuge at 10,000*g* for 10 min at 4°C to pellet unbroken cells.
4. Transfer supernatant and incubate at 30°C for 3 min (until solution is cloudy).
5. Centrifuge at 1300*g* for 10 min at room temperature.
6. Transfer aqueous phase (top phase) but keep detergent phase at room temperature).
7. Add *X* vol of triton X-114 to aqueous phase:

$$\text{X vol of triton X-114} = \frac{\text{Vol of Aq phase}}{\text{2X concentration of triton in \%}}$$

8. Shake well then incubate 3 min at 30°C (until cloudy) then slowly pour over transfer detergent phase.
9. Centrifuge at 1300*g* for 10 min at room temperature.
10. Remove aqueous phase.
11. Precipitate hydrophobic protein (detergent phase) with 10X volume of acetone. Place at –20°C for 1 h or overnight.
12. Pellet proteins by centrifugation at max speed for 10 min at 4°C.
13. Resuspend air-dried pellet of proteins in IEF solution (7 *M* urea, 2 *M* thiourea, 4% CHAPS, 1% DTT) then precipitate protein again with 10 vol of acetone. Place 5–10 min at –20°C.
14. Pellet proteins by centrifugation at max speed for 10 min at 20°C. Then place air-dried pellet in IEF solution to dissolve protein.

3.4. Gel Electrophoresis

1. Samples containing 100-1000 mg of protein can be used to rehydrate immobilized pH gradient strips (*see* **Note 4**).
2. First dimension electrophoresis is performed using the following program: 200-V rapid ramp for 1 h, 500-V rapid ramp for 1 h, 5000-V linear ramp for 5 h, and 5000-V focusing for 80,000 V · h for a total of about 95,000 V · h (*see* **Note 5**).
3. Prior to second dimension electrophoresis, proteins on IPG strips are reduced (1% DTT) and alkylated (4% iodoacetamide) in SDS equilibration buffer (6 *M* urea, 30% glycerol, 2% SDS, and 50 m*M* Tris-HCl, pH 8. 8) for 15 min in each solution.
4. The second dimension is perform on a 1-mm thick polyacrylamide gel at a constant 24–30 m per gel (*see* **Note 6**).

3.5. Gel Staining

3.5.1. Sypro Ruby

1. After second dimension electrophoresis, separated proteins are fixed in the gel for at least 30 min using a 10% methanol/7% acetic acid solution.

2. Gels are then place in sypro ruby overnight with gentle agitation.
3. Gels are washed twice for 30 min in a 10% methanol/7% acetic acid solution.
4. Scan gels with Fluor-S MultiImager (BioRad).

3.5.2. Silver Stain

1. After second dimension electrophoresis, separated proteins are fixed in the gel for at least 30 min using a 50% ethanol/5% acetic acid solution (can be left overnight).
2. Wash gel for 10 min in 50% ethanol.
3. Wash gel twice for 10 min in water (*see* **Note 7**).
4. Place gel in 0. 02% sodium thiosulphate for 5 min.
5. Wash twice for 5 min in water.
6. Place gel in 0. 1% silver nitrate solution for at least 30 min.
7. Rinse gel with small amount of water for a few seconds.
8. Add a small amount of developer solution containing 0. 04% formalin (formaldehyde) and 2% sodium carbonate monohydrate (e.g., 500 μL of 37% formalin in 500 μL of 2% sodium carbonate monohydrate). Swirl briefly and discard.
9. Add more developer and shake slowly until spots appear (may take between 30 s to 5 min depending on amount of protein loaded).
10. Discard developer and place gel in stop solution (5% acetic acid) for at least 5 min.
11. Store in 1% acetic acid.

3.6. Silver Destain and In-Gel Digestion

1. Excised spots are placed in a 96-well plate with pierced well bottom.
2. Protein spots are processed using a Progest automated digestion unit. Briefly, the procedure involved spot destaining with potassium ferricyanide solution (15 m*M* potassium ferricyanide/50 m*M* sodium thiosulfate). Gel pieces are then rinsed three times with water and then shrunk with acetonitrile.
3. The gel pieces are rehydrated with 10–20 μL of trypsin solution (0.01 μg/mL in 50 m*M* ammonium carbonate) and 20–40 μL of 50 m*M* ammonium carbonate is added to the gel pieces prior to overnight incubation (37°C) with trypsin. (*See* **Note 8.**)
4. Peptide fragments are recovered by adding organic solution (30–50 μL of 50% methanol/5% acetic acid) to the gel pieces.
5. Peptide extracts are then evaporated to dryness on a Savant preconcentrator.

3.7. Device Fabrication

1. Chip design.
2. Glass plates are cleaned ultrasonically in detergent (5% Sparkleen, Fisher Scientific) methanol, acetone, and deionized water in an ultrasonic bath in a Class 100 clean room environment. Glass plates are cleaned ultrasonically in acetone. After drying, the plates are immersed in a piranha solution (H_2SO_4/H_2O_2 3/1 vol) for 30 min and thoroughly wash with deionized water.

3. Evaporatively deposit a metal mask, normally consisting of 200-Å Cr and 1000-Å Au onto glass plates under vacuum (<10^{-6} torr).
4. Remove trace organics using H_2SO_4/H_2O_2.
5. Cr and Au masked glass plates are coated with positive photoresist (1.4-μm-thick) using a Solitec photoresist coater-developer (1.5 s at 500 rpm, 3.6 s at 3000 rpm) and then soft-baked at 110°C for 5 min (30 min).
6. Perform photomask layout on a Princess CAD system. A Quintel contact mask aligner is used to expose the photoresist and Microposit 354 (Shipley, Newton, MA) is used as the developer.
7. Bake at 120°C for 30 min and etch away the metal layer with aqua regia and a commercial Cr etch (KTI Chemicals, Sunnyside, CA). The metal layers are etched with commercial gold etch solution and Cr etch solution.
8. Do not remove the remaining metal layer.
9. Etch the exposed glass in a slowly stirred mixture of concentrated $HF/HNO_3/H_2O$ (20:14:66) or a commercial buffered oxide etch (BOE 10:1, Olin-Hunt, NJ).
10. Monitor the channel depth during etching with an Alphastep profilometer (Tencor Ind., Mountain View, CA).
11. Remove the photoresist with acetone and metal masks.
12. Clean the etched glass plate and the cover plate as in **step 2.** In addition to acetone and piranha wash (30 min) the etched glass plates and cover plates are washed with water. Mount glass plates in a wafer frame with blue plastic film, scrub with a sponge soaked in dilute detergent and pressure wash using a pressure cleaning station (we used Model 2066, MicroAutomation, Fremont, CA). The detergent cleaning process is repeated three times, with at least 3 min of scrubbing each time. The final high-pressure washing is followed by a drying step in the pressure washer.
13. Align two plates under microscope and bond them thermally in a Model 6-525 programmable furnace (J. M. Ney Co., Yucaipa, CA). The temperature program is as follows: 40°C/min to 550°C for 30 min; 20°C/min to 610 °C for 30 min; 20°C/min to 635°C for 30 min; and 10°C/min to 650°C for 6 h. Cool to room temperature.

    For 0211 glass the program is:

    10°C/min to 440°C for 30 min; 2°C/min to 473°C for 30 min; 2°C/min to 592°C for 6 h; and 40°C/min to 473°C for 30 min. Cool to room temperature naturally.
14. Repeat bonding cycle if necessary and add weights over poorly bonded regions.

3.8. Fabrication of a Low-Dead-Volume Connector

1. Crystal Bond (having a melting point in the range 0–80°C) is dyed by mixing 1–2 drops of black ink from a Staedtler Lumocolor permanent pen into approx 1–2 mL of melted Crystal Bond to aid visualization.
2. The device is placed on a hot plate (80°C) filling the end of the channel adjacent the edge surface of the microchip with a Crystal Bond to prevent substantial penetration of glass chips into the capillary channel.

3. The device is cut perpendicularly to the separation channel with a diamond saw and the face is sanded with 220- and 600-grit silicon carbide abrasive paper.
4. Drill a hole into the edge surface of the microchip and align with the capillary channel.
5. Heat the chip to 80°C while concurrently applying a vacuum to the capillary channel exit to remove the Crystal Bond.
6. Insert an electrospray emitter into the hole and bond it to the microfluidic device with Crystal Bond.

3.9. BCQ Coating

1. Rinse with 1 *M* NaOH at 25 psi for 1 h.
2. Rinse with water at 25 PSI for 1 h.
3. Rinse with methanol at 25 psi for 1 h.
4. Rinse with silane reagent solution overnight (12 h). The silane solution contains 5 mL methanol, 25 μL glacial acetic acid and 25 μL silane reagent.
5. Rinse with methanol for 1 h.
6. Rinse with water for 1 h.
7. Rinse with BCQ solution overnight. The BCQ solution contains 5 mL water, 100 μL BCQ, 10 μL TEMED, and 70 μL 15% (w/v) ammonium persulfate.
8. Rinse with water for 1 h.
9. Blow dry with nitrogen.

3.10. Chip-CE–nES–MS

1. Small plastic pipet tubes are inserted through small holes made in the center of septa (Thermogreen LB-1 from Supelco or Septa 77 from Chromatographic Specialties) for buffer B and waste A reservoirs. Teflon tubes (180 μm id) are inserted in the center of septa for well D and used to seal a capillary transfer line (*see* **Fig. 1**).
2. In this configuration the chip lays on the Teflon support and a Plexiglas top is used to compress the septa to provide an airtight seal between the sample/buffer reservoirs and the chip device.
3. A 0. 1 *M* formic acid solution is used as the separation background electrolyte.
4. Mass spectrometric experiments are conducted using a Q-Star quadrupole/time-of-flight instrument (AB/Sciex, Concord, ON, Canada). The instrument is optimized by infusing a solution of 1 μg/mL of angiotensin I at a flow rate of 0. 2 μL/min through well E using a Harvard syringe pump.
5. The electrokinetic injection, sample separation, and chip cleaning steps are shown in **Fig. 1B.**
6. During the separation, this peptide solution is introduced at a flow rate of 50 nL/min and the $[M+3H]^{3+}$ and $[M+2H]^{2+}$ ions of angiotensin I are used as internal mass markers for accurate mass measurements.
7. Tandem mass spectra are obtained using the Q-Star and collisional activation of selected precursor ions is obtained using nitrogen as a target gas at collision ener-

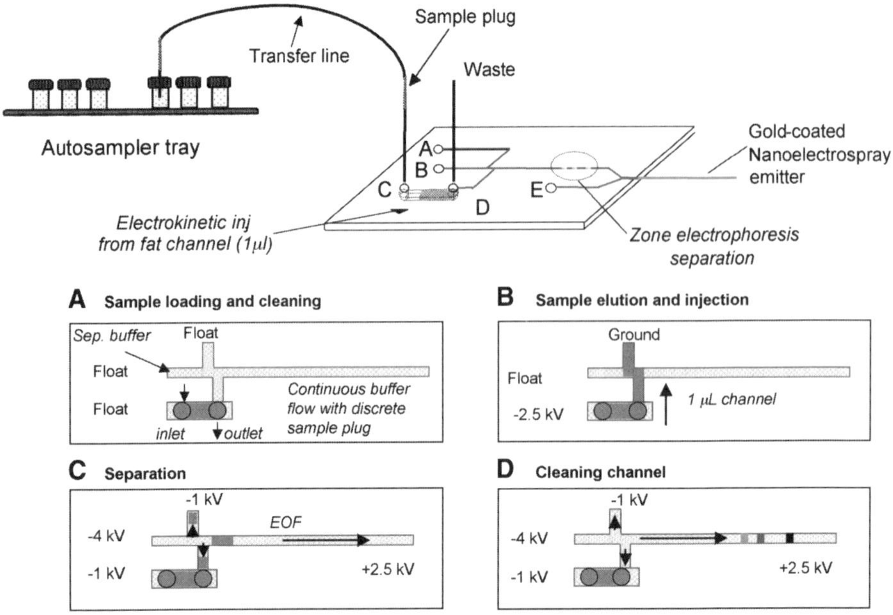

Fig. 1. Components of the autosampler chip-CE–QqTOF interface. **(A)** Microfluidic device comprising a large flow channel (800-μm wide, 150-μm deep, and 22-mm long) located between wells C and D. Sample plugs (typically 8–10 μL) are introduced from the autosampler to well C via a capillary transfer line. Well A, B, and E correspond to waste, separation buffer, and internal standard (1 μg/mL angiotensin I in separation buffer). **(B)** sequential steps involving sample loading and cleaning, desorption, injection, and separation.

gies of typically 50–90 eV (laboratory frame of reference). Fragment ions formed in the RF-only quadrupole are recorded by the time-of-flight mass analyzer at a rate of 1 s/spectrum.

3.11. Sequential Chip-CE–nES–MS (see *Note 9; Figs. 2 and 3).*

1. The separation and waste reservoirs are filled with 30 μL of running buffer (*see* **Fig. 1**).
2. A custom made 96-well plate autosampler with a sealed pressurized inlet is used to introduce a 8–10 μL sample plug to the chip device via a 40-cm length of transfer capillary (360 μm od, 50 μm id) *(23)*.
3. The first two rows of the plate are filled with wash and organic buffers while the remaining 72 wells are filled with reconstituted protein digests.

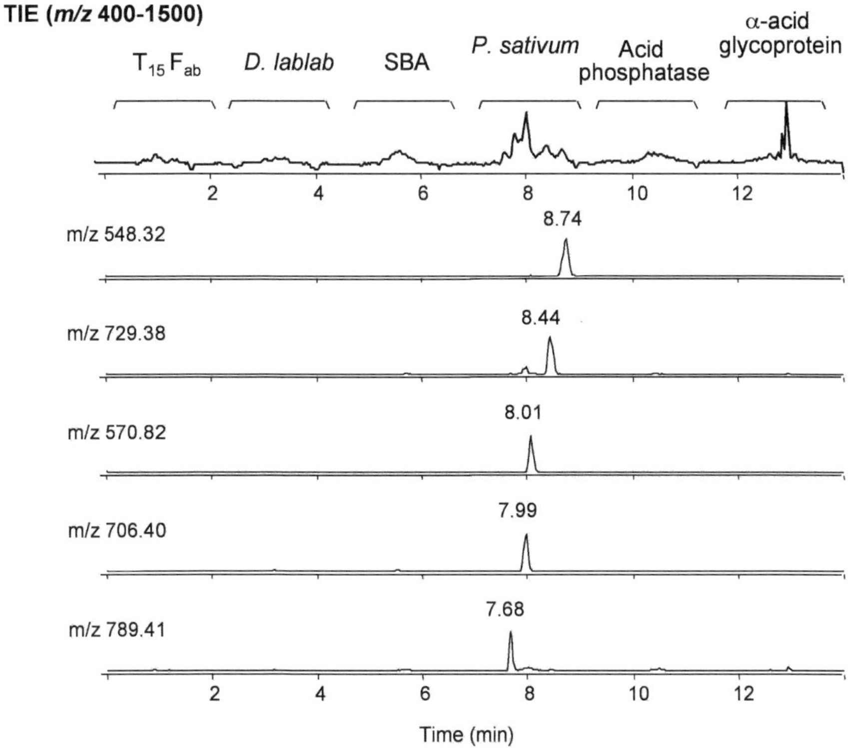

Fig. 2. Sequential analyses of six protein digests using the autosampler chip-CE–QqTOF. Total ion electropherogram, (TIE; (*m/z* 400–1500) and reconstructed ion electropherogram, RIE for the tryptic peptides of *Pisum sativum* lectin. T_{15} F_{ab}: F_{ab} segment of T_{15} antibody, *Dolichos lablab,* SBA and *P. Sativum* correspond to lectins from *D. lablab,* soyabean agglutinin, and *Pisum sativum,* respectively.

3.12. Adsorption Preconcentration Chip-CE–nES–MS (see *Note 10* and *Figs. 4* and *5*)

1. The reverse-phase slurry is prepared by suspending 10 mg C_{18} beads in 50 μL of methanol prior to packing on the chip (*see* **Fig. 1**).
2. A mixed bed composed of 2 mm of 40-μm beads (Waters) and approx 18 mm of 5-μm Poros particles (Applied Biosystems) is filled in sequence into the large channel.
3. A 0. 1 *M* formic acid solution is used for separation background electrolyte.
4. A custom made 96-well plate autosampler with a sealed pressurized inlet is used to introduce 8–10 mL of a sample plug to the chip device via a 40-cm length of transfer capillary (360 μm od, 50 μm id).
5. The first two rows of the plate are filled with wash and organic buffers while the remaining 72 wells are filled with reconstituted protein digests.

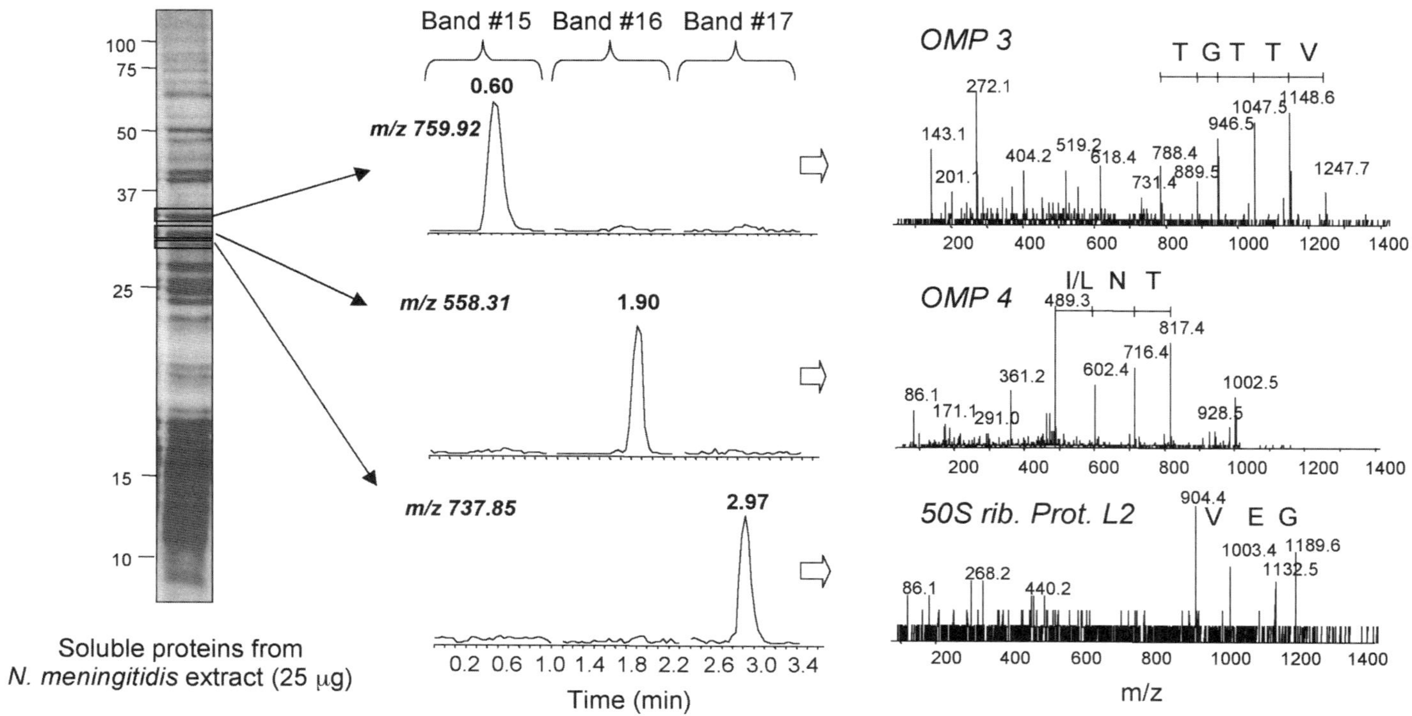

Fig. 3. 1D gel electrophoresis separation of soluble proteins from *N. meningitidis* immunotype L-3 (left panel). Sequential sample loading and electrokinetic injection of gel isolated protein digests from bands #15, 16, and 17. The center panel presents the RIE profile for *m/z* 759. 92, 558. 31, and 737. 85. The right panel presents the corresponding tandem mass spectra of each peptide as indicated.

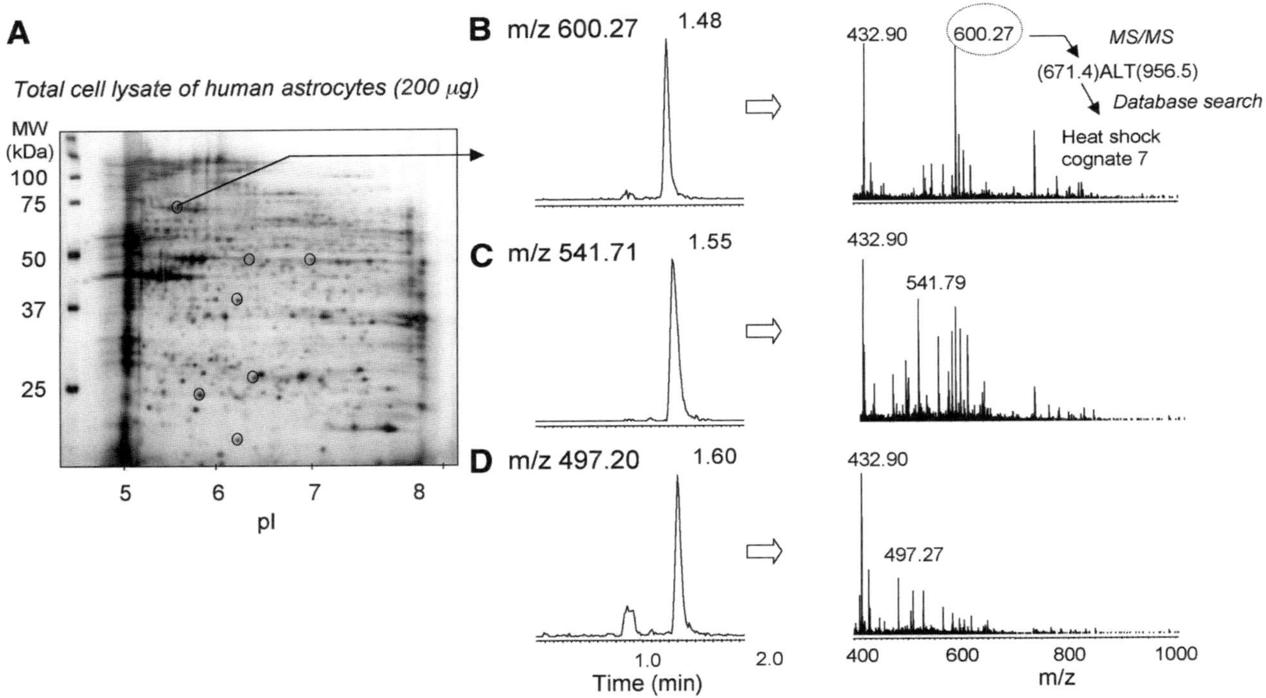

Fig. 4. Identification of trace-level proteins isolated from human astrocytes. 2D gel separation of a total cell lysate (200 μg) and chip-CE–nES–MS analyses of the spot indicated (**A**). (The reconstructed ion electropherograms for doubly protonated tryptic peptide ions at *m/z* 600. 27 (**B**) **5**41. 71 (**C**) and 497. 20 (**D**) are presented with their corresponding extracted mass spectra for the excised spot indicated. M_r and p*I* denote molecular mass (kDa) and isoelectric point, respectively. Sequential loading of peptide samples followed by 200 nL of 1% formic acid in 75% acetonitrile. Electrokinetic injection and separation: –4. 5 kV applied to well D (*see* **Fig. 1**) while other wells are floated. A voltage of +2. 2 kV was applied to the nES emitter.

3.13. Selection of Phosphopeptides Using a Chip-CE–nES–MS (see Note 10)

1. IMAC beads are first washed using 1% acetic acid in 10% aqueous acetonitrile and conditioned with 3 × 10 μL of an aqueous solution of 100 m*M* ferric chloride.
2. The beads are washed again with 1% acetic acid in 10% aqueous acetonitrile, suspended in 50 μL of the same solution and applied to the large packing channel.
3. The IMAC bed is conditioned with 20 μL 0. 1 *M* formic acid before loading the sample on the chip.
4. Prior to sample loading (20 μL) on the chip, the sample is acidified by reconstituting the in-gel digest in 5% acetic acid for proper binding of phosphopeptides to IMAC affinity column.
5. Approximately 20 μL of the digest sample is loaded onto the Fe (III) IMAC beads by pressure at 10 psi for 5 min.
6. The IMAC bed is washed with 50 μL deionized water and 20 μL 0. 1 *M* formic acid.
7. The selected peptides are eluted from the beads using 2% ammonium hydroxide.
8. In order to prevent bubble formation between the acidic buffer and the ammonium hydroxide, a short plug of deionized water is injected before and after the elution buffer.
9. The peptides are eluted from the IMAC bed and electrokinectically injected into the separation channel.
10. A short plug of deionized water (200 nL) is intercalated between the acidic and basic buffers to obtain sample stacking and to minimize bubble formation. In all cases, a constant voltage of –4. 5 kV is applied to well D (**Fig. 1**) during the electrokinetic injection while other wells are floated.

3.14. Database Searching

1. Accurate peptide masses are determined using internal standardization on the Q-Star instrument and are transferred to database search engine.
2. The list of peptide masses was searched against a nonredundant protein sequence database from NCBI or SwissProt.
3. Parameters for all searches assumed that masses corresponded to tryptic peptides and that cysteine residues were converted to S-(carbamidomethyl)cysteine.
4. All peptide masses are considered monoisotopic and the maximum deviation between the calculated and measured masses was set to <10 ppm (*see* **Fig. 3**).
5. Alternatively, the search can be conducted using peptide sequence tag approach in which the precise molecular mass of a given tryptic peptide plus the fragment ion *m/z* values derived from the MS/MS spectrum are used to retrieve potential protein candidates (*see* **Fig. 4**).
6. In situations where no match was obtained from either peptide mass fingerprinting or sequence tags, sequence segments of at least six amino acids were subjected to Blast search at the NCBI web site (www. ncbi. nln. nih. gov).

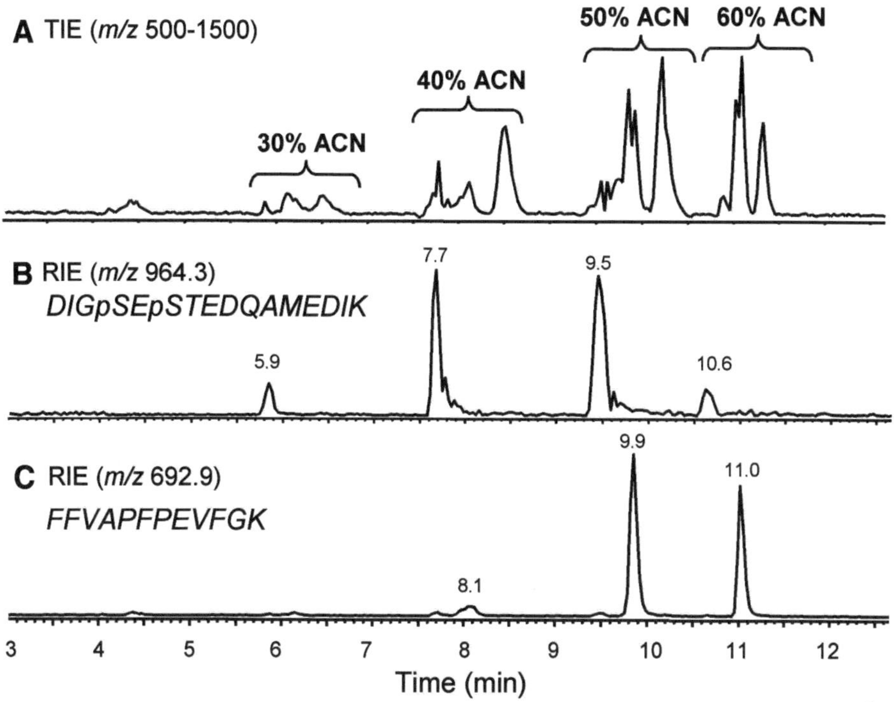

Fig. 5. Enhancement of sensitivity for the analysis of protein digests using adsorption preconcentration and electrophoresis into a microfluidic device. Tryptic digest from alpha-casein (5 pmol) was loaded onto the device. After washing with separation buffer (0.1 *M* formic acid), different elution buffers were sequentially injected to provide a step elution of increasingly hydrophobic tryptic peptides. (**A**) TIE profile for the elution buffers containing 0.1 *M* formic acid and 30%, 40%, 50%, and 60% acetonitrile, respectively. (**B**) RIE profile for *m/z* 964.3. (**C**) RIE profile for *m/z* 692.9.

4. Notes

1. The *N. meningitidis* immunotype L3 strain 406Y (NRCC 4030) was from NRC's Institute for Biological Sciences (Ottawa, Ontario, Canada) bacterial culture collection.
2. Growth and manipulation of *N. meningitidis* bacteria need to be conducted under level II containment to ensure proper biosafety. Once bacteria are killed by stirring cells with a phenol containing solution (bacterial killing solution) proteins can be extracted using the precautions normally followed in the analytical chemistry laboratory.
3. Brief sonication can be use to solubilize membrane, a step often necessary to break down bacterial cell walls. Sonicate until solution becomes clear.

4. Protein load differs for each size of strips available (7, 11, and 17 cm). Follow manufacturer instructions regarding maximum load.
5. Total V·h needs to be adjusted depending on the nature of the sample and the amount of protein loaded on a strip. This protocol is usually adequate for 17-cm strips and protein load up to 600 μg. Higher amounts of protein can also be separated using increased number of total V·h while smaller strips require less focusing V·h.
6. A different concentration is used for the second dimension gel depending on targeted proteins. To better separate high-molecular-weight proteins, gel of 7 or 10% acrylamide is used. To resolve lower-molecular-weight proteins, use 12 or 15% gels. Gradient gels of 7. 5–15% will permit a good visualization of all the proteins.
7. If gels are already stained with Sypro Ruby, start at this step.
8. Since evaporating the samples might result in lost of protein, digestion of 4 h instead of overnight can be done, followed by direct recovery of the peptides in the extract without adding any organic solvents.
9. The modular microsystem shown in **Fig. 1** provides sequential injection and separation of peptide standards and tryptic digests with a throughput of up to 30 samples/h with less than 3% sample carryover. Replicate injections of peptide mixtures (such as Leu–enkephaline and Glu–Fibrinopeptide B) provided reproducibility of migration time with RSD less than 2.3%, whereas RSD values of 3.7–11.8% are observed on peak height. When coupled to nES-MS this system provides detection of submicromolar protein digests (<7 fmol/inj.) using a quadrupole/time-of-flight instrument. An example of application for the analysis of excised bands from an SDS-PAGE gel of *N. meningitidis* membrane proteins is shown in **Fig. 3.**
10. The large channel of this microfluidic device **(Fig. 1)** provides a convenient platform to integrate C_{18} reverse-phase packing or other type of affinity media such as IMAC beads, thus enabling affinity selection of target peptides prior to electrophoretic separation and MS analyses on a quadrupole/time-of-flight instrument. Sequential injection, preconcentration, and separation of peptide standards and tryptic digests are achieved with a throughput of up to 12 samples/h and a concentration detection limit of approx 5 n*M* (25 fmol on chip). When using C_{18} preconcentration packing this device provides reproducible migration time with RSD values of 1.2–1.8% and RSD values ranging from 9.2–11.8% for peak heights. The application of this device for trace level protein identification is demonstrated for 2D gel spots obtained from protein extracts of human astrocytes using on-line MS/MS **(Fig. 4).** Improvement of sensitivity for the analysis of tryptic digests can be achieved using adsorption preconcentration prior to electrophoretic separation on the microfluidic device. Reverse-phase C18 beads are immobilized in the large channel and adsorbed analytes are subsequently separated by applying discrete plugs of increasing organic buffer ranging from 30–60% acetonitrile (0.1% formic acid) as shown in **Fig. 5.**

References

1. S. R. Pennington, M. R. Wilkins, D. F. Hochstrasser, and M. J. Dunn (1997) Proteome analysis: from protein characterization to biological function. *Trends in Cell Biol.* **7,** 168–173
2. Tyers, M. and Mann, M. (2003) From genomics to proteomics. *Nature* **422,** 193–197.
3. Rabilloud, T. (2002) Two-dimensional gel electrophoresis in proteomics: still old, old fashion, but still climbs up the mountains. *Proteomics* **2,** 3–10.
4. Nishihara, J. C. and Champion, K. M. (2002) Quantitative evaluation of proteins in one- and two-dimensional polyacrylamide gels using a fluorescent stain. *Electrophoresis* **23,** 2203–2215.
5. Gorg, A., Obermaier, C., Boguth, G et al. (2000) The current state of two-dimensional electrophoresis with immobilized pH gradients. *Electrophoresis* **21,** 1037–1053.
6. Mann, M. and Wilm, M. (1994) Error tolerant identification of peptide in sequence database by peptide sequence tag. *Anal. Chem.* **66,** 4390–4399.
7. Eng, J. K., McCormack, A. L., and Yates, J. R. (1994) An approach to correlate MS/MS data to amino acid sequence in a protein database. *J. Am. Soc. Mass Spectrom.* **5,** 976–989.
8. Perkins, D. N., Pappin, D. J., Creasy, D. M., and Cottrell, J. S. (1999) Probability-based protein identification by searching databases using mass spectrometry data. *Electrophoresis* **20,** 3551–3567.
9. Link, A. J., Eng, J., Schieltz, D. M., et al. (1999) Direct analysis of protein complexes using mass spectrometry. *Nat. Biotechnol.* **17,** 676–682.
10. Washburn, M. P., Wolters, and D., Yates III, J. R. (2001) Large scale analysis of the yeast proteome by multidimensional protein identification technology. *Nat. Biotechnol.* **19,** 242–247.
11. Davis, M. T. and Lee, T. D. (1998) Rapid protein identification using a microscale electrospray LC/MS system on an ion trap mass spectrometer. *J. Am. Soc. Mass Spectrom.* **9,** 194–201.
12. Moseley A., Blackburn K., Burkhart W., Vissers, H., and Bordoli, R. Proc. 48th ASMS Conference 2000, Long Beach CA.
13. Xue, Q., Foret, F, Dunayevskiy, Y. M., Zavracky, P. M., McGruer, N. E., and Karger, B. L. (1997) Multichannel microchip electrospray mass spectrometry. *Anal. Chem.* **69,** 426–430.
14. Ramsey, R. S., and Ramsey, J. M. (1997) Generating electrospray from microchip devices using electroosmotic pumping. *Anal. Chem.* **69,** 1174–1178.
15. Figeys, D., Ning, Y., and Aebersold, R. (1997) A microfabricated device for rapid protein identification by microelectrospray ion trap mass spectrometry. *Anal. Chem.* **69,** 3153–3160.
16. Bings, N. H., Wang, C., Skinner, C. D., et al. (1999) Microfluidic devices connected to glass capillaries with minimal dead volume. *Anal. Chem.* **71,** 3292–3296.
17. Li, J., Wang, C., Kelly, J. F., Harrison, J. D., and Thibault, P. (2000) Rapid and sensitive separation of trace level protein digests using microfabricated device

coupled to a high performance quadrupole/time-of-flight mass spectrometer. *Electrophoresis* **21,** 198–210.

18. Little, D. P., Cornish, T. J., O'Donell, M. J., Braun A., Cotter, R. J., and Koster, H. (1997) MALDI on a chip: analysis of arrays of low-femtomole to subfemtomole quantities of synthetic oligonucleotides and DNA diagnostic products dispensed by a piezoelectric pipet. *Anal. Chem.* **69,** 4540–4546.
19. Ekstrom, S., Onnerfjord, P., Nilsson, J., Bengtsson, M., Laurell, T., and Marko-Varga, G. (2000) Integrated microanalytical technology enabling rapid and automated protein identification. *Anal. Chem.* **72,** 286–293.
20. Figeys, D. and Aebersold, R. (1998) Nanoflow solventgradient delivery from a microfabricated device for protein identifications by electrospray ionization mass spectrometry. *Anal. Chem.* **70,** 3721–3727.
21. Li, J., LeRiche, T., Tremblay, T.-L., et al. (2002) Application of microfluidic devices to Proteomics research: Identification of trace-level protein digests and affinity capture of target peptides. *Mol. Cell. Proteomics*, **1,** 157–168.
22. Kim, J.-S. and Knapp, D. R. (2001) Miniaturized multichannel electrospray ionization emitters on poly (dimethylsiloxane) microfluidic devices. *Electrophoresis* **22,** 3993–3999.
23. Bonneil, E., Li, J., Tremblay, T.-L., Bergeron, J. J., and Thibault, P. (2002) Integration of solid phase extraction membranes for sample multiplexing; application to rapid protein identification from gel-isolated protein extracts. *Electrophoresis* **23,** 3589–3598.

Index